Inorganic and Physical Chemistry

AN INTEGRATED APPROACH

SI Units Edition

Inorganic
and Physical Chemistry
An Integrated Approach

SI Units Edition

by

R. S. LOWRIE, M.A., D.PHIL., A.R.I.C.

Head of Science Department, Oxford School

and

H. J. CAMPBELL-FERGUSON, M.A., PH.D.

Head of Chemistry Department, Malvern College

WHEATON

A Division of Pergamon Press

A. Wheaton & Company Limited
A Division of Pergamon Press
Hennock Road, Exeter EX2 8RP

Pergamon Press Ltd
Headington Hill Hall, Oxford OX3 0BW

Pergamon Press Inc.
Maxwell House, Fairview Park, Elmsford, New York 10523

Pergamon of Canada Ltd
75 The East Mall, Toronto, Ontario M8Z 2L9

Pergamon Press (Australia) Pty Ltd
19a Boundary Street, Rushcutters Bay, N.S.W. 2011

Pergamon Press GmbH
6242 Kronberg/Taunus, Pferdstrasse 1, Frankfurt-am-Main
West Germany

First edition 1969
Second (SI units) edition 1971
Reprinted 1972
Reprinted 1973
Reprinted 1977

Library of Congress Catalog Card No. 76-142793

Printed in Great Britain by A. Wheaton & Co. Ltd, Exeter

08 016630 X (net)
08 016631 8 (non-net)

Contents

Preface

This textbook is written for A-level, O.N.C. and H.N.C. students using modern syllabuses. The book is not intended to be a comprehensive text and the factual content, especially in the Inorganic chapters, has been kept to a minimum. On the other hand, the underlying principles have been covered in rather more depth than is customary in such books.

In planning this book we have adopted the following general principles:

(1) Detailed factual information has been excluded on the grounds that there are plenty of reference books and traditional comprehensive texts already in existence.

(2) At the end of each chapter, study questions have been provided. The purpose of these is not merely to test the reader's knowledge, but also to promote understanding and to stimulate further thought.

(3) Answers and notes on the study questions, as well as further notes on the text, have been included in a separate publication intended for teachers and more advanced students.

(4) Numerous diagrams and flow sheets have been included in the text.

(5) Summaries at the ends of the Inorganic chapters reinforce the main points.

Our warmest thanks are due to Mr. C. Nicholls, Dr. T. E. Rogers, Dr. T. A. G. Silk, Mr. J. C. Simmons, Mr. R. C. B. Smith and Mr. A. Spiers for reading parts of the manuscript, and especially to Dr. C. S. G. Phillips of Merton College, Oxford, who read the whole manuscript in draft and made many valuable suggestions. We would also like to thank Mrs. E. M. Harvey for patiently retyping the whole manuscript. The responsibility for any errors that remain rests entirely with the authors.

April 1968.

R. S. L.
H. J. C. F.

Preface to Second Edition

Since the publication of the First Edition in 1969, there has been an increasing demand for an edition in SI units. This has necessitated the rewriting of certain numerical questions and short sections of text, but otherwise the book remains substantially unaltered.

June 1971.

R. S. L.
H. J. C. F.

To the Reader

THE chapters are arranged in such an order that they can be taken in sequence. Chapters 12 and 13 may be omitted at a first reading and Chapter 17 may, if desired, be taken earlier. Chapters 1 to 9 are basic to the book as a whole.

The Inorganic chapters, Chapters 20 to 26, may be taken in any order, but the printed sequence will be found to be the most logical.

You will find that the Study Questions vary in difficulty. They are designed to help you to understand concepts rather than to test your knowledge of the book. Some of them you will find very easy; others, especially those marked with a dagger (†), will require rather more thought and sometimes more background reading.

Acknowledgements

FIGURE 9.2 is reproduced with permission from a paper in the *Transactions of the Faraday Society* by Pedley, Skinner and Chernick, **53,** 1612 (1957), and Fig. 9.3 from a paper in the *Journal of Research of the National Bureau of Standards* by F. D. Rossini, **4,** 313 (1930). Figure 9.7 is redrawn with permission from *Senior Science for High School Students, Part 2, Chemistry*, ed. H. Messel, Pergamon, 1966, and Fig. 4.3 is redrawn with permission from the Unilever Educational Booklet *The Physics of Chemical Structure*. The periodic table is reproduced with permission from Mond Nickel.

The data used have been quoted from a variety of sources which it would be difficult to acknowledge individually. We would particularly acknowledge our debt to the *Chemical Data Book* published by the University of New South Wales, and to W. M. Latimer's *Oxidation Potentials*, 2nd ed., Prentice-Hall, 1952.

The following G.C.E. Examining Boards have kindly given us leave to quote A-level questions: Oxford and Cambridge, Oxford Local, Southern Universities, Welsh, Associated Examining Board and Cambridge Local. To them our thanks are also due.

CHAPTER 1

Periodicity

1.1 Introduction

This chapter deals with the important concept of **periodicity** in chemistry. There are now, in 1968, one hundred and four known chemical elements, and with the development of new ways of synthesizing elements this number is bound to continue increasing. Their study would present quite a formidable problem but for the fortunate fact that they can be grouped together in "families" with similar or related properties. Ever since about 1869, when it first became clearly understood by Lothar Meyer and Mendeléeff, the concept of periodicity has been an important unifying factor in the study of inorganic chemistry. Much confusion existed at that time, however, due to the existence of undiscovered elements and incorrect atomic weights, and we shall therefore take up the study of periodicity and the periodic table from the twentieth-century standpoint.

1.2 The atomic nucleus

1909 was a landmark in the development of ideas on the inner structure of the atom. Up to this time it was known that atoms were exceedingly small, of the order 0·1 or 0·2 nano-

metres (nm),* but little was known of how the atoms themselves were built up. In 1909 Geiger and Marsden did some experiments to investigate the effect which a very thin sheet of metal foil had on a beam of fast moving, positively charged particles, called **alpha-particles** (α-particles). These particles were known to be extremely small even compared to the sizes of atoms themselves, and it was hoped that if the metal was thin enough, a few particles might pass through the foil and be scattered in the process.

The result of this experiment was very surprising: despite the fact that the metal foil was several hundred atoms thick, *very nearly all the particles passed straight through the foil* without appreciable deflection; the metal foil behaved as if there were almost no obstructing matter there at all. This in itself was very remarkable, but even more interesting was the fact that a very small fraction of the α-particles were deflected considerably or even bounced back. In most of their experiments about one particle in twenty thousand was affected in this way. The metal foil itself never suffered any mechanical damage, and it thus appeared that the atoms themselves were unaffected by the passage of α-particles through their midst.

The experiments of Geiger and Marsden

---

* One nanometre (nm) equals 10^{-9}m. For a note on units, see Appendix IV.

were examined further by Rutherford, who proposed a new theory of the structure of the atom to account for these observations. Rutherford argued that the atom must contain a very small, positively charged **nucleus** where most of the atomic mass was concentrated. Surrounding this nucleus was a number of negatively charged particles of negligible mass, called **electrons** (Fig. 1.1).

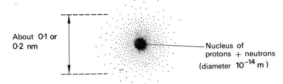

About 0·1 or 0·2 nm

Nucleus of protons + neutrons (diameter 10^{-14} m)

FIG. 1.1. The atom.

Nobody, even today, has ever succeeded in *seeing* inside an atom. It would be fairer to describe Rutherford's picture of it as a *model*. The nuclear model of the atom is something which scientists have found extremely useful in understanding chemistry, and we can safely say today that the existence of the atomic nucleus is a proven fact, even though we shall probably never succeed in "seeing" one. It is quite instructive to see how our model of the atom has changed over the years, and to bear in mind that scientists in the future will very likely be using a different model from us (Fig. 1.2).

1.3 Atomic number

Later experiments have confirmed Rutherford's view about the atomic nucleus. We now know that all atomic nuclei are positively charged, and it has proved possible (largely as a result of work on X-rays by Moseley) to measure the charges on nuclei. It has been found that all atoms have nuclear charges which are an exact whole number multiple of the charge on a hydrogen nucleus.

This observation enables us to extend our model of the atom a stage further, for if nuclear charges are related in this way it is extremely likely that atomic nuclei contain positively charged particles, and that each element is characterized by having a given number of these in its nucleus. We call these particles **protons,** and we refer to the total charge of the nucleus as the **atomic number,** which is defined as the number of protons in the nucleus.

Since the atom as a whole is uncharged, the charge of the protons in the nucleus must be exactly balanced by the negative charge of the electrons which surround it. We may say that:

(a) the positive charge on a proton is equal and opposite to that of the electron, and has been found to equal $1·6 \times 10^{-19}$ coulombs;

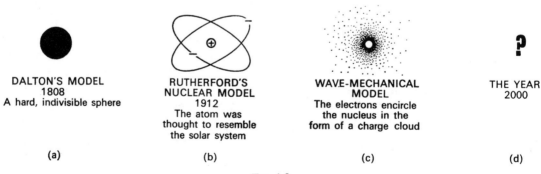

DALTON'S MODEL
1808
A hard, indivisible sphere

(a)

RUTHERFORD'S
NUCLEAR MODEL
1912
The atom was
thought to resemble
the solar system

(b)

WAVE-MECHANICAL
MODEL
The electrons encircle
the nucleus in the
form of a charge cloud

(c)

THE YEAR
2000

(d)

FIG. 1.2.

(b) the atomic number = the number of protons in the nucleus,

= the number of electrons in the neutral atom.

1.4 Mass number

Our model of the atom is still incomplete, for we have not yet taken account of the fact that atomic *masses* do not follow the exact pattern of atomic numbers. To take the simplest case, a helium atom has twice the nuclear charge of a hydrogen atom, and yet its atoms are *four* times greater in mass.

The masses of atoms can be measured in a device called a **mass spectrometer,** which was invented by Aston in 1919 (Fig. 1.3). An element can be made to vaporize into an evacuated tube where it forms separate atoms which are bombarded with electrons. This bombardment knocks one or more electrons off the atoms, so that they have an overall *positive* charge. The atoms are said to have formed **ions,** and the process is called **ionization.** For example, a helium atom in the gaseous state, denoted by the symbol He(g), can lose one or two electrons to form $He^+(g)$ or $He^{2+}(g)$. Chemical equations can be used to represent these processes:

$$He(g) \longrightarrow He^+(g) + e^-$$
$$He(g) \longrightarrow He^{2+}(g) + 2e^-$$

These ions can be accelerated by applying an electric potential of about 200 V, and focused in the form of a beam of charged atoms. If now an electric or magnetic field (or usually a combination of both) is applied at *right angles* to the direction of travel, the particles will be deflected, and will follow a curved path. The more highly charged a particle is, the stronger will be the force acting on it and the greater the deflection. The more massive the ion, the smaller will be the deflection for a given charge, as a result of the relationship *force = mass × acceleration.*

Experiments with the mass spectrometer confirmed a fact that had previously been observed for radioactive elements like uranium, namely that, while all atoms of the same element have the same atomic number, they do not all have the same atomic mass. Neon, for instance, is found to contain atoms which are, almost exactly, 20, 21 and 22 times the mass of the proton. The atomic number of neon is 10.

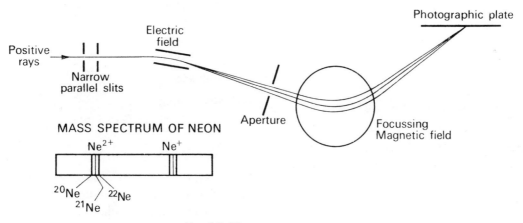

FIG. 1.3. The mass spectrometer.

This observation has been explained by postulating the existence of a third type of particle, called the **neutron,** whose mass is almost exactly the same as that of a proton, but which does not have any electrical charge. Thus atoms of neon always have 10 protons in their nucleus, but they may have either 10, 11 or 12 neutrons to make up the total mass.

Atoms of the same element which differ only in their number of neutrons are called **isotopes,** (from the Greek *isos topos,* same place, meaning that isotopes occupy the same place in the periodic table). The existence of isotopes had been inferred from the study of radioactive materials, but it was not until the invention of the mass spectrometer that they were discovered for stable elements. In fact most elements have isotopes. Hydrogen for instance consists of two isotopes: the majority of hydrogen nuclei have a mass of 1 unit and a charge of $+1$, but a small fraction are "heavy" nuclei with a charge of $+1$ and a mass of 2 units. Heavy hydrogen atoms are known as **deuterium** atoms, and a deuterium nucleus can be regarded as made up of a proton plus a neutron (Figs. 1.4 and 1.5).

The existence of the **fundamental particles,** the electron, proton and neutron, can nowadays be taken as an established fact. Modern physics has revealed a bewildering array of other so-called "fundamental" particles such as mesons and positrons, but their existence does not directly concern the chemist. The fundamental particles which concern us, and the relationships between them, can be summarized thus:

Proton: Charge $= +1$;   Mass $= 1$

Neutron: Charge $= 0$;   Mass $= 1$

Electron: Charge $= -1$;   Mass negligible
(about $1/1840$ unit).

The total mass of a nucleus is called its **mass number,** and it is related to other quantities as follows:

Mass number = number of protons plus number of neutrons in nucleus.

Number of neutrons = mass number minus atomic number.

Single isotopes of an element can be shown by writing the mass number as a superscript, for example ^{20}Ne, ^{21}Ne, ^{22}Ne.

1.5 Atomic weight

The so-called **atomic weight** of an element is the average mass of the isotopes present. For instance neon has an atomic weight of 20·2, which is the weighted average of the masses of its isotopes.

Mass number	20	21	22
Abundance	90·4%	0·6%	9·0%

Weighted average =

$$= \frac{(20 \times 90 \cdot 4) + (21 \times 0 \cdot 6) + (22 \times 9 \cdot 0)}{100}$$

$= 20 \cdot 2 =$ atomic weight.

Fortunately for chemists, the relative proportions of different isotopes in an element do not vary much in nature, so that the atomic weight of an element can be taken as a constant. In

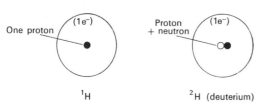

FIG. 1.4. Isotopes of hydrogen.

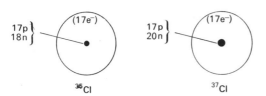

FIG. 1.5. Isotopes of chlorine.

most of the light elements, a single isotope predominates, making the atomic weight an approximate whole number, but with heavier elements many different isotopes are present and the atomic weight is rarely a whole number. Tin for instance has eleven stable isotopes.

By international agreement, all atomic weights are expressed in terms of the isotope of carbon which has a mass number of 12, ^{12}C. Carbon itself contains traces of other isotopes making its atomic weight on this scale 12·01. Practically all atomic weights have now been checked, together with the exact isotopic composition of elements, using the mass spectrometer, for which the "carbon-12" standard is very convenient. A complete table of atomic weights is listed in Appendix 2.

If the mass numbers and atomic numbers of elements are examined carefully, some interesting patterns can be traced. It is rather rare for instance for a nucleus to have an odd number of protons and an odd number of neutrons. The number of neutrons becomes greater than the atomic number as the element gets heavier, making it appear that the very highly charged nuclei need a large number of neutrons to hold them together. Even then the heaviest nuclei tend to disintegrate, and eject particles. This is the phenomenon of radioactivity (Chapter 2).

The atomic number of an element is a more fundamental property than its atomic weight, but nevertheless the atomic weight is of great practical value, since the chemist frequently determines the relative numbers of atoms which take part in a reaction by measuring weight changes. The existence of isotopes was a great source of confusion in nineteenth century attempts to understand chemistry, since an arrangement of the elements in ascending order of their atomic weights is only approximately significant: chemical properties of elements depend upon their atomic number, not upon their atomic weight.

1.6 The periodic law

The properties of the elements are a periodic function of their atomic numbers. This is the periodic law, first stated by Lothar Meyer in 1869, though in fact he stated it for atomic *weights* where it is only approximately true. *Periodic* means repeating regularly after an interval, (cf. the days of the week, which repeat after a *period* of seven days).

The law may be illustrated by plotting a graph of melting points (Fig. 1.6) or boiling points (Fig. 1.7) of the elements against their atomic numbers: despite the discrepancies, a pattern exists. Similar elements occupy corresponding positions on the curve—for instance the **noble gases** (inert gases) have the lowest melting points and boiling points. Immediately following each noble gas in the sequence there is a soft, highly reactive metal known as an **alkali metal**. Immediately preceding each noble gas (except helium) is a highly reactive non-metal known as a **halogen**—fluorine, chlorine, bromine, iodine or astatine. The elements can be arranged in chemical "families" known as **Groups**. There are places in Figs. 1.6 and 1.7 where the periodicity is slightly obscure. Thus the melting and boiling points of the **first row** of elements, lithium to neon, do not correspond all that closely with the **second row**, sodium to argon. Also, the periods are not all of equal length. The first and second rows of elements are termed **short periods.** The third and fourth rows are termed **long periods,** since they contain eighteen elements each instead of eight. Hydrogen and helium do not fit in with this periodic pattern, and in a sense they comprise a complete period on their own.

The periodic law may be demonstrated more convincingly by the volumes occupied by atoms of the elements. Since atoms are inconveniently small, we will compare the volume occupied

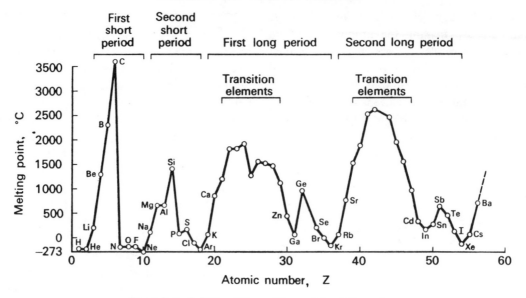

FIG. 1.6. Periodicity of the melting points of elements.

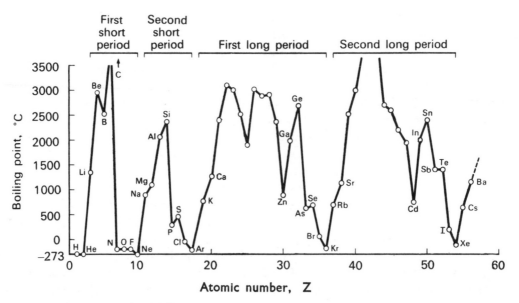

FIG. 1.7. Periodicity of the boiling points of elements.

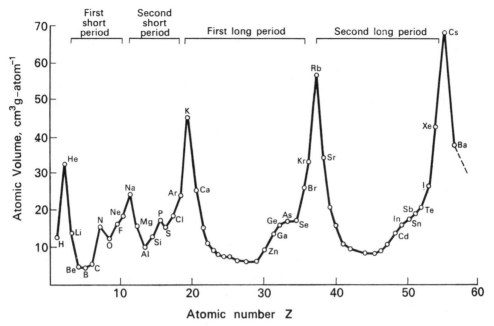

FIG. 1.8. The atomic volumes of the elements.

by *equal numbers* of atoms of all the elements. A convenient number of atoms to take is one **gram-atom,** which is in fact $6 \cdot 023 \times 10^{23}$ atoms:

$$1 \text{ g-atom of hydrogen} = 6 \cdot 023 \times 10^{23} \text{ atoms}$$
$$= 1 \cdot 008 \text{ g of hydrogen.}$$
$$1 \text{ g-atom of carbon} = 6 \cdot 023 \times 10^{23} \text{ atoms}$$
$$= 12 \cdot 01 \text{ g of carbon.}$$
$$1 \text{ g-atom of chlorine} = 6 \cdot 023 \times 10^{23} \text{ atoms}$$
$$= 35 \cdot 45 \text{ g of chlorine,}$$
$$\text{etc.}$$

In each case, one gram-atom of the element is obtained by taking the atomic weight of the element in grams, and it always contains the *same number of atoms*. The number $6 \cdot 023 \times 10^{23}$ is a constant of great importance to the chemist, called the Avogadro number N_A.

The volume of one gram-atom of any element in the solid state is called the **atomic volume.** The atomic volume of an element is a rough measure of the relative size of the atoms. Fig. 1.8 shows a plot of atomic volume of the elements against their atomic numbers. Again a clear periodicity is observed. The alkali metals turn out to have exceptionally large atomic volumes, suggesting that their atoms may be larger than those of neighbouring elements.

Atomic volume fails to take into account the fact that different elements can have their atoms packed together in different ways. In theory a contraction in atomic volume could represent either a decrease in atomic radius, or a closer mode of packing. The atomic radius, which can be taken as half the distance between adjacent nuclei in the element, is plotted against atomic number in Fig. 1.9.

Further examples of periodicity of physical properties will be met in later chapters. Chapter 3 examines the periodic variations in ionization energy, and Chapter 6 investigates the periodicity of binding energy.

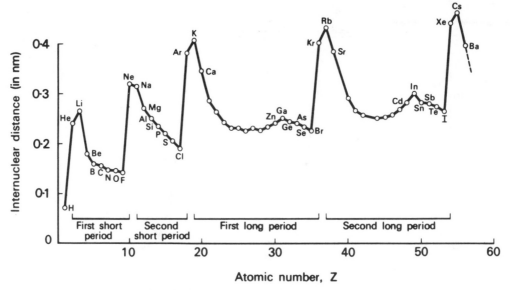

FIG. 1.9. The atomic radii of the elements.

1.7 Chemical periodicity

The periodic law is by no means confined to physical patterns—by far the greatest nineteenth-century contribution to inorganic chemistry was made by the Russian chemist, Mendeléeff, who was primarily concerned with the periodic variation in chemical properties. If an early period in the series of elements is examined, beginning with an alkali metal and ending with a noble gas, the same trend will be seen (Table 1.1).

At a later stage in the book we shall have to be more precise in the use of the word *reactive*.

For the present, we may regard a reactive metal as one which enters vigorously into chemical combination with a non-metal such as oxygen, and a reactive non-metal as one which vigorously attacks metals.

Patterns of periodic behaviour exist not only in the reactivity of elements, but also in the composition of the compounds they form. Figure 1.10 plots the atomic number against the number of gram-atoms of hydrogen which can combine chemically with the element to form a hydride. The noble gases do not combine at all. This is a clear example of periodicity, but unfortunately it breaks down for the long periods. Similar plots can be constructed for oxides of elements (Study Question 13).

TABLE 1.1

very reactive metal	→	metals becoming less reactive	→	element is mid-way between metal and non-metal	→	unreactive non-metals becoming more reactive	→	reactive non-metal	→	noble gas

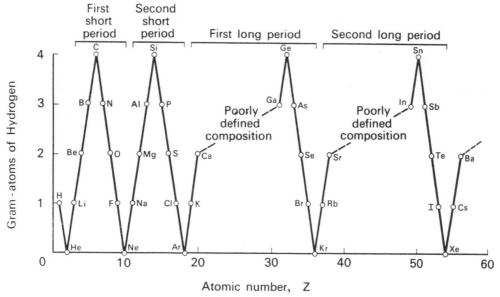

FIG. 1.10. Periodicity shown by the empirical formulae of hydrides.

1.8 The periodic table

The **periodic table** is a convenient way of arranging the elements so that similar elements occur as **Groups** in vertical columns.* Various ways of laying out the periodic table have been used but the so-called "long" form is used throughout this book (see inside cover, also Fig. 1.11). The final chapters of this book each deal with specific areas of the periodic table, and it is useful to be able to refer to the elements concerned by the following terminology:

(a) *Typical and transitional elements.* The elements in the first and second rows (as far as argon) are called **typical elements.** The term **transition element**[†] refers to the block of elements in the centre of the long periods, some-

---

* In this book the word "Group" with a capital G refers to a Group of the periodic table, to distinguish from other, non-specialized, uses of the same word.

† For a more detailed definition of "transition element", see Chapter 26.

---

times known as the *d*-block. The elements of Group IIB, zinc, cadmium and mercury, are *d*-block elements but are not strictly transition metals, since *d*-electrons are not used in bonding. The elements from lanthanum to lutecium and actinium to lawrencium are known as inner transition metals, or *f*-block elements.

(b) "A" *and* "B" *elements.* The elements on the left-hand side of Fig. 1.11 are known as "A" elements, and those on the right as "B" elements.

(c) *s-block, p-block and d-block elements.* The prefixes *s*, *p*, and *d* refer to the energy levels of the electrons within the atom, and it is often convenient to designate blocks of the periodic table by referring to their electronic structures.

A vertical column of elements in the periodic table is known as a Group. A column of transition metals, such as titanium, zirconium and hafnium, is sometimes called a **sub-Group.** The numbering of Groups refers mainly to the typical elements, though it is occasionally useful to refer to the sub-Groups by number as

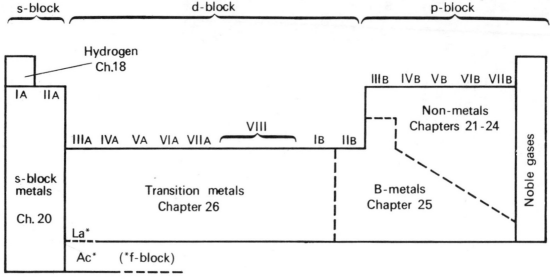

FIG. 1.11. The periodic table (long form) showing the plan of this book.

well. For instance, the Group IIB metals zinc, cadmium and mercury sometimes resemble the Group IIA metals.

1.9 Mendeléeff

The modern form of the periodic table is quite similar to the table introduced by Mendeléeff in 1869. The noble gases (Group 0) had not then been discovered, though this did not obscure the periodicity of the remaining elements. Both Lothar Meyer and Mendeléeff expressed the periodic law in terms of atomic weights.

Mendeléeff's contribution to chemical understanding was remarkable, in that he had the foresight to realize that where discrepancies occurred this was probably due to undiscovered elements. For instance he left a gap in Group IV which was filled only when germanium was discovered in 1885. The properties of germanium were found to be astonishingly close to those which Mendeléeff predicted (Study Question 15).

In one or two cases a pair of elements had to be exchanged in order that their atomic weights might "fit" the periodic law. Tellurium and iodine were such a pair, and these were termed an **inversion.** Inversions are caused by the variation in number of neutrons.

Study Questions

1. What would the results of the Geiger-Marsden experiment have been if matter had been composed of solid spherical atoms rather than particles with nuclei?

2. How is the atomic number related to (a) the number of protons, (b) the number of electrons, (c) the number of neutrons, in a neutral atom?

3. Bromine contains $50 \cdot 53\%$ ^{79}Br and $49 \cdot 47\%$ ^{81}Br.

(a) Calculate the atomic weight of bromine from these figures.

(b) Look up the atomic weight of bromine in a table of atomic weights.

(c) Why do the two values differ? (See also Chapter 2.)

4. Plot the following quantities against atomic number: (a) specific heat, (b) the latent heat of fusion, (c) the latent heat of evaporation, (d) the difference between the melting and boiling point of an element.

  (i) Show how the plots illustrate the principle of periodicity.

  (ii) Comment on any interesting features in the plots.

5. Argon, atomic number 18, has an atomic weight of 39·95, while potassium, atomic number 19, has an atomic weight of only 39·10. How can you account for this? Use a periodic table and a table of atomic weights to discover two more pairs of elements which behave in the same way.

6. The densities of calcium and copper are 1·55 and 8·96 g cm^{-3} respectively. Calculate the atomic volume of each element.

7. Use a periodic table to represent the following in symbols:

  (a) A nucleus with 9 protons and 10 neutrons.

  (b) A nucleus with atomic number 74 and mass number 184.

  (c) Strontium-90.

  (d) A nucleus with 82 protons and 124 neutrons.

  (e) A nucleus with 20 protons and 20 neutrons.

8. Do you think any elements with atomic weights of less than 250 remain to be discovered? Give reasons for your answer.

9. The similarity between calcium, strontium and barium was noted by Döbereiner in 1829, and he called the three elements a triad.

  (a) Find a relationship between the atomic weights of the three elements.

  (b) Can you discover any other properties of the elements that have the same numerical relationship?

  (c) Can you discover any other examples of such triads in a given Group that possess the same atomic weight relationship as calcium, strontium and barium?

10. What are the weights of the following?

(a) 1 g-atom of oxygen.

(b) 0·2 g-atom of calcium.

(c) 3 g-atoms of lead.

(d) 1 atom of hydrogen.

(e) $6·023 \times 10^{23}$ atoms of carbon.

(f) $1·2046 \times 10^{23}$ atoms of uranium.

11. The following elements are each composed of only one stable isotope: F, Na, Al, P, Sc, Mn, Co, As, Y, Nb, Rh, I, Cs, Pr, Tb, Ho, Tm, Ta, Au and Bi.

  (a) How many (i) protons, (ii) neutrons does each isotope contain?

  (b) What have these isotopes in common?

  (c) The element Be is the only other element composed of a single stable isotope. How does it (i) resemble (ii) differ from, the elements above?

†**12.** The mass spectrum of an element X consists of four lines with mass/charge ratios of 34·5, 35·5, 69 and 71 with relative intensities of 3:2:42:28 respectively.

  (a) How many isotopes are present?

  (b) How can you account for four lines?

  (c) What is the approximate atomic weight of X?

†**13.** Use the information in the later chapters in this book to plot the atomic number of the elements against the maximum number of g-atoms of (i) chlorine, (ii) oxygen with which each element will combine.

  (a) How do the plots differ from that obtained for hydrogen? (Fig. 1.11)

  (b) Are the differences sufficient to make you suspect the periodic law?

†**14.** List the ways in which boron and silicon resemble each other. In view of these similarities, justify the fact that they are included in different Groups in the periodic table.

15. The element germanium, which occurs between silicon and tin in Group IV was unknown to Mendeléeff, who attempted to predict its properties. Use the following data to do the same.

Property	Si	Sn
M.p.	1410°C	232°C
B.p.	2680°C	2270°C
Density, g cm^{-3}	2·33	7·30
Oxide m.p.	1700°C	1127°C
Hydride	b.p. –114°C	Decomposes easily
Chlorides	b.p. 57°C	$SnCl_4$ b.p. 113°C
		$SnCl_2$ b.p. 623°C
Fluoride	b.p. –96°C	Sublimes at 700°C

16. Gallium, the element below aluminium in the periodic table, was unknown to Mendeléeff. Aluminium melts at 660° and boils at 2450°C; it has a density of 2·70 g cm^{-3}. It forms an oxide Al_2O_3 (m.p. 2045°C), a chloride Al_2Cl_6 (b.p. 180°C) and a fluoride AlF_3 (m.p.

1040°C). From the properties of aluminium and the trends observed with silicon, germanium and tin (Q.15), predict the properties of gallium.

17. Atomic dimensions are so different from those which we are used to that it is hard for us even to begin to cope with them.

(a) If the nucleus of an atom is the size of a tennis ball, estimate how large the atom itself would be.

(b) Estimate how long you would take to count the atoms present in 1 g of hydrogen.

18. (a) Find out why the symbols for the following elements are not derived from their English names: sodium, potassium, gold, silver, mercury, iron, tin, lead and antimony.

(b) What are the origins of the names rubidium, caesium, chlorine and iodine?

CHAPTER 2

Radioactivity

2.1 Forms of radioactivity

In 1896, Becquerel discovered that some elements have the property of emitting penetrating radiation rather similar to X-rays, which would fog photographic plates even when they were wrapped in black paper. These rays are called **gamma-rays** (γ-rays), and the property is called **radioactivity**. It was later found that high energy particles, either positively or negatively charged, were emitted at the same time, and that the element was itself converted to a new element.

Gamma-rays have extremely short wavelength and very high energy; when they are absorbed by chemical substances, decomposition occurs. The radiation can penetrate solid matter and can have a very damaging effect on living tissues.

Radioactivity is the disintegration of atomic nuclei, with the emission of some of the particles they contain. The factors which determine when a given nucleus is unstable are not yet fully understood, but it may be noted that *all* nuclei with atomic numbers greater than 83 are radioactive.

Two main types of particle are emitted from radioactive nuclei:

(i) **Alpha-particles** (α-particles). These may be regarded as being made up of two protons

and two neutrons, and they are in effect helium nuclei. They have a charge of $+2$ and a mass number of 4, and can be written in symbolic form ^4_2He.

(ii) **Beta-particles** (β-particles). These are simply very high energy electrons, having a charge of -1 unit, and negligible mass.

Since a charged particle is always emitted from a radioactive nucleus, the atomic number must change. For instance radium-226, $^{226}_{88}\text{Ra}$ is alpha-active, emitting γ-rays and α-particles. The atomic number drops by two units, forming the radioactive gas, radon-222, $^{222}_{86}\text{Rn}$. The disintegration may be shown in the form of an equation:

$$^{226}_{88}\text{Ra} \;\rightarrow\; ^4_2\text{He} \;+\; ^{222}_{86}\text{Rn}$$
Radium-226 α-particle   Radon-222

Similarly beta-active nuclei eject an electron (β-particle) from the nucleus, and the atomic number increases by one unit. One of the isotopes of lead, lead-214, is beta-active:

$$^{214}_{82}\text{Pb} \rightarrow {}^0_{-1}\text{e} + {}^{214}_{83}\text{Bi}$$

In β-disintegrations the mass number does not change. The process may be imagined as a nuclear rearrangement in which a neutron changes into a proton and ejects an electron. Both β- and α-emitters usually emit γ-rays as well.

In naturally occurring isotopes of heavy elements like radium, a **radioactive disintegration series** is observed. For instance radium-226 forms radon-222, which forms polonium-218, which forms lead-214. Lead-214, being beta-active, forms bismuth-214, which forms another isotope of polonium, of mass number 214. The process continues in a series of steps of this type until finally a stable species is reached, lead-206. Figure 2.1 illustrates this series, in slightly simplified form.

until Chapter 17, though a simple graph showing the exponential decay of a radioactive element is given in Fig. 2.2.

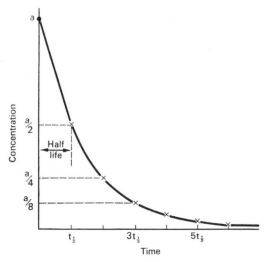

FIG. 2.2. Exponential decay.

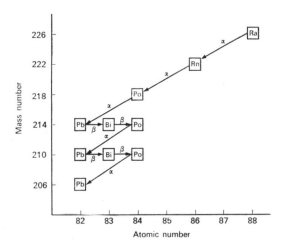

FIG. 2.1. A radioactive decay series.

A peculiar feature of radioactivity is that it is quite independent of the state of chemical combination of the element. The activity of radium for instance is the same whether it is the metal, or its chloride or hydroxide or any other compound. The rate of disintegration is found to be such that after a given time period, known as the **half-life period,** one-half of the nuclei will have disintegrated, no matter how many there were at the beginning of this period. After another half-life period the fraction left will be one-quarter, and so on. This form of disintegration is known as **exponential decay.** The mathematical treatment will be deferred

The masses and charges of alpha- and beta-particles were determined originally in a device analogous to the mass spectrometer. Gamma-rays are not deflected by an electric or magnetic field since they do not have any charge (Fig. 2.3). The **Geiger counter,** and various other electronic devices which make use of the ionizing power of alpha- and beta-particles, can be used for counting the particles and hence estimating the amount of a radioactive element present.

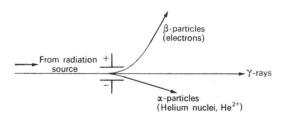

FIG. 2.3.

2.2 Nuclear reactions

The radioactive disintegrations considered above are the simplest form of nuclear reaction, in which a single nucleus decomposes spontaneously. More complex disintegrations can be effected by allowing particles such as α-particles to collide with another nucleus. Rutherford, in 1919, noticed that the normally stable nuclei of nitrogen-14 were disrupted by bombarding them with high energy α-particles. Protons were produced, and the nitrogen nuclei became oxygen-17, a stable isotope of oxygen.

$$\ce{^4_2He + ^{14}_7N -> ^{17}_8O + ^1_1H}$$

α-particle + nitrogen-14 $\rightarrow$ oxygen-17 + proton

Many new isotopes have been synthesized by nuclear reactions of this type. Radioactive isotopes of the lighter elements can be made, by causing an abnormal number of neutrons to be present. Irradiating a stable element with a beam of neutrons may cause it to be converted to an unstable isotope, as in this example with aluminium:

$$\ce{^{27}_{13}Al + ^1_0n -> ^{28}_{13}Al + \gamma};$$

stable   unstable

$$\ce{^{28}_{13}Al -> ^{28}_{14}Si + ^0_{-1}e}$$

stable β-particle

An even more drastic form of nuclear disintegration is **nuclear fission,** which occurs when certain heavy isotopes are bombarded with neutrons. Uranium-235 is the most notorious example of this, since the energy released in its neutron-induced fission was utilized in the atomic bomb. The nucleus splits into two parts, and gives rise to several neutrons in addition. A typical reaction would be:

$$\ce{^{235}_{92}U + ^1_0n -> ^{141}_{56}Ba + ^{92}_{36}Kr + 3^1_0n}$$

three neutrons

Radioactive fission products (cause the "fall-out" from a nuclear bomb)

The energy released in this reaction is 16 000 000 000 joules for every gram-atom of uranium-235.

Since each reacting neutron produces about three new neutrons which can react further, the process becomes a **chain reaction** where each successive stage gives rise to more and more neutrons (Fig. 2.4). In this way a single stray neutron can detonate a whole lump of uranium-235, once the lump exceeds a certain critical mass. To make the reaction yield energy

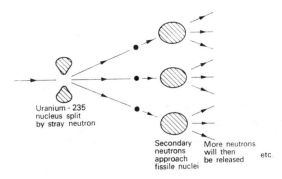

Uranium - 235 nucleus split by stray neutron

Secondary neutrons approach fissile nuclei

More neutrons will then be released   etc.

FIG. 2.4. A chain reaction.

in sufficient quantity for a bomb, it was necessary to separate the fissile isotope, ^{235}U, from the remainder of the uranium, mainly ^{238}U. This was a formidable problem since isotopes are chemically almost identical, and physically very similar. Pure uranium-235 was made during the Second World War by repeated diffusion of the volatile compound uranium hexafluoride, UF_6 (Chapter 8).

A chain reaction of this type can be made to produce energy at a controllable rate, instead of explosively. This is done in the **nuclear reactor** used in a modern nuclear power station. It is necessary to modify the design in such a way that some of the neutrons produced in the reaction are absorbed. In an atomic pile, a chain reaction is set up, and graphite rods can be lowered into the material if necessary to slow down the production of neutrons. The heat produced is used to raise steam to drive electrical generators.

A nuclear reaction which produces even more energy than uranium fission is the **fusion** of two light nuclei, for example:

$$^3_1H + {}^1_1H \rightarrow {}^4_2He;$$
$$\Delta E = -18 \cdot 9 \times 10^{11} \text{ J},$$

where $\Delta E =$ energy change per g-atom of helium.

Reactions of this type are used in the H-bomb: they need an extremely high temperature to make them start—over 1 000 000°C, and a fusion reaction has to be "triggered" by means of an ordinary fission reaction. The energy released from the Sun is produced in this way: hydrogen and helium are the two main elements in the sun (the name helium comes from the Greek word *helios*, meaning sun).

---

†2.3 Energy and mass

There is a small change in mass associated with any energy change, and in the case of nuclear reactions this mass change may be large enough to be detected. The relationship between mass and energy was stated by Einstein:

$$\Delta E = mc^2$$

where $\Delta E =$ change of energy, in joules*
$m =$ change of mass, in kg
$c =$ velocity of light, in m s^{-1}
$(3 \times 10^8 \text{ m s}^{-1})$

Different nuclei have different **nuclear binding energies.** Two effects result from this: (i) there is an energy change associated with any transformation from one nucleus to another (ii) the masses of atoms are not *exact* whole numbers, and there is a change in total mass, for a nuclear reaction. For instance in the above reaction, mass change $= \dfrac{\Delta E}{c^2}$

$$= \left(\frac{18 \cdot 9 \times 10^{11}}{9 \times 10^{16}} \right) \text{kg}$$

$= 2 \cdot 1 \times 10^{-5}$ kg (decrease per gram-equation).

2.4 Synthetic elements

Until the discovery of radioactivity it was not thought possible to change one element into another. The dream of the alchemists was to "transmute" base metals into gold. Transmutation of the elements is now commonplace, though it would be far more expensive to make gold by a nuclear reaction than to mine it! Until quite recently there were gaps in the atomic number series, but these have now been filled by synthesizing the elements in a nuclear reactor. Plutonium, ^{94}Pu, is nowadays made in quite large quantities—it does not occur naturally on Earth.

The periodic table is continually being extended by synthesizing "transuranic" elements, i. e. elements heavier than uranium. A series of inner transition elements has been completed

---

* This symbol Δ means "change of"

with the discovery of element number 103, lawrencium, this series comprising the second set of f-block elements.

2.5 Isotope separation

The separation of a mixture of isotopes of an element presents problems because the properties of isotopes are practically identical in most cases. The extraction of deuterium from ordinary hydrogen is relatively easy because the mass–number ratio is 2 : 1. The properties of deuterium oxide, for instance, are noticeably different from those of ordinary water.

The heavier isotope of an element will give

TABLE 2.1

	Naturally occurring water (mainly ^1H_2O)	Heavy water (D_2O or ^2H_2O)
Melting point	0·0°C	3·8°C
Boiling point	100·0°C	101·4°C
Density	0·997 g cm^{-3}	1·105 g cm^{-3}

rise to compounds of greater density and molecular weight, and separation by diffusion is a possibility. In the case of deuterium, however, use is made of the fact that deuterium gas is evolved more slowly than light hydrogen during electrolysis. In countries like Norway, where hydro-electric power is cheap, high purity deuterium oxide is made by the continual electrolysis of dilute sodium hydroxide solution—the heavier isotope remains behind as part of the solution.

The separation of uranium-235 from uranium-238 is relatively more difficult, since the mass ratio is close. The diffusion of UF_6 has already been mentioned.

Enriched isotopes of most elements are available commercially and are used for various purposes. Such isotopes form conveniently "labelled" atoms. The progress of an element through a living system can be followed by making it radioactive—the working of the thyroid gland, for instance, can be observed by using radioactively labelled iodine. The mechanisms of chemical reactions can often be observed by labelling certain atoms.

Study Questions

1. Write balanced equations for the following nuclear reactions:
   (a) The emission of a β-particle from $^{210}_{82}Pb$.
   (b) The emission of a β-particle from ^{14}C.
   (c) The emission of an α-particle from $^{233}_{92}U$.
   (d) The emission of an α-particle from ^{232}Th.
   (e) The capture of a neutron by ^{238}U followed by the emission of two β-particles consecutively.
   (f) ^{14}N is bombarded with neutrons. The eventual product is ^{14}C.
   (g) ^{65}Zn decays to give ^{65}Cu.

2. Identify A, B, C, and D from the following information.
   (a) $^{239}_{94}Pu + ^1_0n \rightarrow$ A.   (b) $A + ^1_0n \rightarrow$ B.
   (c) $B \rightarrow _{-1}^{0}\beta + C$.   (d) $C \rightarrow ^4_2\alpha + D$.

3. Tritium is the radioactive isotope of hydrogen with a mass number of 3. Would you expect it to be an α- or a β-emitter? Suggest an equation for the nuclear reaction that occurs.

4. (a) The radioactive decay of ^{63}Ni to give ^{63}Cu has a half-life of 120 years. How long will it take for (i) $\frac{3}{4}$, (ii) $\frac{15}{16}$, (iii) $\frac{9}{10}$, of the nickel to change into copper?
   (b) How would this copper differ from the naturally occurring element?

5. (a) Why is helium found in certain radioactive minerals?
   (b) Why is the atomic weight of lead often dependent on the source from which it is obtained?

6. (a) Why is the fallout from a uranium nuclear fission bomb itself radioactive?
   (b) Find out why strontium-90 is a particularly dangerous product of nuclear bombs.

7. The isotopic masses of ^4_2He, ^7_3Li and ^1_1H are respectively 4·00390, 7·01818 and 1·00812. Calculate ΔE for the following fusion reaction:
$$^7_3Li + ^1_1H \rightarrow 2^4_2He.$$
(Velocity of light $= 3 \times 10^8$ m s^{-1}.)

CHAPTER 3

Energy levels and bond formation

3.1 Energy units

Chemistry is largely concerned with the interplay between matter and *energy*, and in this chapter we shall be examining the energy states, or **energy levels,** of electrons in atoms.

In the mass spectrometer, atomic masses are found by ionizing an atom, that is, by knocking one or more electrons off it. A certain minimum quantity of energy is required to do this, known as the **ionization energy** of the atom. To remove the first electron from a gaseous helium atom, it is necessary to strike it with an electron which has been accelerated through a potential of at least 24·58 volts. Thus we say that the first ionization energy of helium is 24·58 **electron-volts.** To remove the second electron from helium will require more energy, since it must be removed from an ion which is already positively charged. The second ionization energy of helium is 54·40 electron-volts.

Many of the energy changes with which chemists are concerned are heat changes, and heat is generally measured in *joules* (J). One joule is defined as the energy converted as work when the point of application of a force of one newton (N) is moved through a distance of 1 metre. In terms of heat, it is found that 4180 J, or 4·18 kJ, are required to raise the temperature of 1 kg of water through one kelvin degree (1 K).

In this book energies are quoted in either electron-volts or kilojoules. Appendix 4 gives a conversion table which can be used if data are required in other units.

1 electron-volt is equivalent to an energy change of approximately 96·5 kJ g-atom^{-1} (see Study Question 12). Thus an element with an ionization energy of 10 electron-volts will require 965 kJ to convert 1 g-atom of it into singly charged positive ions.

$$He(g) \rightarrow He^+(g) + e^-;$$
$$\Delta E = +2370 \text{ kJ g-atom}^{-1}$$
$$\text{or} \quad +24\cdot6 \text{ electron-volts per atom}$$

The symbol Δ is used to denote "change of" a quantity, so in this case ΔE denotes change of energy. The plus sign indicates that the change represents an increase in energy. The helium ion is in a state of higher energy than the uncharged atom (Fig. 3.1).

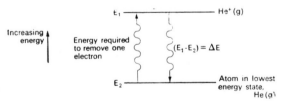

Fig. 3.1. An energy diagram.

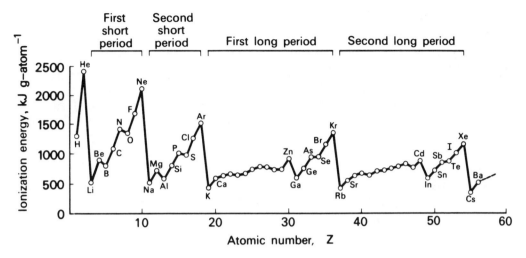

FIG. 3.2. The first ionization energies of the elements.

It might be expected that the first ionization energies of atoms would increase steadily with increasing nuclear charge, but this is not found to be the case. True, the first ionization energy of helium is greater than that of hydrogen, yet that of lithium is *less* than that of helium. In fact ionization energies show a periodicity just like the properties mentioned in Chapter 1 (Fig. 3.2).

3.2 Energy levels

In order to explain the behaviour of electrons in atoms, it is postulated that they can only exist in certain energy states or **energy levels**. Evidence for this comes from two main sources:

(a) *Spectra*. If the gaseous atoms of an element are raised to a high state of energy, they will lose this energy by emitting it as radiation. However, the emission spectrum thus obtained does not consist of a continuous band of frequencies, but rather a series of single frequencies, each corresponding to a fixed energy value. The energy change, ΔE, is related to frequency

v by the expression

$$\Delta E = E_1 - E_2 = hv$$

where h is a constant called Planck's constant.

Thus it would appear that atoms can only lose their energy in fixed quantities or "packets". The loss of energy is attributed to one or more electrons in the atom dropping from a state of high energy E_1 to a lower energy level E_2 (Fig. 3.3).

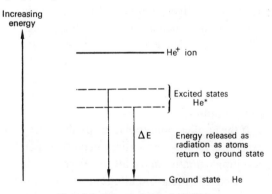

FIG. 3.3. Energy levels in helium.

The values of the energy levels in an atom can be determined by measuring the frequencies of light emitted when the "excited" atom returns to a lower energy state, for instance, the minimum energy state known as the **ground state**. The measurement of frequencies is done by splitting up the emitted light into a spectrum. This can be done either with a prism or a diffraction grating, and the result is known as an **emission spectrum**. The hydrogen atom possesses a relatively simple emission spectrum, due to the fact that it only has one electron. The spectra of some of the heavier elements are far more complex.

The emission spectrum of an element is a form of fingerprint which can be used for identification purposes, in the procedure known as **spectroscopic analysis**. It is interesting to note that this is not a new technique—it was invented over a hundred years ago by Bunsen—and rubidium (meaning red) and caesium (meaning blue) were discovered in this way.

(b) *Excitation potentials*. If an electron of low energy collides with an atom it will undergo an elastic collision. That is to say it will simply bounce off, and due to its small mass it will not transmit any energy to the atom. If the energy of the electron is increased by applying a greater potential, there comes a critical point, known as **the excitation potential** where the collision with atoms becomes inelastic: the atom absorbs some of the electron's energy and is said to be in an **excited state**. The excitation potential is characteristic of the atom concerned, indicating that the atom can only absorb energy in this fixed amount.

If the applied potential is gradually increased, several excitation potentials may be observed, resulting finally in another electron being knocked off the atom altogether (which is the process of ionization described in Chapter 1 for the mass spectrometer) (Fig. 3.4).

The excitation potentials of helium or argon can be investigated in the laboratory with a suitable circuit.

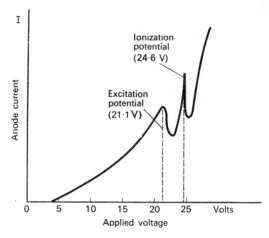

FIG. 3.4. Ionization of helium.

To summarize:

(i) Electrons in atoms and molecules can only exist with certain fixed energy values: they are said to occupy energy levels.

(ii) Electrons can only jump from one level to another by emission or absorption of the necessary amount of energy. This energy, ΔE, is fixed by the value of the energy levels, i.e. energy is emitted in discrete packets or **quanta**.

(iii) $\Delta E = h\nu$.

3.3 Measurement of ionization energy

Where the element can be converted directly into monatomic gas and introduced into a vacuum tube, its ionization energy can be determined in two main ways:

(a) The gas is irradiated with ultra-violet radiation of gradually increasing frequency until the energy, $h\nu$ is sufficient to ionize it and make it conduct electricity.

(b) The gas is bombarded with electrons accelerated through a potential V which is gradually increased until ionization occurs and the gas itself conducts. The ionization energy is thus $96 \cdot 5V$ kJ g-atom^{-1}.

These methods are only applicable when the gas is *monatomic*, (consists of single atoms). It is applicable for instance to the noble gases, and to alkali metal vapours, but not to a gas like oxygen which consists of *diatomic* molecules, $O_2(g)$. In other cases, ionization energies can be calculated from the emission spectra of the elements.

3.4 Magnetic properties of atoms

It is found that all substances can be broadly classified into two types on the basis of their magnetic properties:

(a) **diamagnetic**, i.e: weakly repelled by a magnet; and

(b) **paramagnetic**, i.e: weakly attracted by a magnet.

Substances such as iron and cobalt, which are strongly attracted, are called *ferromagnetic* but this property is limited to relatively few substances.

It is found that all diamagnetic substances contain an even number of electrons in their molecules, and this is the most common class. When a substance is paramagnetic this indicates at least one *unpaired* electron. Paramagnetic substances are much less common, and the property generally results from the presence of an odd number of electrons in the molecule.

Non-ferromagnetic substances require a very powerful magnetic field to produce a measurable force. The number of unpaired electrons in a substance can be measured by placing a sample in a **magnetic balance** (Gouy balance, Fig. 3.5) and measuring the force exerted by a given field.

The association between number of electrons and magnetic behaviour is explained by postulating that electrons possess the property of **spin**. A single electron may be imagined as a spinning charge which behaves as a tiny magnet.*

When two electrons come together they sometimes pair off in such a way that their spins cancel each other out. A molecule made up entirely of paired electrons will have no overall spin, and will therefore be diamagnetic.

3.5 The first row elements

The fact that helium has a higher first ionization energy than hydrogen is easily explained by the higher nuclear charge which helium possesses. The low ionization energy of lithium is explained by postulating that the removed electron is in a higher energy level (Fig. 3.6).

If this postulate is correct, we should expect the second ionization energy of lithium to be rather high. The ion $Li^+(g)$ is **isoelectronic** with (has the same number of electrons as) the helium atom $He(g)$. We should expect the

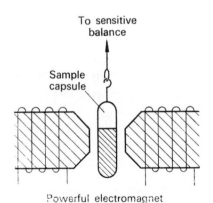

To sensitive balance

Sample capsule

Powerful electromagnet

FIG. 3.5. The Gouy balance.

---

* We have previously likened an electron to a charge-cloud (Chapter 1). This cloud must not be thought of as spinning in the normal sense. The term "spin" is a convenient one to apply to a charge which also has magnetic properties, but it should not be taken too literally.

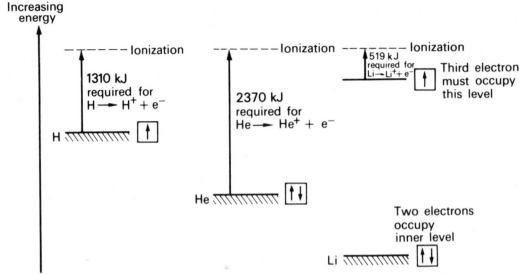

FIG. 3.6.

ionization energy of Li^+ to be greater than that of He on account of the positive charge.

$$Li^+ \rightarrow Li^{2+} + e^-;$$
$$\Delta E = 7300 \text{ kJ g-atom}^{-1}$$
$$\text{or} \quad 75{\cdot}6 \text{ electron-volts per atom.}$$

The periodic variation of first ionization energy with increasing atomic number is explained in terms of two basic postulates:

(a) All the electrons in an atom occupy energy levels: in ionization the electrons which are highest in energy are removed from the atom first.

(b) There is a limit to the number of electrons a given energy level can contain. For instance, the lowest energy level in Fig. 3.6 can only hold two electrons, with their spins paired as indicated by the arrows ⇅.

In the elements as far as neon there are three energy levels:

(a) The lowest energy level, labelled $1s$, which can hold two electrons.

(b) The next energy level, labelled $2s$, which can also hold two electrons only.

(c) Above the $2s$, but not greatly exceeding it in energy, we have the $2p$, which can hold six electrons, that is, three electron pairs. It is convenient to imagine that the electrons can be placed into "boxes" known as **orbitals** each able to hold *two* electrons with paired spins.*

An energy level within an atom is designated by:

(i) **Principal quantum number.** $2s$ and $2p$ both have a principal quantum number of 2. An atom can never have more than eight electrons with a principal quantum number of two.

(ii) **The letters s, p,** etc. This letter shows how many orbitals are available. For "s" electrons there is only one orbital of a given principal quantum number. For "p" electrons there are three orbitals. We shall later meet "d" electrons, where there are five orbitals, and "f" electrons where there are seven.

The energy levels in the hydrogen atom itself are very simple because *all* the states of a given quantum number have the same energy,

---

* The rule which states that not more than two electrons can be placed in the same orbital is known as the Pauli exclusion principle.

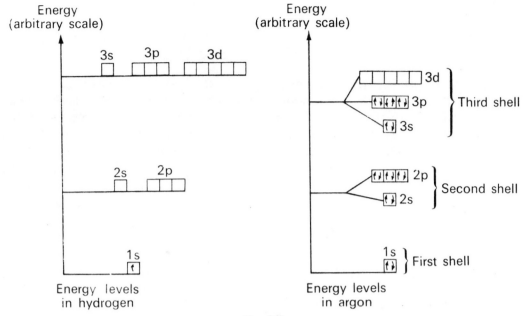

Fig. 3.7.

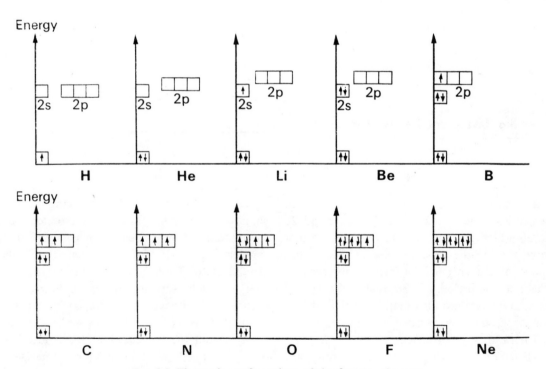

Fig. 3.8. Electronic configurations of the first ten elements.

no matter whether they are *s*, *p*, or *d*. This is only true for hydrogen however, and the levels become more complicated in larger atoms (Fig. 3.7).

The filling up of orbitals with electrons is governed primarily by the fact that electrons repel one another. This is well illustrated by the way in which the electrons are built up in the atoms as far as neon, which is summarized in Fig. 3.8.

We are now in a position to examine in detail the variations in first ionization energy which take place among the first row elements. Fig. 3.2 should be compared with Fig. 3.8.

(a) There is a general rise in ionization energy from lithium to neon, attributable to the change in nuclear charge.

(b) The dip observed in the plot at boron is due to the fact that the ionized electron in boron is lost from a higher energy level.

(c) The dip in the plot at oxygen can be explained by assuming that the electron lost in ionization is one of the paired ones. Since electrons repel one another, one of the paired electrons is lost more readily.

3.6 The second row elements

Fig. 3.2 shows that the second row elements show a very similar trend in ionization energies to the first row. This can be attributed to the existence of further energy levels, 3*s* and 3*p*, which are exactly analogous to the 2*s* and 2*p* levels which govern the behaviour of the first row. One of the consequences of this is that many of the chemical properties observed in the first row period are repeated in the second row. But there are also many chemical differences, some of which can be attributed to the existence of a group of energy levels, 3*d* (Fig. 3.7). Other factors such as size of atoms are also important.

3.7 Ways of representing electronic configurations of atoms

The **electronic configuration** (or electronic structure) of an atom or ion is the arrangement of its electrons within the energy levels. A shorthand notation is used in order to avoid drawing an energy level diagram for every atom. The number of electrons in a given energy level is shown by a small superscript numeral. For instance, five 2*p* electrons are denoted by the symbol $2p^5$. The ground state electronic configurations of the elements as far as argon are represented in Table 3.1.

A group of energy levels with the same principal quantum number is frequently referred to as a **shell**. For instance the electrons in an aluminium atom occupy three shells. The innermost shell, of principal quantum number 1, holds two electrons. The next shell, of principal quantum number 2, holds eight electrons. These two shells are **completed shells**. The shell of principal quantum number 3 is incomplete since it only contains three electrons. If we include the 3*d* orbitals, shell number 3 can hold $(2+6+10) = 18$ electrons.

3.8 Arrangement of electrons in space

We have so far avoided referring to the shapes of the 'orbits' followed by the electrons surrounding atomic nuclei. Consider first the simplest case—that of the hydrogen atom in its ground state. There is a tiny nucleus with a charge of $+1$ "surrounded" by an electron of charge -1 and negligible mass. Bohr, in 1913, postulated that the electron followed a circular orbit around the nucleus, but this model has been abandoned since it did not give a consistent interpretation of the observed properties of

TABLE 3.1

Element	Symbol	Complete electronic configuration			Abbreviated form		
Hydrogen	H	$1s^1$			1		
Helium	He	$1s^2$			2		
Lithium	Li	$1s^2$;	$2s^1$		2,	1	
Beryllium	Be	$1s^2$;	$2s^2$		2,	2	
Boron	B	$1s^2$;	$2s^2$,	$2p^1$	2,	3	
Carbon	C	$1s^2$;	$2s^2$,	$2p^2$	2,	4	
Nitrogen	N	$1s^2$;	$2s^2$,	$2p^3$	2,	5	
Oxygen	O	$1s^2$;	$2s^2$,	$2p^4$	2,	6	
Fluorine	F	$1s^2$;	$2s^2$,	$2p^5$	2,	7	
Neon	Ne	$1s^2$;	$2s^2$,	$2p^6$	2,	8	
Sodium	Na	$1s^2$; $2s^2$, $2p^6$	$3s^1$		2,	8,	1
Magnesium	Mg		$3s^2$		2,	8,	2
Aluminium	Al		$3s^2$,	$3p^1$	2,	8,	3
Silicon	Si		$3s^2$,	$3p^2$	2,	8,	4
Phosphorus	P	(neon "core")	$3s^2$,	$3p^3$	2,	8,	5
Sulphur	S		$3s^2$,	$3p^4$	2,	8,	6
Chlorine	Cl		$3s^2$,	$3p^5$	2,	8,	7
Argon	Ar		$3s^2$,	$3p^6$	2,	8,	8

hydrogen. We now think of the hydrogen electron as a **charge cloud.** Instead of visualizing it as a point charge we imagine it as a "smeared-out" charge. Its distribution around the nucleus is spherically symmetrical but the greatest concentration of charge is nearest the nucleus. Figure 1.2(c) is an attempt to represent the charge cloud picture of such an electron. Such a charge-cloud shows the shape of an orbital. Figure 3.9 shows charge cloud distributions for helium, neon and argon.

It is misleading to think of an electron purely as a "particle" of charge—a charge-cloud is a much better description. In fact the rules which appear to govern the shapes of electron charge-clouds are found to be very similar to those which govern wave motions. A beam of electrons is found to be remarkably similar to a beam of light—it has a wavelength and can be made to undergo interference and diffraction just like a light beam.

Despite the fact that electrons seem to have a split personality, they are still governed by the ordinary laws of electrostatics. Electrons repel one another and are attracted by positive charges such as nuclei.

3.9 Bonds between atoms

Although it is important to understand the behaviour of single atoms, chemistry is mainly concerned with the forces which hold atoms together. Single atoms are rather unusual—the

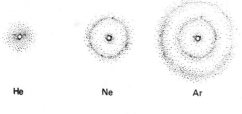

He　　　　Ne　　　　Ar

FIG. 3.9.

only common examples at room temperature are the noble gases (Group 0).

Atoms generally have a tendency to join together as a consequence of the electrostatic forces between electrons and protons, but the noble gases have completely filled energy levels and their atoms have no tendency to join to one another.

Many common gaseous elements form diatomic molecules, for instance $H_2(g)$, $O_2(g)$, $N_2(g)$, $Cl_2(g)$ and $F_2(g)$. Considerable energy has to be put into these substances to disrupt their molecules into atoms, and we therefore say that a **bond** exists between the atoms. This bond is often called a **covalent** bond. The properties of a bond are well illustrated by the simplest molecule of all, that which exists in hydrogen gas.

In hydrogen gas the distance between adjacent nuclei is always found to be the same, 0·074 nm. This distance is called the **bond length.** It is found that 435 kilojoules of energy are required to split up, or **dissociate**, one gram-molecule* of hydrogen gas into atoms. In equation form:

$$H_2(g) \longrightarrow 2H(g)$$

| one gram-molecule of hydrogen molecules; that is, 2·016 g | two gram-atoms of hydrogen atoms; that is, 2·016 g |

$$\Delta E = +435 \text{ kJ}$$

(positive sign indicates energy put into system)

Two hydrogen atoms will combine to form a molecule by taking up an arrangement in space where their electrons form a pair with opposed spins between the nuclei. A simple

---

* The term gram-molecule is used here in the same way as gram-atom in Chapter 1. A gram-molecule of H_2 hydrogen is N_A molecules of H_2, and has a mass of 2·016 g. This concept will be developed further in Chapter 7.

electrostatic argument shows that the attractions will outweigh the repulsions, assuming the electrons to be point charges (Fig. 3.10).

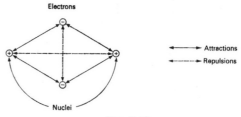

FIG. 3.10.

Taking into account the charge-cloud nature of electrons, they are imagined to form a cloud which has its greatest density on the axis between the nuclei (Fig. 3.11). In hydrogen atoms, the electrons occupied $1s$ **atomic orbitals.** In the hydrogen molecule they have combined together by pairing their spins, to occupy an energy level in the molecule, known as a **molecular orbital.** Hydrogen molecules are diamagnetic (their electron spins cancel out) indicating that the molecular orbital contains a pair of electrons.

The fact that molecules possess energy levels just like atoms explains why helium cannot form a molecule He₂. There are now four electrons to be placed in energy boxes, and

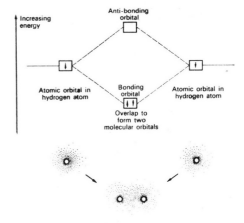

FIG. 3.11.

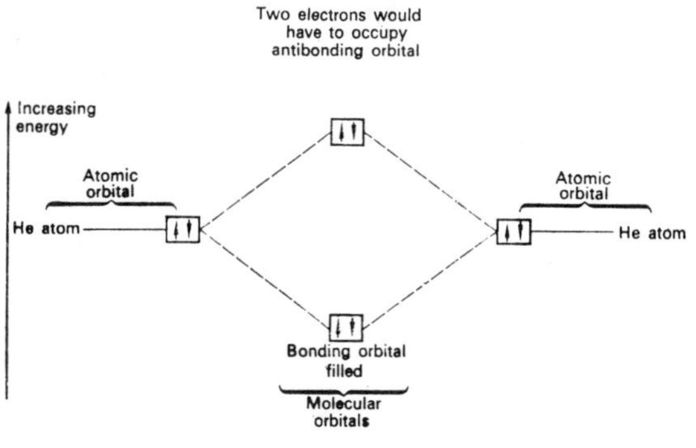

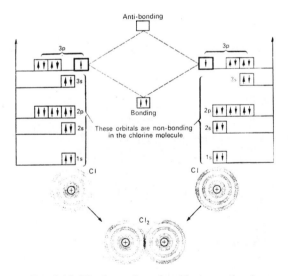

Two electrons would
have to occupy
antibonding orbital

FIG. 3.12.

this would mean placing two of them in a box of higher energy and this would involve a net loss in energy (Fig. 3.12). Since helium atoms cannot attain lower energy by combining they remain separate.

A very similar argument explains why the elements of Group VIIB, the halogens, form diatomic molecules. The single atoms all have one "vacancy" in their uppermost p energy level. Two atoms combine by electron-sharing and a bond is formed in a similar way to hydrogen (Fig. 3.13).

FIG. 3.13. The formation of a chlorine molecule.

3.10 Shorthand ways of representing bonds

It is important always to think of a bond as resulting from the electrostatic forces between electrons and nuclei, and the energy gained by placing electrons in energy levels. Nevertheless, it is rather cumbersome to use energy level diagrams for pictorial representations. A simple notation in which electrons are shown by dots is often helpful. For example:

$$H \cdot + \cdot H \longrightarrow H : H \quad \text{or} \quad H—H$$

This is just a convenient way of counting electrons and energy levels—it is *not* a picture of what a molecule is supposed to be like. It ignores the fact that electrons are clouds of charge, not point charges, and also the fact that molecules are three-dimensional, and not flat like the paper they have to be drawn on.

In these diagrams, only the electrons in the uppermost, partially-filled energy shell are shown. This is known as the **valence shell.** For instance, in chlorine the inner electrons are hardly affected by the formation of a bond as they are deeply buried in the atom (Fig. 3.13).

Where we are interested in the overall structure and shape of a molecule, rather than its energy levels, an even simpler notation is used.

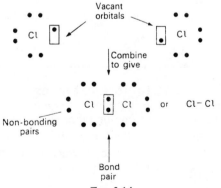

FIG. 3.14.

Here the electron-pairs which form bonds are each represented by a line, and non-bonding electron pairs are omitted (Fig. 3.14).

The bond between two nitrogen atoms in $N_2(g)$ is interesting; nitrogen is an extremely unreactive substance because considerable energy is required to separate the atoms. The dissociation energy of nitrogen into atoms is very high, 945 kJ per gram-molecule. Nitrogen atoms have three unpaired electrons, and it is thought that these all pair up when two atoms come together.

$$:N: + :N: \rightarrow :N::N: \; (N \equiv N)$$

unpaired   three bond
electrons   pairs

We can imagine nitrogen atoms to be "welded" together by three electron-pair bonds. The bond length is short, only 0·1097 nm, which suggests a strong attraction between the electrons and the nuclei.

The oxygen molecule, $O_2(g)$, raises some fascinating problems. Its dissociation energy, 498 kJ per gram-molecule, and its bond length, 0·121 nm, are not exceptional, but oxygen is found to be paramagnetic despite the fact that it contains an even number of electrons! A test-tube containing liquid oxygen is readily attracted to a powerful magnet, indicating unpaired electrons. We have seen that this is what happens in atoms, so it ought not to be surprising that it happens in molecules as well. Paramagnetic molecules are, however, rather unusual.

3.11 Ionic substances

The removal of one or more electrons from an atom results in the formation of a positive ion. Negative ions can be formed in a similar fashion by the addition of electrons to an atom. A fluorine atom will quite readily form a negative ion by filling its $2p$ 'vacancy' with a paired electron.

$$[:\ddot{F}\cdot] + e^- \rightarrow [:\ddot{F}:]^- \;;$$

atom   negative
(gaseous)   ion

$$\Delta E = -350 \text{ kJ}$$

The energy released is called the **electron affinity** of fluorine. (The electron affinity is numerically equal to the ionization energy of the ion $F^-(g)$, i.e. in the above example the E.A. of fluorine is $+350$ kJ.) Compare this energy with the ionization energy of an ionizable metal like potassium:

$$K\cdot(g) \rightarrow K^+(g) + e^- \;;$$

gaseous   gaseous
atom   ion

$$\Delta E = +418 \text{ kJ g-atom}^{-1}$$

Comparing these two equations it might be expected that if a potassium and a fluorine atom are placed together, electron transfer would occur.

$$K\cdot + :\ddot{F}\cdot \rightarrow K^+ + F^- \;;$$

$$\Delta E = -350 + 418$$
$$= +68 \text{ kJ}$$

Energy is released when a pair of opposite charges are brought together, making the overall transfer of an electron from potassium to fluorine, to form an ion-pairs, an energy-releasing process:

$$K^+ + F^- \rightarrow K^+F^- \;;$$

ion-pair

$$\Delta E = -585 \text{ kJ}$$

$$\therefore K\cdot(g) + F\cdot(g) \rightarrow K^+F^- \text{ (ion-pair)};$$

$$\Delta E = -517 \text{ kJ}$$

The compound potassium fluoride has properties which suggest that it is made up of ions. If it is heated until molten it may be **electro-**

lysed: an electric current can be passed through the melt by the movement of ions. Positive potassium ions move to the negative electrode where they become metallic potassium, and negative fluoride ions move to the anode where they lose electrons and form atoms. The atoms immediately combine to form molecules of fluorine gas, F_2.

Study Questions

1. Suggest why the ionization energies of the alkali metals (Group IA) decrease with increasing atomic number.

2. The first ionization energy of beryllium is greater than that of lithium, but the second ionization energy of beryllium is very much less than the second ionization energy of lithium. Explain these facts.

3. Plot the atomic number of the elements as far as argon against their second ionization energies. Compare your plot with Fig. 3.1.

4. $O(g)+e^- \rightarrow O^-(g)$; $\Delta E = -214$ kJ
$O^-(g)+e^- \rightarrow O^{2-}(g)$; $\Delta E = +920$ kJ.

Account for the difference in these two values.

5. What are the ground state electronic configurations of the following?

(a) a neon atom; (b) a lithium atom;
(c) a lithium cation, $Li^+(g)$;
(d) a fluorine atom; (e) a fluoride anion, $F^-(g)$.

6. To which elements do the following electronic configurations apply?

(a) $1s^2$; $2s^2$. (b) $1s^2$; $2s^2p^6$; $3s^2p^5$.
(c) $1s^2$; $2s^2p^6$; $3s^2p^6$; $4s^1$. (d) $1s^2$; $2s^2p^6$; $3s^2p^6d^8$; $4s^2$.

7. Which of the following would you expect to be paramagnetic? Ne, NO, NO_2, N_2O_4, ClO_2, ClO_3, and Cl_2O_6.

8. (a) What have O^{2-}, F^-, Na^+, Mg^{2+}, and Al^{3+} in common?
(b) How would the sizes of these ions compare?

9. In a transition series, the radii of ions with the same charge decrease as the atomic number increases. Suggest an explanation for this behaviour.

10. (a) The first excitation potential of sodium is 2·1 V. Use the expression $\Delta E = h\nu$ to calculate the wavelength of radiation to which this corresponds. In what part of the spectrum does this radiation occur?

(b) The first excitation potential of argon is 13·0 V. What is the frequency and wavelength of the radiation to which this corresponds? (1 electron-volt $= 1·6 \times 10^{-19}$J; $h = 6·62 \times 10^{-34}$ J s; $c = 3 \times 10^8$ m s^{-1})

11. Will a neon light obey Ohm's law? Why do you suppose that a neon light is red?

12. (a) Convert the following to kcal mole^{-1}:
(i) 1 eV; (ii) 100 J mole^{-1}.
(b) Convert the following to eV:
(i) 1 kcal mole^{-1}; (ii) 1 J mole^{-1}.
(c) Convert the following to kJ mole^{-1}:
(i) 1 eV; (ii) 100 kcal mole^{-1}.

CHAPTER 4

Methods of determining structure

4.1 Physical properties and structure

In Chapter 1 some of the aspects of atomic structure were considered, and Chapter 3 dealt with some of the forces which hold atoms together. This chapter is concerned with the different types of structure which matter can have as a result of the bonds between atoms, and the influence which structure has on physical properties.

In a solid the atoms are packed together in a rigid form; generally some regularity of pattern is observable in the packing, and the atoms are said to form a **lattice.** When a solid melts, this rigidity disappears and a slight increase in volume generally takes place. The liquid state is difficult to study experimentally since its structure is so complex, but it is known that the regular lattice structure of the solid breaks up due to vibration of the atoms. The atoms are still close together, though **translation** (that is, movement of particles from one place to another), and rotation, can take place freely. With sufficient energy, some of the molecules of a liquid can escape altogether from the surface—this is **evaporation.** In a gas or vapour the molecules are considerably further apart than in a liquid or solid. Attraction between molecules is relatively slight, and can often be ignored altogether. Gases are compressible as a result of the large amount of space between the molecules, whereas liquids and solids are only very slightly compressible—they are referred to as **condensed phases.**

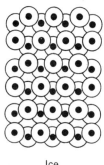

Ice
(Regular lattice)

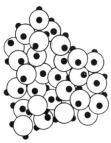

Water
(some disorder)

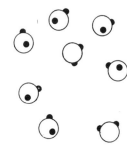

Steam
(Extreme disorder)

Fig. 4.1.

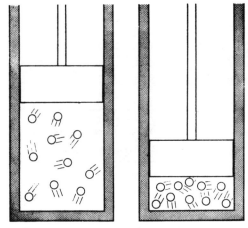

FIG. 4.2.

4.2 Methods of determining the structure of solids

The chief way of determining the positions of atoms in a solid lattice is by **X-ray-analysis**, first devised by von Laue in 1912. X-rays resemble light insofar as they are an electro-magnetic radiation, but their wavelength is much shorter. A typical wavelength for visible light would be 589 nm (the wavelength of the yellow light emitted by sodium vapour in a flame), whereas X-rays have a wavelength in the region of 0·1 nm. Individual atoms are too small to diffract visible light, but if a beam of X-rays of single wavelength, that is a **monochromatic** beam,[*] is directed at a regular crystal, the crystal will act as a diffraction grating. For a given angle of incidence, X-rays will be reflected most strongly in certain preferred directions dependingupon the distance between successive layers of the atoms in the crystal. If the angle of incidence is varied, an analysis of the *diffraction pattern* produced by the crystal will enable all the interatomic distances to be computed.

If the central crystal is rotated slowly, strong "reflections" will be obtained at certain angles. Often a finely-ground powder will produce the same effect. Here the result would be that of superimposing a large number of pictures taken whilst slowly rotating a single crystal, since the powder crystals will all be oriented in different directions. The angle of reflection of X-rays does not *equal* the angle of incidence, except occasionally by chance. A simple arrangement for the examination of X-ray diffraction patterns is shown schematically in Fig. 4.3.

The analysis of X-ray patterns can be a very laborious mathematical chore, but now this can be overcome by the use of the computer.

Figure 4.1 shows the structure that is thought to exist in ice, water and steam. Water is rather an unusual substance as it expands on solidification—a factor which causes pipes to burst and roads to be damaged in frosty weather. Figure 4.2 illustrates the fact that a gas is compressible: the pressure of a gas is due to the bombardment of the container walls by molecules, and the more highly compressed the gas is, the greater will be the number of molecules striking a given area and the greater the pressure.

Two kinds of forces have to be considered in order to understand the structure of materials:

(a) Relatively large forces hold atoms together to form molecules: these we have already referred to as bonds. The electrostatic forces holding ions together in a lattice are of similar magnitude, and so are the forces which hold atoms together in a metal.

(b) Relatively weak forces act between separate molecules, especially where these molecules are small as in gases like carbon dioxide, CO_2, and methane CH_4. Weak forces are even observed in the noble gases where the "molecules" are in fact single atoms.

The precise nature of the forces in (a) and (b) will be examined in more detail later.

---

[*] Of one "colour" only, by analogy with visible light.

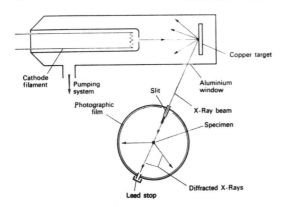

FIG. 4.3. X-ray analysis.

Some spectacular achievements have resulted from the use of computers to work through the calculations. For instance the Nobel Prize for Chemistry was awarded to Professor Dorothy Hodgkin in 1965, for the work she and her research team did in determining the complete structure of Vitamin B_{12}, a highly complex substance of molecular formula $C_{63}H_{90}O_{14}N_{14}PCo$.

A detailed study of how X-ray diffraction patterns are obtained and analysed is beyond the scope of this book, being part of the science of crystallography.* In theory it can tell us all we wish to know about the positions of .*nuclei* in a lattice, but in practice the method has its limitations. In particular hydrogen atoms are difficult to locate in the presence of other, larger nuclei on account of their small nuclear charge. X-ray analysis does not normally give any information about the location of electrons: it tells us about structure but *not* about bonding.

A supplementary technique to X-ray diffraction is **neutron diffraction**. A beam of neutrons is found to behave as if it possessed a definite wavelength, and to give diffraction patterns. (Electrons also have properties which are wave-

---

* For a more detailed description see J. E. Spice, *Chemical Binding and Structure*, chapter 11. Pergamon Press, 1964.

like, and this wave–particle duality is common to all the particles from which matter is built). Neutron diffraction can often enable hydrogen atoms to be located in cases where X-ray analysis fails.

4.3 Methods of determining the shapes of molecules

(a) CLASSICAL METHODS

The problem of determining the shape of a molecule has to be tackled on several fronts. The "classical" methods of determining the structures of molecules do not give details of bond lengths and bond angles, although a general idea of molecular shape can often be obtained. The classical approach includes the following:

(i) The molecular weight is determined, either by gas density measurements (Chapter 7) or from measurement of colligative properties (Chapter 15), and the composition by weight is found. These two pieces of information lead to the *molecular formula* of the compound.

(ii) The way in which the atoms are joined together to form molecules is inferred from

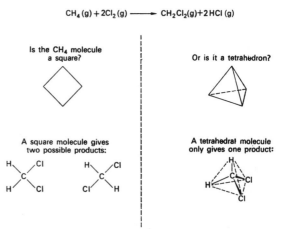

FIG. 4.4.

chemical reactions. We assume the structure of the molecule to be the one which explains its reactions and other properties most consistently. In many cases there is not enough information to decide between different alternatives by classical methods alone, but in other cases the classical arguments have proved to be triumphs of deductive reasoning. Figure 4.4 illustrates how chemists were able to deduce that the carbon atom in methane is tetrahedral rather than planar. A planar formula would have led to two *different* molecules both with the formula CH_2Cl_2 when methane was reacted with chlorine, but only one substance of this formula has ever been prepared. Similarly Kekulé was able to deduce the structure of benzene to be cyclic in 1865, long before the use of physical methods for structure determination.

(b) DIPOLE MOMENTS

In cases where the molecule is simple, or where other features of its structure are already known, determination of its **dipole moment** may enable its exact shape to be decided upon.

Any bond between two unequal atoms possesses an *electrical dipole*, due to the fact that the electron pair is attracted more strongly to one nucleus than to the other. The atom which attracts the electrons more strongly is said to be more **electronegative,** and it will acquire a small excess negative charge $\delta-$ and leave behind a small excess positive charge on the other atom (denoted by $\delta+$). For instance, the hydrogen chloride molecule, HCl, has a dipole moment because chlorine is more electronegative than hydrogen.

$$H—Cl$$

$$\delta+ \quad \delta-$$

$$+ \rightarrow -$$

Direction of dipole moment

The more electronegative elements are those on the right-hand side, and near the top, of the periodic table, the most electronegative of all being fluorine. It is often possible to predict which direction a dipole moment is going to act along a given bond. A molecule composed of two identical atoms, such as H_2, Cl_2, or F_2, will have no dipole moment.

With *triatomic* (three-atom) molecules the measurement of dipole moment often enables us to distinguish between alternative shapes. For instance a consideration of the properties of hydrogen and oxygen atoms, and the reactions of water, does not tell us what *shape* the water molecule is, apart from saying that the oxygen atom is between two hydrogen atoms. There are two possibilities. The molecule might either be **linear** or **bent.**

If H_2O is a linear molecule we shall expect it to have no dipole moment, since the electrical dipoles will cancel one another out. If on the other hand the molecule is bent we shall expect a dipole moment in the direction illustrated:

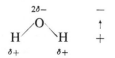

Direction of dipole moment

$$\delta- \quad 2\delta+ \quad \delta-$$
$$O=C=O$$

No resultant dipole moment

In fact measurements on water show it to have quite a large dipole moment, therefore it must be bent. Unfortunately the measurement cannot tell us *how* bent.

Again consider carbon dioxide, CO_2. Measurements show this to have zero dipole moment. The molecule must therefore be *linear* with the carbon atom in the middle—any other arrangement of atoms would produce a dipole moment.

Provided the substance is a liquid, a very simple laboratory experiment will determine

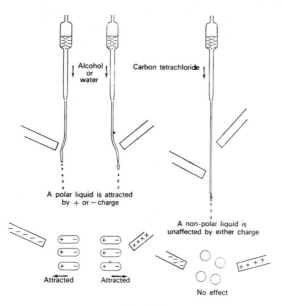

Alcohol or water

Carbon tetrachloride

A polar liquid is attracted by + or − charge

A non-polar liquid is unaffected by either charge

Attracted   Attracted   No effect

FIG. 4.5.

whether or not it has a dipole moment. Allow the liquid to flow as a fine jet from a pipette: **non-polar** substances (substances whose molecules have no dipole moment) are unaffected by bringing a charged rod near the jet, whereas **polar** substances (those with a dipole moment) will be attracted to the rod (Fig. 4.5). It does not matter whether the rod is positive or negative—a dipole is attracted to either charge. The substance under test should be absolutely dry. Since water itself has a high dipole moment, it will affect the properties of other substances it contaminates.

This simple laboratory test cannot be used for quantitative measurements. These are done by finding how the **dielectric constant** of the substance varies with temperature. Dielectric constant is the constant k in the expression for the force of attraction or repulsion between two charges Q_1 and Q_2 a distance r apart, when placed in the given medium:

$$\text{Force} = \frac{Q_1 Q_2}{kr^2}$$

Direct electrostatic measurements are not practicable for measuring k; an electronic method is used instead.[*]

[†] (c) INFRA-RED SPECTROSCOPY

A molecule which can vibrate in such a way that its dipole moment alters while vibrating, can absorb electromagnetic radiation. The energy absorbed corresponds to radiation in the infra-red region of the electromagnetic spectrum ($\Delta E = h\nu$, Chapter 3). Simple molecules have infra-red absorption spectra which depend upon their shapes. The carbon dioxide molecule illustrates this. Although the molecule itself has no dipole moment, being linear, a dipole moment will appear if the molecule (a) bends as in Fig. 4.6a, or (b) stretches unsymmetrically as in Fig. 4.6b. Each of these modes of vibration will absorb infra-red energy of characteristic frequency. Note that the third possible mode of vibration (c), that of symmetric stretching of the bonds, cannot produce a dipole moment and will therefore not lead to absorption of infra-red radiation.

The water molecule is bent, and three possible modes of vibration can be expected to lead to the absorption of infra-red radiation (Fig. 4.7), since all modes lead to a variation in the dipole moment.

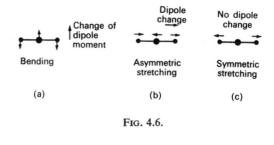

Change of dipole moment

Dipole change

No dipole change

Bending

Asymmetric stretching

Symmetric stretching

(a)   (b)   (c)

FIG. 4.6.

---

[*] For a detailed description of dipole moment measurements, see Mansel Davies, *Some Electrical and Optical Aspects of Molecular Behaviour*, Pergamon Press, 1965.

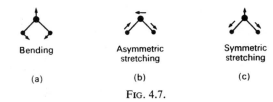

Bending

(a)

Asymmetric
stretching

(b)

Symmetric
stretching

(c)

Fig. 4.7.

We have chosen only simple cases, but analysis of the infra-red spectrum can be applied to polyatomic molecules, and information about their structure inferred. Other spectroscopic methods can be used in conjunction with it (such as the Raman spectrum, which will not be dealt with in this book).

In large polyatomic molecules it is found that each bond in the molecule has what approximates to its own characteristic "stretching frequency" and each bond vibrates more or less independently of the others in the molecule. For this reason measurements of infra-red absorption spectra are very important in the analysis of organic substances. The presence of absorption bands of particular frequencies gives information about which bonds are likely to be present in the molecule. Measurements of this sort can be applied in two ways:

(i) If the structure of the molecules is unknown, it can be deduced by comparing its absorption spectrum with those of substances with known structures.

(ii) If the structure of the molecule is known, its characteristic absorption frequencies can be used to identify it in an unknown mixture, and to measure its concentration.

† (d) OTHER TECHNIQUES

Various other methods are available for the determination of the structure of molecules. X-ray analysis, mentioned above for the determination of macromolecular structures, is

suitable also for determining the shape and mode of packing of molecules in molecular solids. The information can, however, be misleading if a substance undergoes a change of structure when it is melted or vaporized. For instance phosphorus(V) chloride (phosphorus pentachloride) has a vapour density corresponding to molecules of formula PCl_5, yet the solid is shown by X-ray analysis to be composed of separate structural units corresponding to the formulae $[PCl_4]^+$ and $[PCl_6]^-$. The structure of some molecules in the gaseous state can be determined by **electron diffraction:** a vapour will not diffract X-rays but it will diffract a beam of electrons, which behaves as if it has wave properties.

Information about the environment of atoms, particularly hydrogen atoms, can be derived from the **nuclear magnetic resonance** spectrum of the compound. An atomic nucleus, like an electron, possesses the property of spin and can therefore behave as a tiny magnet, orienting itself in a magnetic field. The amount of energy absorbed when the orientation of a nucleus changes in a magnetic field is affected by the electrons surrounding the nucleus. For instance in a molecule of ethanol, $CH_3.CH_2.OH$ the pair of hydrogen nuclei on the $-CH_2-$ group will absorb at a different frequency from the three on the CH_3- group, on account of their different environments; the hydrogen atom in the $-OH$ group will absorb at yet another frequency (Fig. 4.8).

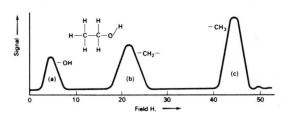

Fig. 4.8. Nuclear magnetic resonance.

Study Questions

1. What can you deduce from the following?

(a) CH_4 has no dipole moment, but CH_2Cl_2 has.

(b) BF_3 has no dipole moment, but NF_3 has.

(c) CO_2 has no dipole moment, but SO_2 has.

(d) PF_5 has no dipole moment, but IF_5 has.

2. How would *you* distinguish between the two compounds:

$$\underset{H}{\overset{Cl}{\diagdown}}C=C\underset{H}{\overset{Cl}{\diagup}} \quad \text{and} \quad \underset{H}{\overset{Cl}{\diagdown}}C=C\underset{Cl}{\overset{H}{\diagup}} \ ?$$

3. A substance of empirical formula $C_1H_{10}O$ had a molecular weight of 74. I.R. spectroscopy showed that none of the hydrogen atoms was attached to an oxygen atom, and N.M.R. spectroscopy showed two types of hydrogen atom, these being in the ratio of $3:2$. Draw the structure of the substance.

CHAPTER 5

Structure of elements

5.1 The structure of non-metals

The elements in the top right-hand region of the periodic table are non-metals. With the exception of graphite, they are poor conductors of electricity. We can conclude from this that the electrons are not mobile, but must be firmly held in place, and that there are no ions present in the structure. Graphite is interesting in that it does conduct electricity better than any other non-metal, albeit rather badly. Graphite shows some of the physical properties of a metal, and its structure provides some clues about the electronic structure present in metals.

Amongst the non-metals some very striking differences in physical properties are found, pointing to profound differences in structure. Carbon for instance, is almost impossible to melt, yet its neighbour in the periodic table, nitrogen, is extremely difficult to liquefy! Often a single element can show more than one structural form. Phosphorus can be obtained as a volatile "waxy" translucent solid, known as white phosphorus or yellow phosphorus, or it can exist as a hard, dark red, rather involatile solid called red phosphorus. There is also a form called black phosphorus. Yellow phosphorus is a violently reactive substance which inflames spontaneously in the air, yet red phosphorus is fairly unreactive. Two or

more distinct structural forms of the same element are called **allotropes,** and the property is called **allotropy.**

5.2 The noble gases

The noble gases represent a simple structural system, and their melting and boiling points are exceptionally low (Fig. 5.1). Helium cannot be solidified at normal atmospheric pressure no matter how much it is cooled.

The solid noble gases consist of atoms which are packed together as closely as possible. Each atom has twelve near neighbours, which is the maximum possible. It is easy to see why atoms form close-packed lattices by doing a simple experiment with a two-dimensional analogy, the *bubble raft*. If a uniformly sized stream of bubbles is allowed to collect on the surface of soap solution in a petri dish, the bubbles will be seen to adhere to one another, most of them being surrounded by six neighbours. This is close packing in two dimensions.

The number of near neighbours which an atom possesses is known as the **co-ordination number.** There are in fact *two* distinct ways in which atoms can arrange themselves in a close-packed lattice of co-ordination number 12, **cubic close-packing** (C.C.P.) and **hexagonal**

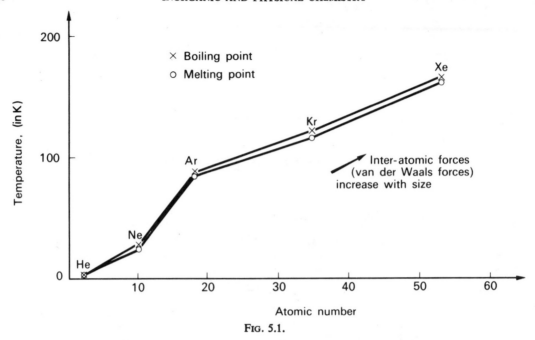

FIG. 5.1.

close-packing (H.C.P.) It is not easy to see the difference in these two forms of packing without constructing a three-dimensional model, for instance with polystyrene spheres. Figure 5.2 shows how this may be done. Join spheres together with pipe-cleaners or cement, to make the layers 1, 2, and 3 in the diagram. Now place layer 2 on top of 1. It will then be found that layer 3 can be added in two different ways, while still contacting the central atom in 2. If 3 is added so that it exactly corresponds to 1, the result will be a H.C.P. arrangement. It is easy to see that twelve spheres now touch the central sphere of layer 2. If layer 3 is rotated through 60 degrees and replaced, it will still be close-packed, but the resultant structure will have a cubic symmetry. The cubic symmetry may not be obvious at first sight but it may be brought out by constructing the C.C.P. form in a different way and comparing it. Construct three layers as in Fig. 5.3 and arrange these one above the other in order. If the resultant struc-

ture is tilted, it will be seen to correspond exactly with the form shown in Fig. 5.3.

Noble gas atoms do not join together to form molecules, on account of their already complete energy levels (Chapter 3). Relatively little energy is required to separate the solids into atoms, a typical value being that for argon:

$$Ar(c) \longrightarrow Ar(g);$$
$$\Delta H = 7 \cdot 4 \text{ kJ g-atom}^{-1}.$$

The forces holding the atoms together are known as **van der Waals forces.** These are electrical in nature, even though the atoms themselves have no overall charge. Suppose that a helium atom underwent a momentary distortion of its electron cloud, making it behave as an electrical *dipole* with positive and negative charge slightly separated. This dipole would then *induce* a separation of charge in an adjacent atom, and the two resultant dipoles would then be attracting one another (Fig. 5.4). The induced dipole can induce further

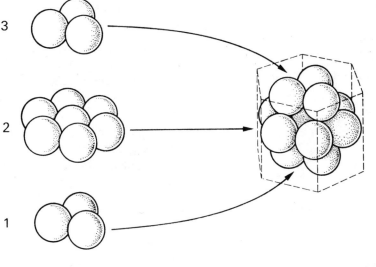

Construct three
layers like this ....

........and the resultant
close-packing has
hexagonal symmetry

FIG. 5.2. Hexagonal close-packing.

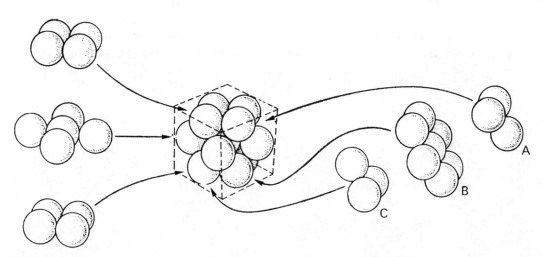

Construct two
squares and a
cross like this

.and the resultant
close packing has
cubic symmetry

the same structure can be made
by packing hexagonal layers in
the order ABC as shown

FIG. 5.3. Cubic close-packing.

dipoles in other atoms and the overall effect will be attraction between all the atoms in solid helium. These dipoles are not permanent: it would be more correct to regard them as continuously oscillating in direction.

 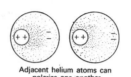

Symmetrical helium atom

Adjacent helium atoms can polarize one another

FIG. 5.4. Van der Waals force in a noble gas.

5.3 The structure of metals

Metals resemble the solid noble gases in structure in that they have high co-ordination numbers. Most metals have co-ordination numbers of twelve or eight. There the resemblance ends, for the solid noble gases are electrical insulators whereas metals conduct electricity freely. An electric current is a flow of charge, and it would appear that a metal structure contains a system of mobile electrons which can pass from atom to atom without disturbing the regular arrangement of the lattice. In addition, metals generally have the following properties:

(a) They are good conductors of heat and electricity. Electrical conductivity decreases as the temperature is raised.

(b) They are opaque, and are good reflectors of light.

(c) They are **malleable,** i.e. they can be flattened by rolling or hammering.

(d) They are **ductile,** i.e. they can be drawn into wire.

An element which has only some of these properties is not a metal; for instance iodine has a shiny "metallic" appearance but does not conduct electricity and does not have a closely packed structure; it is therefore a non-metal. To be classed as a metal, an element must have an electrical conductivity which decreases as the temperature is raised.

Aluminium is a typical example of a metal with a H.C.P. structure, while copper, silver and gold have C.C.P. structures.

The alkali metals (Li, Na, K, Rb and Cs) and some transition metals (such as Fe, Cr, Mo and W) have a more open structure with a co-ordination number of eight. The arrangement is known as **body-centred cubic** (Fig. 5.5).

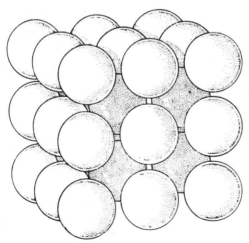 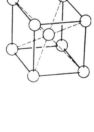

FIG. 5.5. Body-centred cubic structure.

The more open mode of packing accounts in part for the lightness of the alkali metals, though a more important factor is their high atomic radius.

Observations of a bubble raft show many other features which simulate the arrangement of atoms in metals. There are places where bubbles are out of position, which may be called dislocations. There are groups of regularly arranged bubbles, separated from other groups by lines along which the pattern fails to correspond. This situation occurs in metals, the boundaries being called **grain boundaries** (Fig. 5.6).

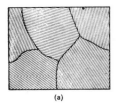 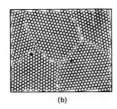

FIG. 5.6. (a) The appearance of a metal surface with grain boundaries. (b) "Grain boundaries" on a bubble raft.

In view of the fact that metals contain atoms so regularly arranged, it is surprising at first sight that they do not appear to be crystalline, but all solid metals are in fact crystalline when viewed under magnification. The crystalline appearance may have been obscured by polishing of the surface. You have probably seen zinc crystals without realizing it—the surface of a "galvanized" bucket is patterned due to molten zinc having crystallized out on cooling. Metals such as lead and silver can be made to form beautiful crystals by displacing their ions from solution, in reactions such as:

$$Zn(c) + Pb^{2+}(aq) \rightarrow Pb(c) + Zn^{2+}(aq).$$

Even sodium and potassium form crystals when the molten metal is cooled.

A close-packed metal lattice is readily rearranged, and this accounts for the malleability and ductility of metals.

A well-known characteristic of metals is the way in which their physical properties can be modified by mechanical stress or heat treatment. Metals seem to vary in the way in which they respond: some types of steel become softer on bending, while copper becomes hard and brittle when repeatedly deformed—a phenomenon known as **work hardening.** Mixtures of metal atoms—alloys—often have special properties which the pure elements themselves lack.

Work hardening occurs when mechanical deformation shifts the grain boundaries in such a way as to produce interlocking grains of great mechanical rigidity. Heat treatment also modifies the grain structure by removing many of the defects present in the lattice. A metal is often found to be less deformable when it is impure—it is thought that the impurity atoms hinder the rearrangement of the grain boundaries and dislocations.

5.4 Bonding in metals

Figure 5.7 shows the heat of atomization of some metals plotted against the number of electrons in their valence shell, and it will be seen that for a given period there is an approximately linear relationship. The heat of atomization is the energy required to convert 1 g-atom of the solid into atoms in the vapour state. This suggests that it is the valence electrons in a metal which somehow hold the atoms together—the more electrons the stronger the bond between atoms.

A metal lattice is thought of as containing positive ions held together by a mobile "electron gas". As an approximation sodium can be thought of as an array of Na^+ ions welded together by an equal number of electrons. Magnesium has a higher heat of atomization because it contains Mg^{2+} ions held together by

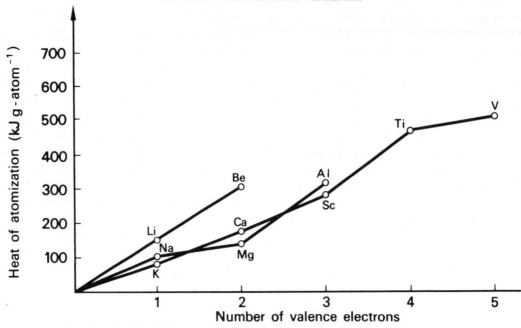

FIG. 5.7. The relation between number of valence electrons and binding energy.

TABLE 5.1

METALLIC RADII (nm)

Li 0·157	Be 0·112										
Na 0·191	Mg 0·160	Al 0·143									
K 0·235	Ca 0·197	Sc 0·164	Ti 0·147	V 0·135	Cr 0·129	Mn 0·137	Fe 0·126	Co 0·125	Ni 0·125	Cu 0·128	Zn 0·137

twice as many electrons. Aluminium has Al^{3+} ions held together by three times the number of electrons.

The observed densities of metals are in accordance with this model. Table 5.1 gives figures for what might be called *metallic radii*, that is half the internuclear distance between near neighbours in a metal lattice, or the radius which the atoms would have if we assume them to be hard spheres in contact.

Where ions are isoelectronic, for example Na^+, Mg^{2+} and Al^{3+}, the ion with the highest nuclear charge will be the smallest (Fig. 5.8).

The weakly-bonded alkali metals, Group IA, are therefore the least dense metals in their respective periods. Lithium will even float on the oil in which it is customarily stored, and a large lump of lithium is so light that it feels as if it is hollow.

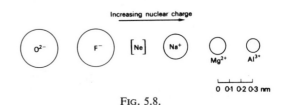

FIG. 5.8.

The conductivities of the alkali metals, of both heat and electricity, are extremely high. Conductivity is the result of the electrons being mobile and we can imagine that the electrons of Group IA metals are more mobile because the energy levels are less "crowded". This is an over-simplified picture, but it helps us to see why sodium, with only one mobile electron per atom is a better conductor of heat and electricity than calcium. Sodium is in fact used in nuclear reactors: being both easily melted and a good conductor of heat it can be used for heat transfer. It has also recently been proposed as a conductor to replace copper in special electricity cables.

5.5 Structure of non-metals

Excluding the noble gases which do not form bonds, the structure of non-metals is the result of electron-pair bonds forming between atoms. Two factors determine the number of bonds formed:

(i) The number of electrons in the valence shell (section 3.10).
(ii) The number of electrons which the valence shell can accommodate altogether.

A useful empirical rule is that if there are N electrons in the valence shell, the atoms in non-metallic elements form $8 - N$ bonds with their neighbours. In the sections which follow, some non-metallic structures will be dealt with in more detail, providing illustrations of the rule.

5.6 Diamond and graphite

Carbon shows two distinct allotropes, diamond and graphite. It is readily shown by chemical tests, or by using the mass spectro-

TABLE 5.2

	Diamond	Graphite
M.p. (°C)	—	3730
B.p. (°C)	—	3830
Density g cm^{-3}	3·51	2·26
Internuclear distance	0·154 nm	0·142 nm (0·335 nm between layers)

meter, that they are the same element, and differ only in structure. Diamond is denser than graphite. Their physical properties are compared in Table 5.2.

Figure 5.9 shows the structure of diamond—each carbon atom has *four* near neighbours, i.e. carbon has a co-ordination number of four in diamond.

Diamond is an electrical insulator, its electrons being rigidly held between atoms, with four electron-pair bonds per atom. The whole solid is a macromolecule of enormous rigidity and hardness. It is quite unlike a metal in physical characteristics. Apart from its insulating properties, it is transparent, and if attempts are made to bend a crystal or alter its shape it fractures in specific directions called **cleavage planes.** A cleavage plane is found to lie in the direction which cuts across the least

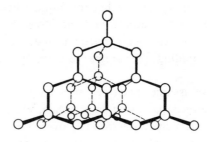

FIG. 5.9. The structure of diamond.

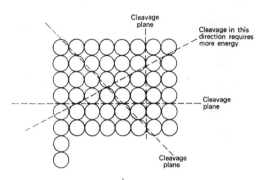

FIG. 5.10.

number of bonds. It requires energy to fracture the crystal, and the crystal will come apart in the direction which requires least energy (Fig. 5.10).

Graphite is quite unlike diamond. It looks "metallic" although this can be deceptive, and many non-metallic substances can be mistaken for metals just because they happen to be shiny in appearance. A more convincing metallic property is that it conducts electricity. In other respects graphite is quite unlike a metal. Structurally it has a co-ordination number of three, and the atoms are arranged in hexagonal sheets, corresponding to a lattice rather like wire netting (Fig. 5.11).

Graphite is a much softer substance than diamond—indeed it is used as a solid lubricant, whereas diamond is used as an abrasive. Clearly there are weaknesses in the structure. The distance between adjacent sheets of carbon atoms is much greater than the distance between bonded carbon atoms, and the forces holding adjacent sheets together are much weaker than those holding the atoms together within a sheet.

If we assume that three electron-pair bonds are formed per atom in graphite, then each atom has a spare electron left over. Two properties of graphite must be explained: first,

that it *conducts electricity* and second, that it is a diamagnetic substance. The spare electrons must be paired off in some way, otherwise graphite would be attracted by a magnet. However, the spare electrons must be free to wander about the lattice, rather than being rigidly held between pairs of atoms. These spare electrons are **delocalized,** just as they are in a metal. Electron pairs which remain in the vicinity of one atom, or in the space between a pair of atoms, are said to be localized. Graphite is a relatively poor conductor of electricity because the electrons can only flow freely within a given sheet of atoms—they cannot readily hop from one sheet to another. Metals are better conductors than graphite because their electrons are mobile in three dimensions.

So-called *amorphous carbon* is not really a third allotrope. A detailed examination of its structure shows it to be microcrystalline graphite. However, since its surface area is so very large for its overall mass it has very different physical properties from ordinary graphite, and it is often treated as a third allotrope. Special forms of carbon, such as carbon fibre and vitreous carbon, have also been prepared.

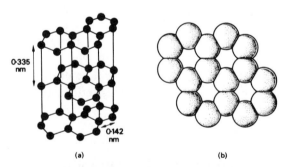

(a)   (b)

FIG. 5.11. The structure of graphite (a) showing that the distance between the atoms in different layers is greater than between adjacent atoms in a layer; (b) showing the hexagonal arrangement of one layer.

5.7 Nitrogen and phosphorus

The very great contrast between these two elements is in a sense a violation of the periodic law. Phosphorus is much more reactive, and its two common forms have structures which are different from that of nitrogen. Both elements can form three covalent bonds per atom, but they form them in quite different ways. Nitrogen, as we saw in Chapter 3, has a co-ordination number of *one*, that is, it forms diatomic molecules containing triple bonds. Phosphorus solves its energy-level problem in quite a different way. It appears that double and triple bonds of the type met with in nitrogen do not form at all easily, for phosphorus allotropes generally have a co-ordination number of three. The simplest way of explaining this is to postulate three electron-pair bonds and one lone pair per atom. Figure 5.12 illustrates the energy levels in an isolated phosphorus atom.

In white and black phosphorus a given phosphorus atom is connected to its three neighbours in a pyramidal manner, leaving the lone pair of electrons projecting away from the pyramid (Fig. 5.12).

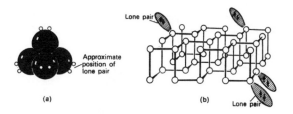

FIG. 5.13. (a) White phosphorus. (b) Black phosphorus: for clarity, only a few lone pairs are shown.

White phosphorus melts very easily (m.p. 44°C), and vaporizes as tetrahedral clusters of four atoms, that is, P_4 molecules (Fig. 5.13(a)). These molecules are present in the solid state. It is not surprising that white phosphorus is "waxy" since the forces between its molecules are about the same as those in a hydrocarbon wax. Red phosphorus on the other hand exists as some sort of giant structure, which is less volatile, and similar to the structure of black phosphorus shown in Fig. 5.13(b). Notice that in each case the environment of a given phosphorus atom corresponds to that shown in Fig. 5.12.

When describing the energy levels in phosphorus molecules the $3d$ levels must be taken into account. At this stage it will suffice for us to treat the bonds between phosphorus atoms as if they were simple electron-pairs, but this is a slight over-simplification. In fact it is the very existence of available d-orbitals in phosphorus, arsenic, antimony and bismuth which probably accounts for the fact that these elements are so different from nitrogen. There are many differences between first row (lithium to neon) and second row (sodium to argon) elements which can be attributed to the non-existence of a $2d$ sub-shell in the first row elements.

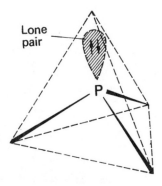

FIG. 5.12. The environment of a phosphorus atom.

5.8 Oxygen and sulphur

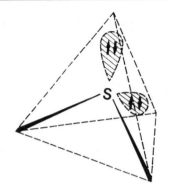

Fig. 5.14. The environment of a sulphur atom.

The structural relationship between nitrogen and phosphorus is repeated here between oxygen and sulphur. Oxygen has a co-ordination number of *one* (but see ozone, below) while sulphur has a co-ordination number of *two*. Sulphur can be expected to have two electron-pair bonds and two lone pairs (Fig. 5.14).

Given this environment for individual sulphur atoms, all sorts of structures are theoretically possible, and so it is not surprising that many allotropes of sulphur exist. The commonest basic units are:

(i) a puckered ring of eight atoms, S_8 (Fig. 5.15);

(ii) a long zig-zag chain of many atoms, made by S_8 rings opening out and the ends joining up.

The S_8 rings can pack together in various crystalline forms, two common ones being *rhombic sulphur*, stable below 96°C, and *monoclinic sulphur*, stable above this temperature. The temperature of change is called the **transition temperature**—it is analogous to a melting point or boiling point, but the changes from one phase to another, being between two solids, takes place more slowly.

Changes between two solid phases are bound to be slow, because translational motion of molecules is severely restricted. Once a "nucleus" is established on which the new structure can build up, transition from one crystalline form to another can take place quite rapidly.

If rhombic or monoclinic crystals are heated to just above their melting point, S_8 molecules persist in the liquid state. Further heating disrupts the rings, and the molecules *polymerize*, that is, many molecules join together to form a long chain. In the same way that small ethylene molecules C_2H_4 can be polymerized to form the plastic "polythene", *plastic sulphur* is produced when the strongly heated liquid is rapidly cooled by pouring into cold water. Unfortunately plastic sulphur has little practical use—articles manufactured from it would rapidly disintegrate at room temperature as the rings closed up again forming microcrystalline rhombic sulphur!

If sulphur is slowly heated all sorts of changes take place in the viscosity and colour of the liquid. On boiling, a dark red vapour is produced containing various molecules, S_8, S_4 and S_2, the proportion of each depending on the temperature.

Oxygen, as we normally know it, exists as diatomic molecules, O_2. The absorption of ultra-violet radiation in the upper atmosphere

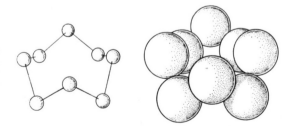

Fig. 5.15. S_8 rings which occur in rhombic and monoclinic sulphur.

dissociates many of these molecules and *atomic oxygen* is produced.

$$O_2(g) \xrightarrow{h\nu} 2O(g);$$

$$\Delta E = +498 \text{ kJ}.$$

The energy for this reaction can alternatively be provided by passing an electric discharge. At room temperature and normal atmospheric pressure, atomic oxygen is too reactive to persist, and some O_3 molecules are produced:

$$O(g) + O_2(g) \rightarrow O_3(g).$$

O_3 is called *ozone*. It has a characteristic smell, which can often be detected in the region of electrical equipment. Contrary to popular belief it is far from health-giving, and, being highly reactive, is harmful to living tissues in high concentrations. The "ozone" smell at the sea-side can generally be attributed to rotting sea-weed!

5.9 Borderline elements

Among the elements two distinct structural types can be distinguished: on the one hand metals of high co-ordination number, and on the other non-metals of low co-ordination number. Near the centre of the *p*-block of the periodic table there exist many elements with intermediate properties and structures. For instance, arsenic, antimony and bismuth have layer structures which are fairly close-packed, but with only three near neighbours, and the structure becomes more "metallic" down Group V. This is in accordance with the $8-N$ rule, and would suggest that covalent bonds form between atoms. However, the elements conduct electricity and their electrons are therefore mobile. The truth is that the bonding is intermediate between metallic and covalent.

A similar state of affairs exists with the elements of Group IVB, which exhibits a changeover in structure and physical properties with germanium on the borderline. Borderline elements like germanium are **semiconductors**.

TABLE 5.3

SOME PROPERTIES OF THE GROUP IVB ELEMENTS

Element	Melting point	Bond length (nm)	Electrical conductivity	Remarks
Carbon	3730°C (graphite)	0·154	non-conductor ⎫	These data refer to dia-
Silicon	1420°C	0·234	semi-conductor ⎬	mond-like forms of the
Germanium	937°C	0·244	semi-conductor ⎭	elements
Tin	232°C	0·280 (grey) 0·302 (white)	semi-conductor (grey) conductor (white)	Grey form with diamond structure. Changes to white metallic form at 13°C
Lead	327°C	0·350 (metallic)	conductor	Diamond-like form does not exist

Study Questions

1. How are the properties of the following elements related to their structures: (a) diamond, (b) graphite, (c) plastic sulphur, (d) copper, (e) tin?

2. In each of the following sets, arrange the elements in order of increasing melting and boiling points:

(a) Ar, Br, Cr.

(b) Na, Ne, Ni.

(c) Cl_2, Br_2, I_2.

(d) Li, Na, K.

(e) K, Ca, Sc, Ti.

Explain any trends you observe.

3. Give examples of elements which have the following co-ordination numbers: 1, 2, 3, 4, 8 and 12.

4. Place the elements W, X, Y and Z as accurately as you can in the periodic table:

(a) W is a red solid containing puckered rings of atoms.

(b) Solid X has a close-packed structure and does not conduct electricity.

(c) Y has a body-centred cubic structure and a density of 0·53 g cm^{-3}.

(d) Z consists of puckered sheets in which each atom is bonded to three neighbours, the sheets being held together by metallic bonding.

CHAPTER 6

Structure of compounds

THIS chapter is concerned with the relationship between physical properties of compounds and their structure. This is illustrated by the hydrocarbons which, although they contain only two elements, exhibit a wide range of properties. Other compounds possessing different types of structure are then examined.

6.1 Structure of hydrocarbons

All hydrocarbons contain C—H bonds. All except methane also contain bonds linking carbon atoms together. Methane, the simplest hydrocarbon, is at room temperature, a gas made of CH_4 molecules.

Figure 6.1 shows that an isolated carbon atom has vacant energy levels. Four electrons are quite readily accommodated. One way of

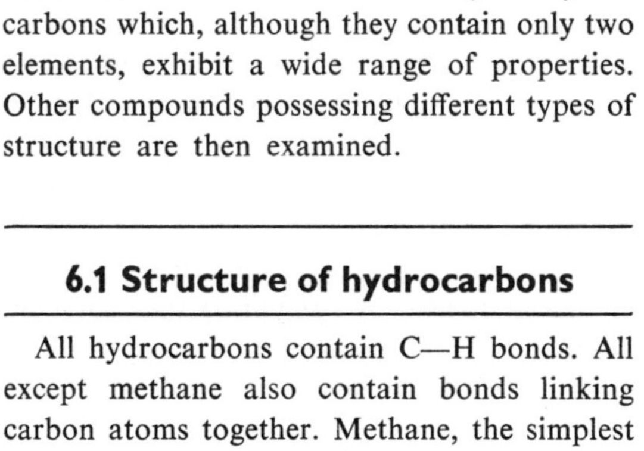

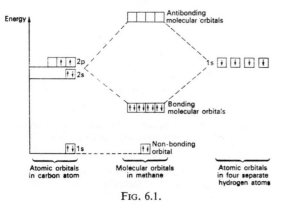

FIG. 6.1.

achieving this is by forming four **electron-pair** bonds with hydrogen atoms. The $2s$ and $2p$ atomic orbitals in carbon, together with the four $1s$ atomic orbitals in the four hydrogen atoms, have now become eight molecular orbitals in the CH_4 molecule, four of which are low energy (bonding) and four high energy (antibonding). Figure 6.2 shows the same process in electron-dot notation.

$$\text{H} \cdot \overset{\textstyle\cdot}{\underset{\textstyle\cdot}{\text{C}}} \cdot \text{H} \longrightarrow \overset{\textstyle\text{H}}{\underset{\textstyle\ddot{\text{H}}}{\text{H} \!:\! \ddot{\text{C}} \!:\! \text{H}}} \quad \text{or} \quad \begin{array}{c} \text{H} \\ | \\ \text{H}\!-\!\text{C}\!-\!\text{H} \\ | \\ \text{H} \end{array}$$

Fig. 6.2.

The electron pairs in methane repel one another, and so the molecule, instead of being a flat square, has the shape of a tetrahedron (triangular pyramid). Figure 6.3 shows various ways in which the solid three dimensional shape of a methane molecules can be represented on a flat piece of paper, but the best way to visualize it is by making a model.

Methane is a gas which is liquefied and solidified only by considerable cooling (m.p. $-182°C$, b.p. $-160°C$). Very little energy is required to separate methane molecules from one another—we say that intermolecular attraction is small. ΔH denotes change in heat energy, and the positive sign denotes heat added to the system.

49

Tetrahedron

The four bonds in methane are directed to the corners of a tetrahedron, with carbon at the centre

The shape can be shown by drawing the bonds in "perspective"

"Space-filling" picture

Electron-cloud picture

FIG. 6.3.

$$CH_4(l) \longrightarrow CH_4(g);$$

one gram-molecule (16 g) of liquid methane

one gram-molecule of methane gas

$$\Delta H = +9\cdot22 \text{ kJ at } -160°C$$

In contrast to this, an enormous amount of energy would be needed to split up the same number of molecules into separate atoms of carbon and hydrogen. So much energy is needed that it cannot be done in practice, though the energy is calculated to be:

$$CH_4(g) \longrightarrow C(g) + 4H(g);$$

one gram-molecule of methane gas

carbon vapour (atoms)

hydrogen in the form of separate atoms

$$\Delta H = +1660 \text{ kJ}$$

Carbon atoms can also form electron-pair bonds with each other, and a whole range of molecules can be built up from carbon and hydrogen atoms, called **hydrocarbons.** Methane is the simplest possible hydrocarbon. Another simple one is ethane. The molecule consists of two carbon and six hydrogen atoms and its

chemical formula is written C_2H_6. Similarly the gas propane has a chain of three carbon atoms, and a formula C_3H_8.

A whole series of substances is known, the molecules of which consist of chains of carbon atoms joined by electron-pair bonds. If n is the number of carbon atoms, there will be $2n+2$ hydrogen atoms. The carbon atoms may be either linked up in "straight" chains* or in branched chains. A wide range of these hydrocarbons can be extracted from **petroleum,** the crude oil obtained from an oil well.

The size and shape of molecules have a considerable effect on their physical properties. If the boiling point of a hydrocarbon is plotted against number of carbon atoms, a gradual rise is noted (Fig. 6.4) as the force of attraction between molecules become larger. This is illustrated by comparing the heat required to vaporize 1 gram-molecule of each of the hydro-carbons (Table 6.1).

The effect of structure on physical properties is well illustrated by comparing the character-istics of the main products obtained by distilling crude petroleum at an oil refinery (Table 6.2).

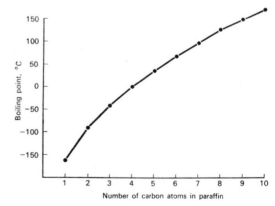

Number of carbon atoms in paraffin

FIG. 6.4. Force of attraction between hydrocarbon molecules increases with chain length.

---

* All carbon chains are bound to be bent since the bonds are tetrahedral. "Straight" here means un-branched.

TABLE 6.1

Formula	Name	Heat required to vaporize 1 gram-molecule (kJ)	Boiling point (°C)
CH_4	Methane	9·2	−160
C_2H_6	Ethane	13·8	−88
C_3H_8	Propane	18·4	−41·4
C_4H_{10}	n-Butane	22·2	+0·8

TABLE 6.2

Product	Number of carbon atoms	Properties
"Calor" gas	3 or 4	Gas, liquefiable under pressure
Motor fuel	about 8	Liquid, easily evaporates
Diesel oil	about 14	Liquid, does not evaporate easily
Lubricating oil	about 18	Thick liquid, does not evaporate
Pitch (road tar)	20 or more	Semi-solid black substance

The larger molecules from substances which are thick and treacly in the liquid state, due to the molecules becoming tangled up. Translational motion is difficult, and we say the liquid is *viscous*, or has a high viscosity.

If a hydrocarbon has very long chains the result is a more rigid solid structure. *Polyethylene* (polythene) $(C_2H_4)_x$ is an example of this.

ethylene gas   solid polythene

Rubber affords another illustration of the connection between structure and physical

(a) Single rubber molecule (diagrammatic)

(b) Tangled rubber molecules before stretching

(c) Rubber molecules when stretched

FIG. 6.5.

properties. Natural rubber is a hydrocarbon which consists of long zig-zag chains of carbon atoms tangled together. When rubber is stretched some straightening out of molecules takes place (Fig. 6.5). By studying the structure of the molecules of natural rubber, chemists have been able to develop the far superior synthetic rubbers which are now commonly used.

6.2 Principles underlying shapes of molecules

Although the shapes of different molecules are very varied, they are determined by quite simple principles, which may be understood by considering the electrostatic repulsions which exist between electrons.

It has already been noted that the four electron-pair bonds of a carbon atom are directed towards the four corners of a regular tetrahedron. This shape is in fact the one which we

FIG. 6.6. Four balloons tied together assume a tetrahedral shape.

A very large number of molecules and ions are found to be not planar but pyramidal, such as ammonia NH_3, and the sulphite ion SO_3^{2-}. The reason for this difference becomes apparent upon examining the electronic configurations of the molecules concerned (Fig. 6.8). It is found that molecules of the type AX_3 are

(i) planar, if A does not have any **lone pairs;**

(ii) pyramidal, if A has one lone pair in addition to its **bond pairs.** As far as shape is

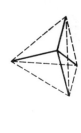

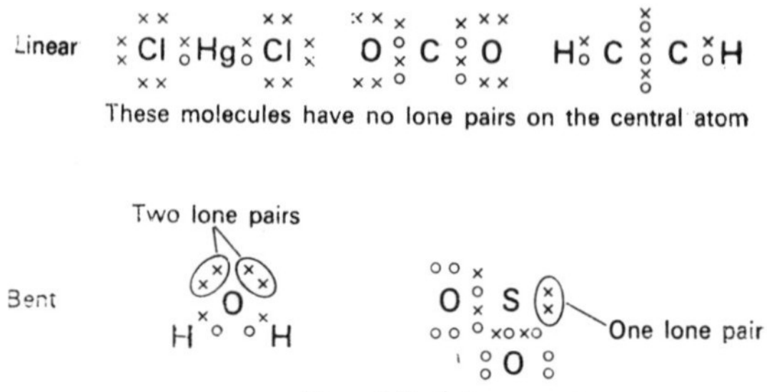

These molecules have no lone pairs on the central atom

would *expect*, assuming the bonds to repel one another.

A simple model, using toy balloons, can be used to demonstrate the effect of repulsion between bonds. If four long balloons are attached in the manner shown in Fig. 6.6 they will automatically assume a tetrahedral shape, making angles of approximately 109° with one another. Note that a square planar shape, although symmetrical, has angles of only 90° between adjacent bonds (balloons), and that this configuration is therefore not the most stable. A similar balloon experiment would suggest that three bonds will form a flat triangular molecule might result (Fig. 6.7). A number of molecules and ions, for instance boron trifluoride BF_3, and sulphur trioxide SO_3, do indeed have this shape.

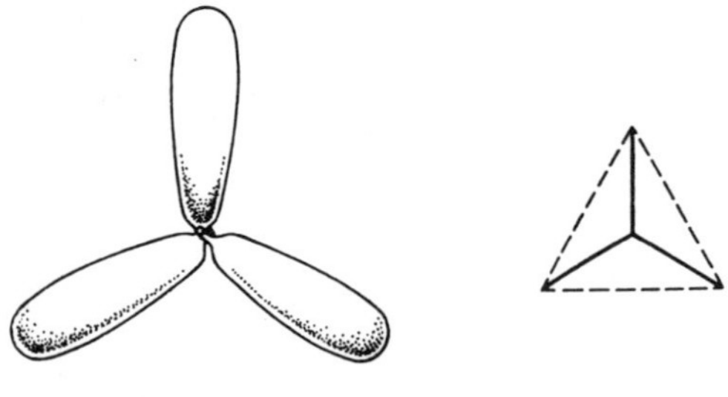

FIG. 6.8. (a)

BF₃ No lone pairs: planar

NH₃ pyramidal

SO₃ No lone pairs: planar

SO_3^{2-} The two extra electrons now form a lone pair making a pyramidal shape

FIG. 6.7. Three balloons tied together assume a planar shape.

AX_3 Planar when lone pairs absent

AX_3 Pyramidal when one lone pair present

FIG. 6.8. (b)

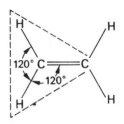

FIG. 6.9.

FIG. 6.10. The formation of an octahedral structure illustrated with balloons.

concerned it does not matter whether the bond A—X is a single bond (that is, one using only one electron pair) or a double bond (using two electron pairs). Ethylene, for instance, is a planar molecule, with bond angles of approximately 120° (Fig. 6.9).

The shapes of molecules of the type AX_2 can be similarly explained. Our balloon model suggests that such a molecule will be linear, but this is only found to be true in fact for molecules where A has no lone pairs, such as mercury(II) chloride, $HgCl_2$, and carbon dioxide, CO_2. When A has one or more lone pairs the molecule is found to be *bent*, as in water, H_2O, hydrogen sulphide, H_2S, or sulphur dioxide, SO_2.

There are surprisingly few instances where a consideration of the number of lone-pairs and bond pairs on the central atom fails to give a correct prediction of the molecular shape. All that is necessary is to write down the electronic structure correctly, which can be done by observing the following rules:

(a) The valence shell in hydrogen can only accommodate 2 electrons.

(b) First row elements cannot accommodate more than eight electrons in their valence shell.

(c) Later elements can accommodate more than eight electrons in their valence shell.

The rules of electron pair repulsion can be applied to compounds of the type AX_5, AX_6,

and even AX_7. Type AX_6 is quite simple to understand, and sulphur hexafluoride, SF_6, provides an example. A preliminary experiment with balloons predicts an octahedral shape, with exactly 90° between adjacent bonds, and this is in fact correct (Fig. 6.10).

The balloon experiments are generally too rough and ready to give a reliable answer for AX_5 and AX_7. Phosphorus(V) chloride (phosphorus pentachloride), PCl_5, has molecules shaped like a trigonal bipyramid, while iodine(VII) fluoride (iodine heptafluoride), IF_7, has the shape of a pentagonal bipyramid (Fig. 6.11).

The complex ions of transition metals sometimes have distorted shapes, due to the presence of lone pairs, or lone electrons. For instance the ion tetrammine copper(II), $Cu(NH_3)_4^{2+}$, has the nitrogen atoms arranged around the central copper atom in a plane.

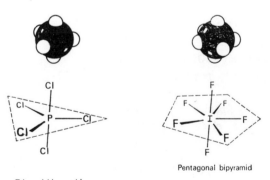

Trigonal bipyramid

Pentagonal bipyramid

FIG. 6.11.

† 6.3 The iso-electronic rule

A consequence of the above rules for molecular shape is that molecules which are **iso-electronic,** that is, have the same number of outer electrons and the same number of atoms, tend to have the same shape. It will be left as an exercise for the student to work out the reasons for the rules which follow, but they are a consequence of the principles laid down in section 6.2:

(a) Species of the type AX_2 containing 16 valence electrons are linear; more than 16 electrons lead to a bent structure.

Linear:

CO_2; number of valence electrons $= 4+6+6 = 16$
N_2O; number of valence electrons $= 5+5+6 = 16$
$BeCl_2$; number of valence electrons $= 2+7+7 = 16$
NO_2^+; number of valence electrons $= 5+6+6-1 = 16$

Bent:

NO_2; number of valence electrons $= 5+6+6 = 17$
NO_2^-; number of valence electrons $= 5+6+6+1 = 18$

(b) Species of the type AX_3 containing 24 valence electrons are planar triangular; more than 24 electrons lead to a pyramidal structure.

Triangular plane: BF_3; $3+(3\times7)$ $= 24$
 BO_3^{3-}; $3+(3\times6)+3 = 24$
 CO_3^{2-}; $4+(3\times6)+2 = 24$
 NO_3^-; $5+(3\times6)+1 = 24$
 SO_3; $6+(3\times6)$ $= 24$

Pyramid: SO_3^{2-}; $6+(3\times6)+2 = 26$
 PCl_3; $5+(3\times7)$ $= 26$

(c) Species of the type AX_4 containing 32 valence electrons are tetrahedral:

 CCl_4; $4+(4\times7)$ $= 32$
 PO_4^{3-}; $5+(4\times6)+3 = 32$
 SO_4^{2-}; $6+(4\times6)+2 = 32$
 ClO_4^-; $7+(4\times6)+1 = 32$

6.4 Giant structures (macromolecules)

Previous sections have dealt with the shapes of small molecules, generally with only one central atom. The same principles of shape govern larger molecules, as it is simply necessary to consider the number of bonds and lone-pairs associated with each atom in turn.

The element carbon, in the form of diamond, has each atom joined to its neighbours by four covalent bonds arranged tetrahedrally. In other words the same principle of electron-pair repulsion applies. The result is an absolutely rigid macromolecule of enormous strength—diamond is one of the hardest substances known to man. Since the whole solid is in effect a single molecule a very large amount of energy is required to vaporize it. Bonds have to be broken to produce carbon vapour, and a temperature of over 3000°C has to be attained.

$$C(c) \longrightarrow C(g);$$

1 g-atom of 1 g-atom of
solid carbon monatomic vapour

$$\Delta H = + 715 \text{ kJ.}$$

If a solid is very difficult to melt, then it must be some sort of giant structure in which the atoms are joined throughout by relatively large forces, and not an assembly of molecules held together by relatively weak forces (van der Waals forces). Rather more evidence about the properties of the solid is needed before concluding that the bonds between atoms are covalent. In many cases the forces are of a different nature, as we shall see in the next section.

6.5 Bonding in solids

This chapter has so far been largely concerned with *structure*, that is the stereochemical arrangement, of molecules. It is also important to consider the question of **bonding,** which is concerned with the *nature* of the forces between atoms. The principal types of bonding which exist can be illustrated with examples:

(a) Bonds between non-metallic atoms are **essentially covalent.** The physical properties of such substances depend upon whether they consist of finite molecules or of macromolecules, but in general they are electrical insulators both as solids and when molten.

(b) Compounds between elements situated on the extreme left and extreme right of the periodic table respectively, such as sodium chloride, are **essentially ionic.** Such substances have structures which appear to be related to the relative sizes of ions, rather than to electron-pair repulsions. They do not conduct electricity in the solid state, but they can be electrolysed when molten.

(c) Compounds between metals are **metallic** in character. By this we mean that they conduct electricity, without decomposition, in the solid state. Metals generally have a lustrous appearance, and are malleable and ductile. The stoichiometry of intermetallic compounds is complicated, and effects such as solid solutions are common. The general term **alloy** can be taken as applying to all systems of two or more metals, whether true compounds exist or not.

Very many compounds cannot be fitted exactly into the three categories defined above.

6.6 Ionic lattices

It was shown in Chapter 3 that if a potassium atom and a fluorine atom are brought together, complete transfer of an electron is energetically favoured. The presence of ions can actually be demonstrated in the molten state by electrolysis, but their presence in the solid can only be inferred from physical properties.

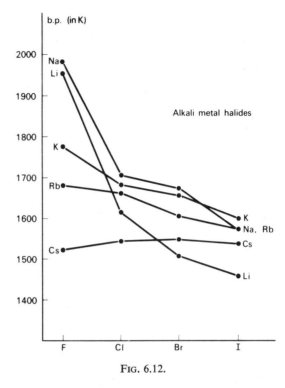

FIG. 6.12.

The halides of the alkali metals provide the simplest possible ionic systems. A few facts about them are summarized below:

(i) They all have high melting and boiling points. This suggests that they are giant lattices, not separate molecules in the solid phase. The melting and boiling points show regular trends related to the position of the elements in the periodic table. Figure 6.12 shows how the boiling points vary.

(ii) They are all soluble in water, the solubility being least for the atoms of lowest atomic weight. Lithium fluoride, which represents the lowest atomic weight combination of all, is only sparingly soluble (0·13 g per 100 g of water at 20°C).

(iii) Their crystals are cubic in shape. X-ray analysis shows that in nearly all cases the coordination number is six. Each alkali metal atom has six halogen atoms as near neighbours, and vice versa—the 6 : 6 structure. Caesium chloride, bromide and iodide are cubic but with a coordination number of eight—the 8 : 8 structure. These structures are not close-packed; a coordination number of 12 would be impossible for ions of opposite charge.

(iv) The solids conduct electricity very badly. Therefore there can be no mobile electrons in the lattice. In fact there is very slight electrical conductance arising from the presence of mobile M^+ ions.

(v) When molten or when dissolved in water the alkali metal halides will all conduct electricity, with electrolysis, i.e. they are electrolytes. This again is not *proof* that ions exist in the solid, and we shall later meet cases where it is more reasonable to suppose that ions exist in the melt but *not* in the solid. Nevertheless the ionic model gives a good interpretation of the behaviour of the alkali metal halides.

(vi) Measurement of internuclear distances between pairs of ions, by X-ray analysis, shows that the ions *behave roughly as if each had its own separate radius*. For instance the internuclear distances have been measured in sodium chloride, sodium bromide, potassium chloride and potassium bromide with the following results:

Salt	Distance between nuclei (nm)	Salt	Distance between nuclei (nm)
NaCl	0·2814	NaBr	0·2981
KCl	0·3139	KBr	0·3293

$$0·2981 - 0·2814 = 0·0167 \text{ nm,}$$
$$0·3293 - 0·3139 = 0·0154 \text{ nm.}$$

The differences, approximately 0·016 nm, can be attributed to the difference in radius between

Cl^- and Br^-.

$$0·3129 - 0·2814 = 0·0325 \text{ nm}$$
$$0·3293 - 0·2981 = 0·0312 \text{ nm}$$

The differences, approximately 0·032 nm, can be attributed to the difference in radius between Na^+ and K^+.

Not even X-ray analysis enables us to measure the radii of *separate* ions, and the discrepancy of the figures shows that the "radius" of an ion appears to vary with its environment. Various methods of calculating single ionic radii are available, and the following figures are generally accepted as the radii of the ions assuming 6 : 6 co-ordination:

Na^+ 0·098 nm	Cl^- 0·181 nm
K^+ 0·133 nm	Br^- 0·196 nm

The 6 : 6 structure is shown in Fig. 6.13 and the 8 : 8 structure in Fig. 6.14.

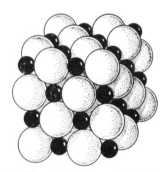

FIG. 6.13. The 6 : 6 (sodium chloride) structure.

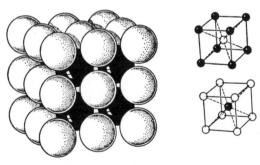

FIG. 6.14. The 8 : 8 (caesium chloride) structure.

Shear stress

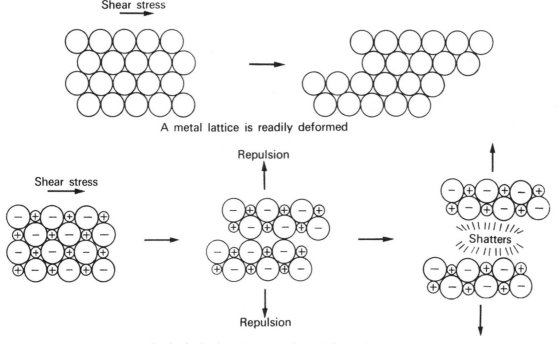

A metal lattice is readily deformed

Shear stress

Repulsion

Repulsion

Shatters

An ionic lattice shatters when deformed

FIG. 6.15.

(vii) The crystals are brittle, and readily fractured along planes of cleavage. This property does not in itself point specifically to the presence of ions, but it does suggest that a close-packed metallic arrangement is absent. The fracture of a crystal is readily interpreted on the ionic model. If adjacent planes of ions are dislocated, repulsion between ions of similar charge will force the layers apart (Fig. 6·15).

The above list gives *some* of the reasons why we believe that a salt like sodium chloride consists of ions. In Chapter 9 we shall test this hypothesis quantitatively, and it will be shown that the ionic model fits the energy data extremely well, and gives a good interpretation of observations such as solubility in water and heat of solution. Other salts, for instance the halides of zinc, do not give such good agreement, suggesting that the ionic model is less

applicable. Their structures are less simple, and cannot be interpreted as being due to the packing together of charged spheres. Some ions are thought to be more readily polarized (distorted) than others, causing the structures to be modified (Fig. 6.16).

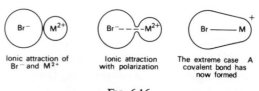

ionic attraction of Br^- and M^{2+}

Ionic attraction with polarization

The extreme case A covalent bond has now formed

FIG. 6.16.

6.7 Electrolysis

The properties of alkali metal halides are best interpreted by assuming that each ion carries a single charge. It is found by experiment that when a molten alkali metal halide is electrolysed, the same number of gram-atoms of element are always liberated provided the total *charge* passed through the cell is the same.

The liberation of one gram-atom of an alkali metal, or of a halogen, always requires 96 490 coulombs of charge. One coulomb is the charge which flows when 1 ampère of current flows for one second. To liberate one gram-atom of an alkali metal requires one ampère for 96 490 seconds, or about $26\frac{1}{2}$ ampère-hours. (One 2-volt accumulator of the type commonly used for laboratory d.c. supplies will yield approximately this amount of charge.) The quantity 96 490 coulombs is called one **faraday.**

If a salt like lead(II) bromide* is electrolysed, *two* faradays of charge are required to liberate one gram-atom of lead. This is interpreted by postulating that the lead ions carry two positive charges whereas the alkali metal ions only carry one:

$$Na^+ \quad + \quad e^- \quad \longrightarrow \quad Na(l)$$
ions in melt   one faraday   one gram-atom
                                        of molten sodium

$$Pb^{2+} \quad + \quad 2e^- \quad \longrightarrow \quad Pb(l)$$
ions in melt   two faradays   one gram-atom
                                        of molten lead.

Quantitative observations on the amounts of substances liberated when an ionic melt or solution is electrolysed, were made by Michael Faraday (1832). Faraday coined the term "electrochemical equivalent" for the mass of an element liberated by 1 coulomb in electrolysis, and framed three laws of electrolysis to describe his observations. The term electrochemical equivalent has now largely fallen into disuse, and we can re-state Faraday's second law in more modern terms.

* The Roman numeral (II) is included in the name because lead does not *always* exist as ions with two positive charges. The numeral indicates that in this case the ion has a charge of +2.

Faraday's laws of electrolysis (modern form):

(1) The mass of a substance liberated at an electrode during electrolysis is directly proportional to the charge which flows through the cell.

(2) The quantity of electricity needed to liberate one gram-atom of an element at an electrode is 96 490 coulombs, or a small whole-number multiple of this amount.

(3) Chemical changes during electrolysis occur at the electrodes, and not within the body of the cell.

Electrolytes, unlike metals, do not obey Ohm's law: a certain minimum potential difference has to be applied to an electrolyte in order for current to flow at all. Below this minimum, loosely termed **discharge potential,** no electrolysis occurs. This suggests that a certain minimum energy has to be attained before the electrolyte will split up into its simpler electrolysis products. We can make a rough estimate of this energy by measuring the discharge potential, after making due allowance for side effects such as the extra voltage needed to overcome the internal resistance of the electrolytic cell.

If one coulomb of charge is raised through a potential of one volt, one *joule* of work is done. To decompose one g-formula of sodium chloride completely into its elements, 96 490 coulombs will be required, at a minimum discharge potential of 4·0 V.

Hence energy required $= 96\ 490 \times 4{\cdot}0$.
$$= 3{\cdot}86 \times 10^5$$
J g-molecule $^{-1}$
$$= 386$$
kJ g-molecule $^{-1}$

This figure is only approximate, since various inaccuracies can occur when using a discharge potential as a measure of energy, but it illustrates the magnitude of the quantities involved.

6.8 Conductivity of electrolytes

The electrical conductivity of an electrolyte is generally less than that of a metal because the current is carried by bodily movement of ions rather than by electrons. Measurements of conductivity can give useful information about the nature of the ions in an electrolyte and the number of ions present in a given solution.

The conductivity of a solution is measured in a cell of known dimensions, called a conductivity cell, in a Kohlrausch bridge circuit (Fig. 6.17). This is similar to a Wheatstone's bridge but with the following modifications:

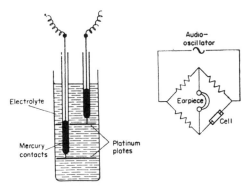

FIG. 6.17. Conductivity cell and Kohlrausch bridge.

(a) a.c. has to be used instead of d.c. to prevent electrolysis taking place. The a.c. source is usually an oscillator.

(b) An ordinary moving coil galvanometer cannot be used for a.c. For rough work a telephone earpiece is used, and the operator balances the bridge for minimum sound. For more accurate work a cathode ray oscilloscope is used.

It is convenient to define a quantity known as the **resistivity** of a substance. This is the electrical resistance, in Ω, of a cube of material of unit side, the potential difference being applied to two opposite faces of the cube. The S.I. unit of resistivity is the ohm metre (Ω m), this being the resistance of a metre cube; however, in the section which follows a more convenient unit is the ohm cm, or the resistance of a centimetre cube.

In a conductivity cell,

$$\text{resistivity} = k \times \text{measured resistance}$$

where k is a constant of proportionality that depends upon the dimensions of the cell and is called the **cell constant.** The cell constant is best determined by measuring the resistance in the cell of an electrolyte of known resistivity, such as potassium chloride solution of known concentration.

The reciprocal of resistivity is called the **conductivity** of the cell.

6.9 Molar conductance

The **molar conductivity** (molar conductance) of a solution is defined as the conductivity, in Ω^{-1} cm^{-1}, multiplied by the volume in cm^3 containing one gram-formula weight of dissolved electrolyte. It is denoted by the symbol Λ. The significance of molar conductivity can best be understood by imagining a cell with plates 1 cm apart, and of such area that they enclose 1 gram-formula of electrolyte (Fig. 6.18).

If an ionic substance dissolves in water completely giving ions, the molar conductivity ought not to alter if the solution is diluted. If the area of the plates is increased as the solution is diluted, so that 1 gram-formula of electrolyte is enclosed throughout, then the same ions are responsible for the conductivity throughout, and they have to conduct across a path of 1 cm.

If this experiment is done, it is found that many electrolytes do in fact give constant values for Λ at a given temperature provided the solution is dilute. At high concentrations Λ decreases somewhat. This can be interpreted by

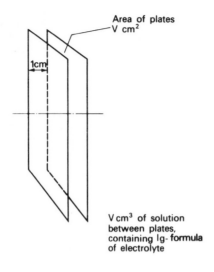

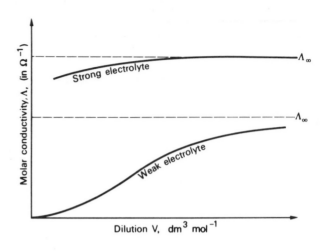

FIG. 6.18.

assuming that in the concentrated solutions the ions are attracting one another and hence impeding each other's progress. In dilute solutions interionic attractions become negligible as the ions are further apart.

The conductivity of an electrolyte increases with temperature, and this is interpreted by imagining that the ions move with greater velocity at the higher temperature.

Some electrolytes show a very marked change in molar conductivity as their concentration is changed, and these are known as *weak electrolytes*. It is supposed that in these cases ionization is incomplete and the degree of ionization increases as the solution is diluted (section 10.9).

where Λ_∞ = the molar conductance at "infinite dilution";

Λ_+ = the ionic conductance of the positive ion, assumed to be singly charged;

Λ_- = the ionic conductance of the negative ion, assumed to be singly charged.[*]

The above relation is called *Kohlrausch's law of ionic conductances*. The data in Table 6.3 illustrates the law.

A given ion has the same ionic conductance, no matter what other ions are present, and the

6.10 Ionic conductance

It is found experimentally that, provided solutions are very dilute, molar conductances are additive in nature, and it appears that the molar conductance can be represented as the sum of the separate conductances of the positive and negative ions.

$$\Lambda_\infty = \Lambda_+ + \Lambda_-$$

TABLE 6.3

Electro-lyte	Λ_∞	Electro-lyte	Λ_∞	Difference
KCl	130·0	NaCl	108·9	21·1
KNO$_3$	126·3	NaNO$_3$	105·2	21·1

[*] The argument will have to be modified if multiple-charged ions are present, since it is assumed that there are equal numbers of each ion present.

Λ_∞ of a given salt is the sum of the ionic conductances of the ions contained in it. Table 6.4 gives some typical values for ionic conductances at 25°C.

The values for the hydrogen and hydroxide ions are rather higher than for other ions, and this is because these ions can move more rapidly through water. Hydrogen ions are thought to attach to water molecules forming species such as H_3O^+, and to move through the electrolyte by a kind of "relay race" mechanism:

TABLE 6.4

Cation	Conductance $\Omega^{-1}\,cm^2$	Anion	Conductance $\Omega^{-1}\,cm^2$
H^+	349·8	OH^-	198·5
K^+	73·5	Br^-	78·4
NH_4^+	73·4	I^-	76·8
Ag^+	61·9	Cl^-	76·3
Na^+	50·1	NO_3^-	71·4
Li^+	38·7	$CH_3CO_2^-$	40·9

Similarly with hydroxide ions:

Most other ions have rather similar values for ionic conductance, suggesting that they move with roughly the same velocity down a given potential gradient. Differences are observed however, and these are not always of the type predicted by considering the simple ionic sizes. It might be imagined for instance that the ionic conductance of Li^+ would be greater than that of Na^+ because the former ion is smaller, but in fact the reverse is found to be true. This observation is explained by assuming that ions in solution attract water molecules electrostatically to themselves, forming a *hydration sheath*. Since the lithium ion produces a more intense electrostatic field at its surface it is hydrated to a greater degree and is therefore less mobile in water. This is one of the reasons why it is important to write $Li^+(aq)$ and $Na^+(aq)$ for the ions in solution: the state symbol (aq) refers to the presence of the hydration sheath. The average number of water molecules associated with an ion in aqueous solution is termed the **hydration number.**

Study Questions

1. Predict the shapes and write down electron dot structures for the following molecules and ions:

(a) BCl_3, CCl_4, NCl_3, PCl_5, SCl_4, SF_6, IF_5.

(b) H_3O^+, AlH_4^-, NH_4^+, PCl_4^+ and PCl_6^-.

2. Suggest three ions that would be isoelectronic with SF_6. What will the shape of the ions be?

†**3.** Predict the shapes of the following three-atom systems: O_3, NOCl, N_3^-, SO_2, ICN and SCN^-.

†**4.** Two shapes are theoretically possible for the ion ICl_4^-. What are these? Which shape is the ion most likely to adopt?

5. (a) The internuclear distance in RbCl is 0·3285 nm. Use this information and the data given in the chapter to predict the internuclear distance in RbBr.

(b) The internuclear distances in NaF and KF are 0·2307 and 0·2664 nm. Predict the internuclear distance in RbF.

6. The melting and boiling points of the chlorides of five consecutive elements are:

Empirical formula of compound	m.p. (°C)	b.p. (°C)
NaCl	800	1465
$MgCl_2$	712	1418
$AlCl_3$	190	180
$SiCl_4$	−68	57
PCl_5	148	164

What can you say about the probable structure and bonding of these compounds? What further experiments would you carry out to confirm your suspicions?

7. The melting point of the Group IVB dioxides are: CO_2, −57°C; under pressure; SiO_2, 1720°C; GeO_2, 1120°C; SnO_2, 1130°C. Comment on the structures of these compounds in the light of these data.

8. Magnesium oxide, MgO, is harder than sodium fluoride, NaF. It also melts at a higher temperature (2640°C as compared with 990°C). How can you explain these facts?

9. Make an estimate of the radii of the ions Si^{4+} and C^{4-}. The co-ordination number in silicon carbide, SiC, is found to be 4. Do you think it likely that silicon carbide contains these ions?

10. Diamond, by virtue of its structure, is a very hard substance. Suggest another substance which might be about as hard as diamond.

11. Copper, silver, and gold are in Group IB of the periodic table. Separate solutions containing ions of these metals were electrolysed in series, and the following results were obtained:

	Copper cathode	Silver cathode	Gold cathode
Weight before electrolysis (g)	1·243	1·163	1·435
Weight after electrolysis (g)	1·370	1·594	1·698

What conclusions can you draw about the magnitude of the charges on the three ions?

12. A current of 0·1 A is passed through a solution of nickel ions for 16 minutes 5 seconds, and 0·0294 g of nickel was deposited on the cathode.

(a) How many g-atoms of nickel were deposited?

(b) How many faradays were passed?

(c) What is the charge on the nickel ion in this solution?

13. A minimum discharge potential of 3·34 V is required to decompose potassium iodide completely into its elements. Assuming that the ions are singly charged, calculate the energy required in kJ g-molecule^{-1}.

†**14.** Polymers called silicones are derived from the reactions of methyl silicon chlorides with water. Hydrogen chloride is eliminated and Si—O—Si links are formed, for instance:

$$2(CH_3)_3SiCl + H_2O \longrightarrow (CH_3)_3Si—O—Si(CH_3)_3.$$

With $(CH_3)_2SiCl_2$ as the reactant, two types of polymer consisting of rings or chains are formed.

(a) Draw the ring and chain structures.

(b) What would the physical properties of the two types of polymer be?

(c) At high temperatures the chains become rings. Why does this limit the use of silicones as high temperature plastics?

(d) What polymer would form from the reaction of CH_3SiCl_3 with water? How would its physical properties differ from those of the ring and chain polymers?

CHAPTER 7

Formulae

7.1 Atoms, molecules and moles

When elements combine together to form a compound, the composition of the compound is denoted by its chemical formula. Two distinct meanings may be attached to this formula, according to the context in which it is used. Taking ammonia, NH_3, as an example:

(i) the formula denotes that a *molecule* of ammonia consists of a unit containing three hydrogen atoms and one nitrogen atom; or

(ii) the formula denotes that the compound ammonia is formed when one *gram-atom* of nitrogen is combined with three *gram-atoms* of hydrogen.

The context makes it perfectly clear which meaning is intended: If we are discussing the structure of the ammonia molecule, then the symbol N denotes one atom of nitrogen in the molecule NH_3. If, on the other hand, we are interested in the heat of a chemical reaction, then the symbols in a chemical equation refer to gram-atom quantities:

$$N_2 + 3H_2 \rightleftharpoons 2NH_3;$$
$$\Delta H = -92 \text{ kJ}$$

This equation means that 92 kJ of heat are evolved when 28 g of nitrogen combine with 6 g of hydrogen, 34 g of ammonia being formed. Here the symbol N does not denote a single atom of nitrogen—what it really denotes is a *gram*-atom. Similarly the formula NH_3 may be called a gram-molecule.

The concept of the gram-atom was introduced in Chapter 1. One gram-atom of any element represents a given number of atoms, namely the Avogadro number, $6 \cdot 023 \times 10^{23}$. One gram-molecule of ammonia similarly represents $6 \cdot 023 \times 10^{23}$ molecules of ammonia.

The term **mole** is used by chemists to denote gram-atoms, gram-molecules, in fact $6 \cdot 023 \times 10^{23}$ particles of anything. Thus a mole of hydrogen gas contains $6 \cdot 023 \times 10^{23}$ molecules, and has a mass of approximately 2 g. A mole of ammonia contains $6 \cdot 023 \times 10^{23}$ molecules of ammonia, with a mass of approximately 17 g. *The formula NH_3 can be used either to denote a molecule of ammonia or a mole of ammonia.*

In a giant lattice the term molecule has no real meaning, and the formula of such a substance should be taken as referring to the mole. Sodium chloride, $NaCl$, does not exist as separate molecules composed of ion pairs (except in the vapour phase), and when we write its formula as $NaCl$ we are stating that one gram-atom of sodium and one gram-atom of chlorine combine together to produce a mole of sodium chloride. The term **gram-formula weight** can also be used to denote a mole of a substance.

7.2 Stoichiometric and non-stoichiometric compounds

A compound whose composition can be represented by a simple whole number formula is said to be **stoichiometric**. Any pure compound which is made up of molecules is bound to be stoichiometric, since the molecules are all identical and atoms cannot be split up.

Giant structures are more often non-stoichiometric. A perfectly crystalline solid such as pure sodium chloride will be stoichiometric since the rules of geometry determine the relative numbers of each kind of particle in the structure. Such perfection is frequently not attained in giant structures however, due to lattice defects. It is meaningless for instance to talk of "pure" silver oxide since a whole range of solids, Ag_xO_y, of continuously variable composition can be made, depending upon the pressure of oxygen gas which is applied. We may at times write Ag_2O as the formula of silver oxide, but this exact ratio of gram-atoms is only attained at one particular oxygen pressure.

Many compounds are non-stoichiometric due to their lattices being imperfect: the oxides and sulphides of many metals, particularly transition metals, often deviate considerably from exact stoichiometry, and the metal hydroxides precipitated by the addition of alkali, $OH^-(aq)$, to a solution of a metal salt have an indefinite amount of water in their composition. Thus for convenience we may write aluminium hydroxide as $Al(OH)_3$, but in practice we should be writing $Al(OH)_3 . x\ H_2O$ where x is not a whole number.

Metal alloys are interesting in that "compounds" of definite composition and stoichiometry can often be identified. The number ratios which occur often seem to be quite arbitrary, but a closer study has revealed that there are rules underlying the formulae of alloys. Typical formulae of inter-metallic compounds are $MgZn_2$, Cu_3Sn, $CuZn$, Cu_5Zn_8 and $CuZn_3$.

In very many cases the deviation from exact stoichiometry of solids is so slight that only very accurate measurements of combining weights can detect it. In the nineteenth century atomic weights were calculated assuming the law of constant composition, first stated in 1802 by Proust. We now realize that the law is not universal and that non-stoichiometric compounds are widespread where giant lattices are involved.

7.3 Determination of the Avogadro number

The **Avogadro number** is the number of particles in a mole of a substance, for instance the number of molecules of CH_4 in 16 g (approximately) of methane, or the number of atoms of carbon in 12·01 g (one gram-atom) of carbon. It is generally given the symbol N_A. The actual value of this number was not known during the nineteenth century, even though the existence of atoms was firmly established. Essentially some method of determining the actual size of atoms was required.

(a) *X-ray method.* In describing X-ray diffraction (Chapter 4) it was stated that in order to measure absolute values of internuclear distances, the actual wavelength of the X-rays used must be known. This can be done by diffracting X-rays by reflection at the surface of a ruled metal diffraction grating. Normally for diffraction to occur the spacing between the rulings on the gratings must be comparable to the wavelength undergoing diffraction. However, if the glancing angle (the complement of the angle of incidence) is made less than about 1°, then a very finely ruled grating will diffract X-rays.

Having established the wavelength of the X-rays on an absolute scale, the same X-rays are diffracted by the adjacent planes of atoms in a crystal. Once the distance between atoms has been established, it is easy to calculate the volume which the atoms occupy, and measurements of density establish the volume occupied by a *mole* of the substance. The following calculation illustrates how this has been used to determine the value of the Avogadro number.

Density of potassium chloride, KCl,
    $= 2 \cdot 01$ g cm^{-3} approximately

Distance between K and Cl nuclei
    $= 3 \cdot 14 \times 10^{-8}$ cm (from X-ray data)

Volume of 1 mole of potassium chloride
    $=$ volume of $(39 \cdot 0 + 35 \cdot 5)$ g of KCl
    $= \dfrac{74 \cdot 5}{2 \cdot 01} = 37 \cdot 1$ cm^3.

Since potassium chloride has a simple cubic structure (Chapter 6), each atom may be regarded as being at the centre of a cube, of side $3 \cdot 14 \times 10^{-8}$ cm.

$\therefore$ volume occupied by each atom
    $= (3 \cdot 14 \times 10^{-8})^3 = 30 \cdot 9 \times 10^{-24}$ cm^3.

$\therefore$ in potassium chloride $\dfrac{1}{30 \cdot 9 \times 10^{-24}}$ atoms occupy 1 cm^3;

$\therefore \dfrac{37 \cdot 1}{30 \cdot 9 \times 10^{-24}}$ atoms occupy $37 \cdot 1$ cm^3 (one mole) $= 1 \cdot 20 \times 10^{24}$ atoms.

However, the Avogadro number is *half* this figure, for it is the number of ion pairs, KCl, which go to make up one mole of potassium chloride. This works out at approximately 6×10^{23}.

We do not need to know the true volume occupied by a single potassium ion or a single chloride ion separately. We know the internuclear K—Cl distance, but we do not need the separate radii of the ions.

(b) *Radioactivity method.* When a substance emits α-particles, helium gas is produced by the reaction:

$$He^{2+} \quad + \quad 2e^- \quad \longrightarrow \quad He(g)$$

| alpha particles | stray electrons | gaseous helium |

In one year, a gram of radium emits $11 \cdot 6 \times 10^{17}$ α-particles. This figure is estimated by counting the particles with a Geiger counter over a suitable time interval, making allowances for other particles emitted by secondary disintegration products of radium. Careful measurement has shown that over the same period of time, one gram of radium gives rise to $7 \cdot 67 \times 10^{-6}$ g of helium.

$7 \cdot 67 \times 10^{-6}$ g of helium contain $11 \cdot 6 \times 10^{17}$ particles

4 g (one mole) of helium contains
    $\dfrac{11 \cdot 6 \times 10^{17} \times 4}{7 \cdot 67 \times 10^{-6}} = 6 \cdot 05 \times 10^{23}$
    molecules mol^{-1}.*

The fact that the figure obtained by radioactivity measurements ties up well with the figure obtained by X-ray diffraction measurements is one of the best pieces of evidence we can obtain for the atomic nature of matter. The figure $6 \cdot 023 \times 10^{23}$ molecules mol^{-1} is the accepted value for N_A, the Avogadro number.

(c) *From the charge on an electron.* Millikan, in 1913, succeeded in determining the charge of an electron by careful observation of oil drops produced by a finely divided spray in an electrostatic field. The apparatus which he employed is shown diagrammatically in Fig. 7.1: the vessel, containing air, was immersed in a thermostat. A finely divided spray of oil was introduced through the atomizer, and some passed through the holes in the upper plate A.

---

* The abbreviation *mol* is used for *mole* when expressing units. Note that *mol* is *not* an abbreviation for "molecule".

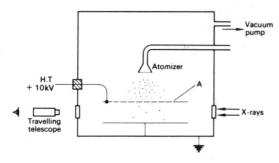

FIG. 7.1. Millikan's apparatus (diagrammatic)
for determining the charge on an electron.

Observations on a single oil drop could be made
with a travelling telescope; when an oil drop
was observed the air was ionized by passing
X-rays in through the window on the right.
The velocity with which the drop fell under
gravity alone was measured, and then an
electric field of about 10 000 V was applied so as
to make the drop move upwards. From these
measurements the magnitude of the charge
carried by the oil drop could be calculated. The
results showed that an alteration in velocity
frequently occurred, due to the oil drop capturing
different numbers of ions on successive occasions.
The values for the charge were found always to
be whole number multiples of about 1.6×10^{-19}
coulombs. This figure was taken to be the charge
on a single electron.

Since one faraday is a mole of electrons, the
Avogadro number is the ratio

$$N_A = \frac{\text{charge of one faraday}}{\text{charge of one electron}} = \frac{96\,490}{1.6 \times 10^{-19}}$$
$$= 6.03 \times 10^{23}$$

There are other methods of determining N_A,
for instance from Brownian motion, or from
observations on scattered light. The fact that
all these methods give approximately the same
value for the charge on an electron is our best
direct evidence for believing in the existence of
atoms.

7.4 Calculations involving changes in weight

Accurate measurements of changes in weight
during a chemical reaction were important
to the historical development of chemistry,
since atomic weights were originally determined
from weight changes in chemical reactions.
Nowadays accurate measurements of weight,
gravimetric measurements, are mainly of
analytical importance, for instance in determining the chemical formula of an unknown pure
substance, or for determining the percentage
composition of a mixture. The methods used
in calculations of this type are described in this
section. Provided that relevant atomic weights
are known, weighing may enable the *stoichiometry* (relative numbers of reacting particles)
of a reaction to be determined.

*Worked Example 1. 0·53 g of iron filings was
placed in an excess of silver nitrate solution.
Silver was precipitated which, after filtering
and drying, was found to weigh 2·05 g. Write
a chemical equation for the reaction.*

We shall assume that silver nitrate solution
consists of the ions $Ag^+(aq)$ and $NO_3^-(aq)$.
All common nitrates are soluble in water, and
therefore $NO_3^-(aq)$ ions are **spectator ions,**
that is, they do not participate in the reaction.

Atomic weight of silver = 107·9
Atomic weight of iron   = 55·8

∴ Number of gram-atoms of iron added

$$= \frac{\text{weight}}{\text{atomic weight}} = \frac{0.53}{55.8} = 0.0095$$

Number of gram-atoms of silver precipitated
$$= \frac{2.05}{107.9} = 0.0188$$

From this it is seen that the ratio

$$\frac{\text{number of gram-atoms of silver}}{\text{number of gram-atoms of iron}}$$
$$= 2, \text{ approximately.}$$

We must write an equation which shows two gram-atoms of silver, on the right-hand side, produced from one gram-atom of iron, on the left-hand side.

That is, Fe → 2 Ag (incomplete).

The silver is produced from silver ions Ag^+, and in order to conserve positive charge in the reaction, iron must produce ions which are positively charged, which we may provisionally write Fe^{n+}.

$$Fe + 2 Ag^+ \longrightarrow 2 Ag + Fe^{n+}.$$

The charges are balanced by making $n = 2$. The gram-atom quantities balance also, and the completed equation, with state symbols, reads:

$$Fe(c) + 2 Ag^+(aq) \longrightarrow 2 Ag(c) + Fe^{2+}(aq).$$

It is important to make some sort of measurement in order to establish a chemical equation, and to establish what substances are actually formed. If it is *known* that the iron dissolves to form iron(II) ions, Fe^{2+}, the equation can be written down straight away. If iron had formed the species iron(III), Fe^{3+}, the stoichiometry would have been different.

Gravimetric measurements are not confined to solids and liquids. Gases are readily weighed if they can be absorbed into a solid or liquid phase. For instance, the carbon dioxide produced in the combustion of carbon compounds can be absorbed in previously weighed soda-lime and the increase in weight noted. Water vapour can similarly be absorbed in silica gel.

Worked Example 2. 5·601 g of a barium salt was heated. Oxygen gas only was evolved, and the residue was found to be barium chloride weighing 3·834 g. Write down the name and formula of the original salt. The barium ion carries two positive charges and the chloride ion one negative charge.

Since the charges must balance in barium chloride, the compound must contain two g-ions of Cl^- for every one of Ba^{2+}. Therefore the formula is $BaCl_2$. Since there was a weight loss, and only oxygen was evolved, the original barium salt must be written $BaCl_2O_x$.

Atomic weight of barium = 137·3
Atomic weight of chlorine = 35·5
Atomic weight of oxygen = 16·0

We need to know the number of *moles* of barium chloride formed. It is convenient to talk about the **molecular weight,** or strictly the gram-formula weight, of barium chloride, namely the sum of the atomic weights $Ba + 2 Cl$.

g-formula weight of barium chloride
$= 137·3 + (2 \times 35·5)$
$= 208·3$

∴ Number of moles of barium chloride
$= \dfrac{3·834}{208·3} = 0·0184$

Number of moles of oxygen atoms evolved
$= \dfrac{\text{loss in weight}}{\text{atomic weight of oxygen}}$
$= \dfrac{5·601 - 3·834}{16·0} = 0·110$

∴ Number of moles of oxygen atoms which would be evolved per mole of barium chloride
$= \dfrac{0·110}{0·0184} = 6.$

The equation may be written:

$$BaCl_2O_x \longrightarrow BaCl_2 + 6[O] \quad \text{(incomplete)}$$
....   one mole   six gram-atoms

Balancing up the oxygen atoms, and taking into account the fact that oxygen is evolved as diatomic molecules, the complete equation, with state symbols, becomes:

$$BaCl_2O_6(c) \longrightarrow BaCl_2(c) + 3 O_2(g)$$

In fact the salt is barium chlorate, and its formula is more correctly written $Ba(ClO_3)_2$, to take into account that it gives chlorate ions $ClO_3^-(aq)$, when dissolved in water.

Measurements involving solely changes in weight can be used to deduce the so-called **empirical formula** of a substance, that is, the relative numbers of gram-atoms of elements which combine. It will not however give the **molecular formula** of a molecular substance. To give an example, measurements on the weights of carbon dioxide and water evolved when benzene is burned lead to the knowledge that benzene contains one gram-atom of carbon for every gram-atom of hydrogen. That is to say,

empirical formula of benzene = CH.

However when the *structure* of benzene is examined, it is found to be a molecular substance. Each molecule contains six carbon atoms arranged in a hexagon with six attached hydrogen atoms. Therefore the molecular formula of benzene is C_6H_6, molecular weight 78. When we talk about one mole of benzene, we mean $6\cdot023\times10^{23}$ molecules of C_6H_6, that is, 78 g.

The empirical formula of an organic substance can be established by a procedure such as **combustion analysis** (Fig. 7.2). The substance

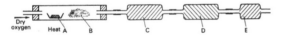

Dry oxygen   Heat A   B   C   D   E

FIG. 7.2. Combustion analysis. A, sample; B, copper oxide as an auxiliary source of oxygen; C, drying agent; D, soda-lime; E, drying agent.

is completely burned in oxygen, forming carbon dioxide and water vapour, which are led through a tube C containing a drying agent such as magnesium perchlorate. After the water has been removed, carbon dioxide is absorbed in another tube D containing soda-lime, a mixture of calcium and sodium hydroxide:

$$CO_2(g)+NaOH(c) \longrightarrow NaHCO_3(c).$$

If its molecular weight is also known, the molecular formula can be calculated. Methods of determining molecular weights are described in Chapters 8 and 15.

Worked Example 3. A hydrocarbon X of molecular weight 128 was subjected to combustion analysis. 64 mg were completely burned in oxygen. The increase in weight of the soda-lime tube was 220 mg, and the absorbed water weighed 36 mg. Deduce the molecular formula of the hydrocarbon.

220 mg of carbon dioxide

$$= \frac{220}{44} = 5 \text{ millimoles (mmol)}$$

36 mg of water

$$= \frac{36}{18} = 2 \text{ mmol}$$

64 mg of X

$$= \frac{64}{128} = 0\cdot5 \text{ mmol}$$

The above ratios can now be expressed in terms of a chemical equation:

$0\cdot5$ mmol X + oxygen $\longrightarrow$
5 mmol CO_2 + 2 mmol H_2O
$\therefore$ 1 mol of X + oxygen $\longrightarrow$
10 mol CO_2 + 4 mol H_2O

$$C_xH_y + \text{oxygen} \longrightarrow 10\,CO_2 + 4\,H_2O$$
$$\therefore \quad x = 10$$
$$y = 8.$$

The completed equation, balancing the oxygen, becomes:

$$C_{10}H_8 + 12\,O_2 \longrightarrow 10\,CO_2 + 4\,H_2O.$$

The hydrocarbon in this problem is probably naphthalene.

7.5 Calculations involving solutions

Reactions which occur in solution are often investigated quantitatively using aqueous (or sometimes non-aqueous) solutions of known molar concentration in the procedure known as **volumetric analysis.** As the name implies, the volumes of solutions are measured, using the pipette and burette, in a **titration**.

The **molarity** of a solution is defined as the number of moles of solute dissolved in a litre (1000 cm³) of solution. A solution containing one mole of sodium hydroxide, NaOH, per litre (that is $23+16+1 = 40$ g dm⁻³* is described as a **molar** solution, written 1 M NaOH. Similarly 4·0 g dm⁻³ of sodium hydroxide is a one-tenth molar solution, written 0·1 M NaOH.

Worked Example 4. Calculate the weight of solute in the following solutions:

(a) *One dm³ two-molar nitric acid, 2 M HNO₃.*

Formula weight = weight of one mole
$$= 1+14+48 = 63 \text{ g}$$

∴ 2 mol weigh $2\times63 = 126$ g

∴ One dm³ of 2 M HNO₃ contains 126 g of nitric acid.

(b) *50 cm³ of 0·5 M sodium carbonate, Na₂CO₃.*

Formula weight $= (2\times23)+12+(3\times16)$
$$= 106 \text{ g}$$

∴ Weight of solute per dm³ $= 0·5\times106$
$$= 53 \text{ g}$$

∴ Weight of solute in 50 cm³

$$= 53 \times \frac{50}{1000} = 2·65 \text{ g}.$$

Worked Example 5. Calculate the molarity of the solution produced by adding water to 50 cm³ of 2 M sulphuric acid, H₂SO₄ to give 200 cm³ of solution.

50 cm³ of 2 M acid contain $2 \times \dfrac{50}{1000}$ moles.

New volume = 200 cm³.

∴ 1dm³ contains $2 \times \dfrac{50}{\cancel{1000}} \times \dfrac{\cancel{1000}}{200}$

$$= 2 \times \frac{50}{200} = 0·5 \text{ mol}$$

∴ The diluted solution is 0·5 M.

This result is easily remembered and understood in the form: volume before dilution × molarity before dilution = volume after dilution × molarity after dilution.

Worked Example 6. To what volume would 500 cm³ of 0·5 M potassium permanganate, KMnO₄, have to be diluted in order to prepare a 0·2 M solution?

Let total volume after dilution = V cm³
Using the above relationship,

$$V \times 0·2 = 0·5 \times 500$$

$$\therefore V = \frac{0·5}{0·2} \times 500 = 1250 \text{ cm}^3.$$

Worked Example 7.

(a) *25 cm³ of 0·1 M sodium carbonate solution, from a pipette, were titrated with 0·2 M hydrochloric acid from a burette. The indicator was phenolphthalein. The burette readings were: initial reading 1·6 cm³, final reading 14·1 cm³. How many moles of hydrochloric acid have reacted with one mole of sodium carbonate?*

Volume of HCl added $= 14·1 - 1·6 = 12·5$ cm³.

∵ Number of moles of HCl added

$$= \frac{12·5}{1000} \times 0·2 = 0·0025 \text{ mole.}$$

Number of moles of Na₂CO₃ used

$$= \frac{25}{1000} \times 0·1 = 0·0025 \text{ mole.}$$

∴ The answer is one mole of hydrochloric acid.

The equation for this reaction may be written:

$$Na_2CO_3(aq)+HCl(aq) \longrightarrow$$
$$NaCl(aq)+NaHCO_3(aq).$$

(b) *The titration was repeated using methyl orange indicator, and this time it was found that twice the volume of acid had to be added to produce a colour change. Explain!*

---

* One litre is equal to 1000 cm³ or 1 dm³. Throughout this book dm³ will be used as an abbreviation for litre.

Since twice the volume of acid was required, it follows that 2 moles of acid were required to react with one mole of sodium carbonate. The equation is:

$$Na_2CO_3(aq) + 2\ HCl(aq) \longrightarrow$$
$$2\ NaCl(aq) + CO_2(aq) + H_2O.$$

This subject is discussed further in section 17.14.

Worked Example 8. The concentration of chloride ion, $Cl^-(aq)$, in a solution can be estimated by titrating it with silver nitrate from a burette. Silver chloride, $AgCl$, is precipitated. The endpoint is detected by adding a few drops of potassium chromate which gives brick-red silver chromate immediately a slight excess of silver ions has been added. Silver nitrate solution, 34 g dm^{-3}, was used to titrate 20 cm^3 portions of a solution of chloride ion. Use the burette readings to estimate the concentration of chloride ion.

Burette readings (cm^3):

Final	(1) 26·0	(2) 26·1	(3) 28·2	(4) 29·4
Initial	0·5	1·1	3·1	4·4
Volume	25·5	25·0	25·1	25·0

Titration (1) is taken as inaccurate, due to overshooting of the end-point. Within the limits of experimental error, the mean of titrations (2) to (4) is 25·0 cm^3.
Equation:

$$Ag^+(aq) + Cl^-(aq) \longrightarrow AgCl(c)$$

i.e. 1 mole reacts with 1 mole

Concentration of silver nitrate

$$= 34\ g\ dm^{-3}$$
$$= \frac{34}{170}\ mol\ dm^{-3} = 0.2\ M.$$

Number of moles of Ag^+ added in titration

$$= 0.2 \times \frac{25.0}{1000}$$

∴ Number of moles of Cl^- present in 20 cm^3 is equal to this.

∴ Number of moles of Cl^- per dm^3 of solution

$$= 0.2 \times \frac{25.0}{1000} \times \frac{1000}{20.0} = 0.25$$

7.6 Calculations involving gas volumes

Quantities of gases are usually measured more conveniently by volume than by weight, owing to their low density. Avogadro, in 1811 noticed that all gases obeyed approximately the same laws relating volume to temperature and pressure, and this observation led to the hypothesis that *equal volumes of gases at the same temperature and pressure contain equal numbers of molecules.* Subsequent observations have proved that this is true to a reasonable approximation, so we now refer to it as **Avogadro's law.** It holds most accurately for gases of very low boiling point.

The volume of 1 mole of a gas, that is, N_A molecules, at 273 K and 760 mm of mercury pressure, is called the **gram-molecular volume.*** It is approximately 22 400 cm^3. 273 K and 760 mm represent standard temperature and pressure **(s.t.p.).** At room temperature 1 mole of any permanent gas will occupy approximately 24 000 cm^3 (24 dm^3). This volume is easily visualized as it is somewhere near to one cubic foot.

Worked Example 9. Hydrogen peroxide decomposes when a catalyst of manganese dioxide powder is added, to give water and oxygen only. Calculate the concentration of hydrogen peroxide solution, in g dm^{-3}, which will give

10 times its own volume of oxygen when decomposed, at s.t.p.

$$H_2O_2(aq) \longrightarrow H_2O + \tfrac{1}{2} O_2(g)$$

In this equation, $\tfrac{1}{2}O_2$ denotes half a *mole* of oxygen, not half a molecule! Half a mole of oxygen $= 11\cdot2$ dm³ at s.t.p. $= 16$ g. From the equation, this quantity of oxygen is given by 1 mole of hydrogen peroxide, $= 34$ g.

For the solution to give ten times its volume of oxygen,

Volume of solution $= \dfrac{11\cdot2}{10}$ dm³

Concentration of solution

$= 34$ g dissolved in $\dfrac{11\cdot2}{10}$ dm³

$= 34 \times \dfrac{10}{11\cdot2}$ g dm⁻³ $= 30\cdot4$ g dm⁻³.

It is generally not convenient in the laboratory to make measurements at s.t.p. In practice, the volume of a gas is measured at whatever temperature and pressure are convenient. The volume that the gas would occupy at s.t.p. is then calculated by applying:

(1) **Boyle's law.** At constant temperature, the volume of a gas is inversely proportional to its pressure.

$pV = $ constant,

where p is the pressure and V the volume.

(2) **Charles' law.** At constant pressure, the volume of a gas is directly proportional to its absolute temperature.

$$V \propto T,$$

where T is measured in degrees Kelvin (K). 0°C is 273 K approximately.

These two laws are conveniently brought together in a single mathematical equation, using a new constant of proportionality, R, termed the **gas constant.**

$$pV = RT.$$

R has the same numerical value for one mole of any permanent gas. For n moles of a gas, it is often convenient to use the form:

$$pV = nRT$$

where V now equals the volume of n moles. If a gas, occupying a volume v_1 at pressure p_1 and temperature T_1 is subjected to altered conditions v_2, p_2 and T_2, it follows that

$$\frac{p_1 v_1}{T_1} = \frac{p_2 v_2}{T_2}.$$

The above relationship is useful for Boyle's law and Charles' law calculations.

Worked Example 10. Calculate the molecular weights of the following gases from the data given.

(a) *2·24 dm³ of an oxide of nitrogen weigh 4·4 g at 273 K and 760 mm Hg.*

Weight of 22·4 dm³ at s.t.p.

$$= \frac{4\cdot4 \times 22\cdot4}{2\cdot24} = 44 \text{ g}$$

∴ Molecular weight $= 44$. (The gas must be N_2O.)

(b) *350 cm³ of an oxide of carbon weigh 0·416 g at 298 K and 745 mm Hg.*

$$\frac{p_1 v_1}{T_1} = \frac{p_2 v_2}{T_2}$$

$$\begin{array}{cc} \uparrow & \uparrow \\ \text{actual} & \text{s.t.p.} \\ \text{conditions} & \text{conditions} \end{array}$$

Substituting,

$$\frac{745 \times 350}{298} = \frac{760 \times v_2}{273}$$

$$\therefore v_2 = \frac{745 \times 350 \times 273}{298 \times 760} = 315 \text{ cm}^3$$

$$\therefore 22\,400 \text{ cm}^3 \text{ weigh } \frac{0\cdot416 \times 22\,400}{315} = 28 \text{ g}$$

∴ Molecular weight $= 28$. (The gas must be CO.)

Measurements of gas volumes are sometimes used to determine molecular formulae of gases, and the composition of gaseous mixtures of known formula. In the case of hydrocarbon gases, the gas is sparked in a eudiometer tube (Fig. 7.3) with excess oxygen. The products are water, whose volume at room temperature is negligible in comparison with the gas volumes involved, and carbon dioxide, which is estimated by introducing potassium hydroxide solution and measuring the decrease in volume.

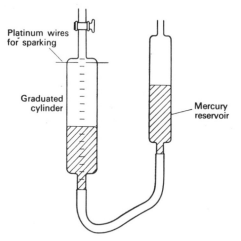

FIG. 7.3. Eudiometer.

The formula of the gas is then determined by applying Avogadro's law, together with **Gay-Lussac's law of combining volumes,** which states that when gases react they do so in volumes which bear a simple ratio to one another, and to the volume of any gaseous products formed, all measurements being made at the same temperature and pressure. Gay-Lussac's law is a consequence of Avogadro's law, as the following worked example shows:

Worked Example 11. 10 cm³ of a gaseous hydrocarbon were exploded with 75 cm³ of oxygen, and the gaseous products occupied 65 cm³ when cool. Potassium hydroxide solu-

tion caused the volume to be reduced to 45 cm³. The residual gas was shown to be oxygen by absorbing it completely in alkaline pyrogallol (an organic reducing agent). Determine the molecular formula of the hydrocarbon.

Let the hydrocarbon be C_xH_y. The equation for its combustion is

$$C_xH_y + \left(x + \frac{y}{4}\right)O_2 \longrightarrow xCO_2 + \left(\frac{y}{2}\right)H_2O$$

Applying Avogadro's law:

$$1 + \left(x + \frac{y}{4}\right) \longrightarrow x + \left(\frac{y}{2}\right)$$

mole   moles   moles   moles

$$\therefore 1 + \left(x + \frac{y}{4}\right) \longrightarrow x + \begin{matrix}\text{negligible}\\\text{volume of}\end{matrix}$$

volume volumes   volumes liquid H_2O

Substituting:

$$10 \text{ cm}^3 + (75-45) \text{ cm}^3 \longrightarrow 20 \text{ cm}^3$$
$$\therefore 1 \text{ cm}^3 + 3 \text{ cm}^3 \qquad \longrightarrow \quad 2 \text{ cm}^3$$
$$\therefore x = 2$$
$$\therefore y/4 = 3-2. \therefore y = 4.$$
$$\therefore \text{ The molecular formula is } C_2H_4.$$

Worked Example 12.   25 cm³ of a mixture of hydrogen, methane and carbon dioxide were exploded with 30 cm³ of oxygen; the total volume decreased to 22·5 cm³. Potassium hydroxide solution caused the volume to be reduced further to 12·5 cm³. The residue was completely absorbed by alkaline pyrogallol. Deduce the volume composition of the mixture.

Let x = number of cm³ of hydrogen,
    y = number of cm³ of methane,
    z = number of cm³ of carbon dioxide
        originally present.

$$H_2 + \tfrac{1}{2}O_2 \longrightarrow H_2O$$

Volumes: x $x/2$ negligible

$$CH_4 + 2O_2 \longrightarrow CO_2 + 2H_2O$$

Volumes: y $2y$ y negligible

Final volume of carbon dioxide after exploding $= y+z$.

$y+z$ = contraction in volume on adding KOH
$$= 22\cdot5 - 12\cdot5$$
$$= 10$$

Original volume of mixture before adding oxygen $\qquad = x+y+z = 25$
$$\therefore x = 15$$

Volume of oxygen used, from the equations
$$= x/2 + 2y = 7\cdot5 + 2y$$

By experiment, volume of oxygen used
$$= 30 - 12\cdot5 = 17\cdot5$$
$$2y = 17\cdot5 - 7\cdot5$$
$$\therefore y = 5$$
$$\therefore z = 10 - y = 5$$

$\therefore$ The gas consisted of 15 cm³ hydrogen, 5 cm³ methane, and 5 cm³ carbon dioxide.

7.7 Molecular weights of gases from density measurements

The relative density (otherwise known as the vapour density) of a gas may be defined as

$$\frac{\text{weight of a given volume of a gas under given conditions}}{\text{weight of the same volume of hydrogen under the same conditions}}$$

By Avogadro's law, it follows that the relative density equals

$$\frac{\text{molecular weight of the gas}}{\text{molecular weight of hydrogen}}$$

$$= \frac{\text{molecular weight}}{2\cdot016} = \frac{\text{molecular weight}}{2},$$

approximately.

Hence the molecular weight of a gas may be determined by weighing. It is not necessary in practice to make direct comparisons with hydrogen, since 1 mole of any gas which is approximately ideal in behaviour occupies 22 400 cm³ at s.t.p.

A rough laboratory determination of the molecular weight of a gas may be obtained by weighing a flask (i) evacuated, and (ii) containing the gas at known temperature and pressure. The weight which will occupy 22 400 cm³ at s.t.p. is then calculated—this is the molecular weight.

Measurement may be done in a flask of known volume, or alternatively in a graduated syringe such as that shown in Fig. 7.4. For instance, the molecular weight of carbon dioxide can be roughly checked by rapidly weighing a piece of solid carbon dioxide and allowing it to evaporate into a syringe.

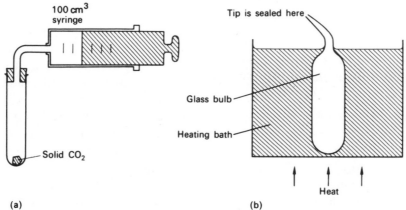

(a)   (b)

FIG. 7.4. (a) Measurement of a gas density with a syringe. (b) Dumas' method.

More accurate determinations were done by Regnault in 1845. Two identical globes fitted with stopcocks were counterpoised on a balance, one having been previously evacuated. The evacuated globe was then filled with hydrogen at a known temperature and pressure, and reweighed. The globe was then filled with the given gas and reweighed. The relative density of the gas, and hence its molecular weight, was then calculated.

The vapour of a volatile liquid may be weighed by **Dumas' method.** The sequence of operations is as follows:

(i) A glass bulb or boiling tube with a drawn off tip is first weighed full of air (W_1 g).

(ii) A few ml of the volatile liquid are introduced by warming the bulb and cooling it with the tip held under the surface of the liquid.

(iii) The bulb is placed in a heating bath to volatilize the liquid completely, until no more vapour issues from the bulb.

(iv) Still keeping the bulb in the heating bath, the tip is sealed off with a small flame.

(v) The bulb is cooled, dried, and weighed (W_2 g).

(vi) The tip of the bulb is broken off under water, causing water to rush in and fill the bulb completely.

(vii) The bulb is dried externally, and reweighed together with the broken off tip (W_3 g).

$$\text{Volume of bulb} = (W_3 - W_1) \text{ cm}^3.$$

The weight of air occupying this volume is calculated, at laboratory temperature and pressure (W_4 g). This is subtracted from W_1 giving the weight the bulb would have had if evacuated. Hence

Weight of vapour in bulb $= W_2 - (W_1 - W_4)$ g. The weight of vapour occupying 22 400 cm^3 at s.t.p. is then calculated: this is the molecular weight.

7.8 Determination of atomic weights

An exact knowledge of atomic weights is of importance to the analyst when finding the formula of a compound or the composition of a mixture by means of gravimetric or volumetric measurements. Nowadays very precise values for atomic weights, expressed on the carbon-12 scale, have been determined by means of the mass spectrometer; the accurate atomic masses of the isotopes of an element, and their relative abundance, are determined (section 1.5).

Before the mass spectrometer was invented chemists had to rely on other, often very ingenious, methods of finding atomic weights. Most of these are now of academic interest only, but one method of outstanding interest is **Cannizzaro's method** (1858). Cannizzaro based his argument upon the hypothesis of Avogadro (which we now know as Avogadro's law, but which at that time was not susceptible to experimental verification). Cannizzaro determined the molecular weight of a series of compounds of an element by assuming that the density of a vapour relative to hydrogen under the same conditions is always one-half of the molecular weight. Cannizzaro then determined the number of grams of the element in one g-molecule of each compound. The series of weights thus obtained were found to be whole number multiples of a weight which he took to be the weight of one g-atom.

The data in Table 7.1 illustrate the argument which enabled Cannizzaro to fix 35·5 as the atomic weight of chlorine:

TABLE 7.1

Substance	Molecular weight	Weight of chlorine in 1 g-molecule
Chlorine	71	$71 \text{ g} = 2 \times 35 \cdot 5 \text{ g}$
Hydrogen chloride	36·5	35·5 g
Mercury(II) chloride	271	$71 \text{ g} = 2 \times 35 \cdot 5 \text{ g}$
Arsenic(III) chloride	181·5	$106 \cdot 5 \text{ g} = 3 \times 35 \cdot 5 \text{ g}$
Phosphorus(III) chloride	138·5	$106 \cdot 5 \text{ g} = 3 \times 35 \cdot 5 \text{ g}$
Iron(III) chloride	325	$213 \text{ g} = 6 \times 35 \cdot 5 \text{ g}$

The argument does not *prove* that the atomic-weight of chlorine is 35·5, but the more compounds of chlorine are taken the more reliable the method becomes. Since no one has ever discovered a compound of chlorine in which one gram-molecule contains *less* than 35·5 g of chlorine, the atomic weight of chlorine may safely be assumed to be the least weight present.

The reasoning is in a sense analogous to that applied to Millikan's oil-drop method for determining the charge on an electron: the measurements there do not *prove* that the unit charges gained or lost by an oil-drop are equal to the charge on an electron, since there is always a possibility that electrons are gained or lost in pairs, or threes. We therefore take the unit amount of charge gained or lost as the most *probable* value for the charge on an electron, and there is plenty of corroborative evidence to support this assumption.

Study Questions

1. What are the weights of the following?

(a) 1 g-atom of carbon.
(b) 5 g-atoms of silicon.
(c) 0·1 g-atom of germanium.
(d) 0·2 g-atom of tin.
(e) 0·5 g-atom of lead.
(f) 1 g-molecule of CO_2.
(g) 1 mole of COS.
(h) 3 moles of $C_6H_{12}O_6$.
(i) 1 mole of $MnCl_2$.
(j) 0·1 mole of $Al_2(SO_4)_3$.

2. How many atoms are there in the following?
(a) One molecule of $(CH_3)_2CO$.
(b) 1 mole of $(CH_3)_2CO$.
(c) 15 molecules of CH_4.
(d) 8 g of CH_4.

3. (a) What is meant by a mole of electrons?
(b) What is (i) the charge, (ii) the mass of a mole of α-particles?

4. The internuclear distance in NaBr is 0·298 nm. The salt crystallizes with the simple cubic (6 : 6) structure. Work out a value for the density of NaBr.

5. Write down the equations for the following reactions:

(a) 0·653 g of zinc dust were added to an excess of copper sulphate ($CuSO_4$) solution. The copper that formed was filtered, dried and found to weigh 0·635 g.
(b) When 0·52 g of chromium was added to excess silver nitrate ($AgNO_3$) solution, 3·234 g of silver were formed.
(c) 20 cm³ of M KOH was titrated with 0·4 M H_2SO_4. The reaction was complete when 25 cm³ of the latter had been added.
(d) 7·17 g of a lead oxide was reduced in a current of hydrogen to give 6·21 g of metallic lead.
(e) 5 cm³ of M KI reacted with exactly 2·5 cm³ of M $Pb(NO_3)_2$ to give a yellow precipitate of lead iodide.

6. 6·98 g of a colourless liquid was heated; decomposition into chlorine and 5·56 g of a white solid, $PbCl_2$, occurred readily.

(a) What is the formula of the liquid?
(b) What structure would you expect the liquid to have?

7. When 2·74 g of barium was burned in oxygen, 3·38 g of a white solid was formed.

(a) Deduce the formula of the white solid.
(b) Is this result consistent with your ideas of the periodic law?

8. 2·6 g of a white crystalline compound was heated to 400°C in a vacuum, when it decomposed into 0·92 g of sodium, and nitrogen. What is the formula of the compound?

9. 0·54 g of aluminium was heated in a stream of chlorine to give 2·67 g of a white product, which sublimed at about 200°C. The whole of the product occupied a volume of 224 cm³ at 760 mm and 273 K.

(a) Calculate the empirical formula of the product.
(b) Calculate the molecular formula of the product.

10. 3·2 g of sulphur was burned in air to give 6·4 g of A. A reacted with oxygen in the presence of a catalyst to give 8·0 g of B. With water B gave first 8·9 g of C, and then 9·8 g of D. Identify A, B, C and D.

†11. 11·90 g of salt containing only potassium, sulphur and oxygen was heated. Only an acidic reducing gas was evolved, and 8·70 g of a white solid remained. The solid was dissolved in an excess of dilute nitric acid and an excess of barium nitrate added: 11·65 g of a white precipitate formed. Only the potassium ions from the original salt now remained in solution.

(a) What is the empirical formula of the salt?
(b) Write down the equation for the thermal decomposition of the salt.

12. A hydrocarbon of molecular weight 92 was subjected to combustion analysis. When 184 mg was completely burnt in oxygen, 616 mg of CO_2 and 144 mg of water were formed. Calculate (a) the empirical formula, and (b) the molecular formula of the hydrocarbon.

13. A compound was found to contain carbon and hydrogen. When 120 mg of it was completely burned in oxygen, the evolved CO_2 was found to weight 176 mg and the water 72 mg. Since all tests for other elements proved negative, the experimenter assumed that the compound also contained oxygen.

(a) Deduce the empirical formula of the compound.
(b) A further experiment showed that the molecular weight of the compound was 180. What is the molecular formula?

14.(a) Calculate the weight of solute in the following:
       (i) 1 dm³ of 2 M NaOH.
       (ii) 25 cm³ of 0·1 M H_2SO_4.
       (iii) 40 cm³ of M $KMnO_4$.
       (iv) 20 cm³ of 0·02 M KI.
       (v) 4 dm³ of M/2 HCl.

(b) To what volume would 250 cm³ of 2 M HCl have to be diluted in order to obtain a 0·1 M solution?
(c) 20 cm³ of 2 M KOH solution was diluted to 500 cm³. What is the molarity after dilution?

15. The concentration of thiocyanate ion, CNS^-, in solution can be estimated using silver nitrate solution, $AgNO_3$(aq.) Silver thiocyanate, AgCNS, is precipitated and the end point can be detected because thiocyanate ions form a deep red complex with Fe^{3+}(aq). A thiocyanate solution was titrated into 25 cm³ of a solution of silver nitrate containing 17 g dm⁻³. Successive burette readings were 23·9, 23·34, 23·36 and 23·32 cm³.

(a) Estimate the thiocyanate ion concentration.
(b) Suggest why the first reading was higher than the others.

16. The gram molecular volume of a gas at s.t.p. is 22 400 cm³. What would be the volume of a mole of an ideal gas at 23°C and 735 mm Hg?

17. The following analytical figures were obtained for certain compounds. Which are stoichiometric and which non-stoichiometric? Write empirical formulae where possible.

(a) 50% S and 50% O.
(b) 38·7% Ti and 61·3% F.
(c) 77·3% Ti and 22·7% O.
(d) 75% Ti and 25% O.
(e) 72·9% Ti and 27·1% O.
(f) 60·9% Fe and 39·1% S.
(g) 76·5% Fe and 23·5% O.
(O = 16, S = 32, Ti = 48, Fe = 56, F = 19.)

18. 15 cm³ of a gaseous hydrocarbon X was exploded with 100 cm³ of oxygen. When cool, the resulting gas mixture occupied 70 cm³; the carbon dioxide was absorbed by potassium hydroxide solution, and the gas volume decreased to 25 cm³. The remaining gas was shown to be oxygen by absorbing it in alkaline pyrogallol. What is the molecular formula of the hydrocarbon?

19. A mixture of ethane, C_2H_6, ethene, C_2H_4, and hydrogen was obtained after an experiment to test a new catalyst. 24 cm³ of this was exploded with 100 cm³ of oxygen. The volume after reaction was 70 cm³. Excess potassium hydroxide solution absorbed 40 cm³ of this and alkaline pyrogallol the remainder. What was the volume composition of the mixture?

CHAPTER 8

Molecular motion

8.1 The kinetic theory of gases

According to Boyle's law (Chapter 7) the volume of a given mass of gas should be inversely proportional to its pressure, but this is only approximately true, even for the so-called "permanent" gases. The extent of the deviation is greatest for those gases which are most easily liquefied. Figure 8.1 shows the variation of pV with pressure for three common gases. For an **ideal gas** there would be no variation—Boyle's law would be exactly obeyed—but a gas like nitrogen is found to be only about half as compressible at 1000 atm as Boyle's law would predict. Carbon dioxide at 40°C is found to be *more*

compressible than predicted; in fact, at room temperature carbon dioxide condenses to the liquid state on applying pressure alone.

The **kinetic theory** was developed mainly in the nineteenth century, and is a good example of the scientist's *model*, devised here in order to rationalize the behaviour of gases. The kinetic theory of gases is based upon the following postulates:

(1) Pressure is due to the bombardment of the walls of the containing vessel by molecules of the gas. Evidence for this comes from **Brownian motion**, for instance in smoke particles or colloidal sulphur. Observation with a microscope shows the particles to be in continual erratic motion, due to molecular bombardment.

(2) The average kinetic energy of the molecules in a gas increases as the temperature is raised. Heat energy is kinetic energy of molecular motion.

(3) Collisions occur between molecules and the walls of the containing vessel. Such collisions are perfectly *elastic*, i.e. no energy is dissipated in any other form. If this were not so, the pressure and temperature of a given volume of gas would gradually drop as the molecules slowed down.

(4) Attraction between molecules is negligible. (It will later be shown that this is only true at low pressures where the distance between gas molecules is large.)

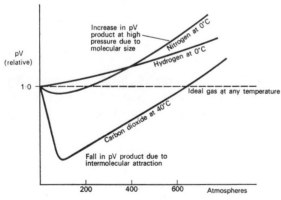

FIG. 8.1.

(5) The molecules are taken as being infinitely small.

This simple kinetic theory leads directly to Boyle's law and Charles' law. As far as the chemist is concerned, one of the most interesting things is that real gases do *not* obey the gas laws. The kinetic theory can help to explain the deviations.

8.2 Diffusion and effusion

One of the strongest supports for the kinetic theory came from the work of the Scottish chemist, Graham, in 1846. If a gas is allowed to escape from its container through a small hole into a vacuum, the rate at which molecules escape will depend upon the rate at which they reach the area represented by the hole. This process is called **effusion.** An analogous process is that of **diffusion,** which is the passage of a gas through a porous partition, such as a porous pot. Diffusion is similar to effusion, with a large number of tiny holes through which the molecules can escape instead of one single hole.

Graham's work on the effusion and diffusion of gases led him to the following law which applies equally well to either process:

The rates of effusion (diffusion) of gases under given conditions are inversely proportional to the square roots of their densities.

Effusion rates are most readily compared by taking two gases and allowing them in turn to pass through the same effusion hole, under identical pressure and temperature conditions, comparing the times for the same volume of each gas to pass through. It is found that:

$$\frac{time_1}{time_2} = \sqrt{\frac{d_1}{d_2}} \quad \text{or} \quad \frac{rate_2}{rate_1} = \sqrt{\frac{d_1}{d_2}}$$

It follows from Avogadro's law that the density, d, of a gas is proportional to its molecular weight, or to the mass m of its molecules; therefore

$$\frac{rate_2}{rate_1} = \sqrt{\frac{m_1}{m_2}} \qquad (1)$$

Does Graham's law support the elementary kinetic model of gases? If we were comparing equal volumes of two different gases under identical pressure and temperature conditions, they would contain equal numbers of molecules. Translated into molecular terms therefore, Graham's law suggests that the *number* of molecules of a gas which can effuse per unit time is inversely proportional to the square root of the molecular mass. But the number of molecules passing through the effusion hole per second is directly proportional to the *mean velocity* with which molecules impinge upon the area of the hole. Graham's law in fact suggests that

$$\frac{\text{mean velocity of molecules of gas}_2}{\text{mean velocity of molecules of gas}_1} = \sqrt{\frac{m_1}{m_2}}$$

Let c_1 = mean velocity of the molecules of gas$_1$

c_2 = mean velocity of the molecules of gas$_2$

Then, $\quad \dfrac{c_2}{c_1} = \sqrt{\dfrac{m_1}{m_2}}$

$$\therefore m_1 c_1^2 = m_2 c_2^2$$
$$\therefore \tfrac{1}{2} m_1 c_1^2 = \tfrac{1}{2} m_2 c_2^2 \qquad (2)$$

Graham's experiments therefore suggest that, if the kinetic theory is true, the mean kinetic energy of the molecules of different gases are the same, at a given temperature.

Diffusion is employed as a method of separating gases or vapours of different molecular weights. It is particularly valuable for separating isotopes, for example, uranium-235 (Chapter 2).

8.3 Boyle's law

Boyle's law is readily explained by the simple kinetic theory. The pressure of a gas is assumed to be due to the bombardment of molecules on the containing vessel. Now,

pressure = force exerted per unit area,
force   = rate of change of momentum in a given direction.

If we compress a gas from a volume V to a volume $V/2$ at constant temperature, it may be assumed that, since the temperature does not change, the mean kinetic energy of the molecules remains the same. However, since the gas is now occupying only half its original volume, a given volume of gas will now contain twice as many molecules. Consequently a given area of surface will now receive twice as many collisions per unit time as it did before. The number of collisions per unit time determines the total rate of change of momentum of the particles when they collide. Therefore

Rate of change of momentum

per unit area $\propto 1/V$
that is, force per unit area $\propto 1/V$
that is, pressure $\propto 1/V$.

†8.4 The pressure exerted by a gas

Consider a gas enclosed in a sphere of radius a (Fig. 8.2). The diagram represents any particle chosen at random, moving with velocity v.

Simple kinetic theory assumes collisions to be perfectly elastic: on this assumption the angle θ at which the particle bounces off the wall equals the angle at which it strikes the wall.

Triangle OAB is isosceles, therefore the particle always meets the surface at the same angle θ.

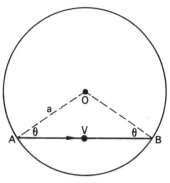

FIG. 8.2.

Time between collisions, for the particle shown

$$= \frac{AB}{v} = \frac{2a \cos \theta}{v}$$

The contribution of this one particle to the pressure

$$= \frac{\text{rate of change of momentum}}{\text{area of surface}}$$

$$= \frac{2mv \cos \theta}{2a \cos \theta} \cdot \frac{v}{4\pi a^2}$$

$$= \frac{mv^2}{4\pi a^3}$$

where m = mass of the particle.

The total pressure, p, = the sum of the pressure due to all the particles present

$$\therefore p = \Sigma \left(\frac{mv^2}{4\pi a^3} \right)$$

The mean square velocity, $\overline{c^2}$, is equal to the sum of the squares of the velocities divided by the number of molecules, n.

$$\overline{c^2} = \frac{1}{n} \Sigma v^2$$

$$\therefore p = \frac{nm\overline{c^2}}{4\pi a^3}$$

The density of the gas, ρ = mass ÷ volume

$$\rho = \frac{mn}{\left(\frac{4\pi a^3}{3} \right)}$$

$$\therefore p = \tfrac{1}{3} \rho \overline{c^2}$$

Also,

$$pV = \tfrac{1}{3} N_A m \overline{c^2}$$

where V = volume of one mole,
$\quad N_A$ = the number of molecules in a mole
$\qquad$ (the Avogadro number).

8.5 Kinetic energy and temperature

The experimental observations of the gas laws lead to the equation

$$pV = RT$$

where V = volume of 1 mole of the gas,
$\quad R$ = the gas constant, which is approximately the same for all gases.

In the previous section we derived the relationship

$$pV = \tfrac{1}{3} N_A m \overline{c^2}$$

where N_A is now the number of molecules in a mole, the Avogadro number.

$$pV = \tfrac{2}{3} \times \tfrac{1}{2} N_A m \overline{c^2}$$

$$= \tfrac{2}{3} \times \text{total kinetic energy of the molecules.}$$

Therefore the kinetic theory can be reconciled with the statement $pV = RT$ by making the simple assumption that
$T \propto$ total kinetic energy of the molecules.

8.6 Avogadro's law and the Avogadro number

The assumption that the mean kinetic energy of the molecules of gases at a given temperature will be the same, is also in accord with Avogadro's law. Let suffixes denote quantities relating to gas$_1$ and gas$_2$:

$$p_1 V_1 = \tfrac{1}{3} n_1 m_1 \overline{c_1^2} \quad \text{and} \quad p_2 V_2 = \tfrac{1}{3} n_2 m_2 \overline{c_2^2}$$

But,

$$\tfrac{1}{2} m_1 \overline{c_1^2} = \tfrac{1}{2} m_2 \overline{c_2^2} \quad (\text{since } T_1 = T_2)$$

Therefore, if we take equal volumes of gases ($V_1 = V_2$) at the same temperature and pressure ($p_1 = p_2$), they will contain equal numbers of molecules, i.e.:

$$n_1 = n_2.$$

8.7 Dalton's law of partial pressures

Dalton (1808) noticed that when two or more gases are mixed together, the total pressure exerted by the gas mixture is equal to the sum of the pressures which each separate gas would exert by occupying the space alone. Dalton coined the term **partial pressure** to describe this property. The partial pressure of a gas in a mixture is the pressure which that gas would exert if it alone occupied the same volume. Dalton's law of partial pressures states that the pressure exerted by a mixture of gases is equal to the sum of the partial pressures of the component gases.

Simple kinetic theory explains this behaviour. The mean kinetic energy of the molecules of a given gas depends solely on the temperature, and is not affected by the presence of different molecules. Hence each gas bombards the walls of the containing vessel independently of each other gas. Intermolecular collisions away from the vessel walls will not have any effect on the pressure.

8.8 The van der Waals equation

The fact that real gases do not obey the relationship $pV = RT$ exactly shows that some, at least, of the assumptions of the simple kinetic theory do not always apply. It is necessary to examine some of these assumptions more closely:

(a) *Collisions between molecules, and collisions with the vessel wall, are taken to be perfectly elastic.* This is reasonable for, if it were not so, the molecules would gradually slow down and finally settle out at the bottom of the vessel.

(b) *The molecules are taken to be infinitely small.* For real gases the validity of this assumption is bound to be questioned. The point is whether the volume occupied by actual molecules is negligible compared to the total volume occupied by the gas. It would be a fair assumption with small molecules at very low pressures. Van der Waals (1873) suggested a modified form of the gas equation in which the volume term V was replaced by a term $(V-b)$ where b represents a constant correction term which allows for the actual volume of the molecules. b becomes important when V becomes small, that is, when the gas is highly compressed.

(c) *Intermolecular attraction is taken to be zero.* This assumption becomes progressively less true the nearer the gas molecules are to one another. Again the attraction term becomes strongest when the gas is highly compressed. Van der Waals proposed modifying the pressure term p in the gas equation. If p is the actual pressure which a gas exerts on the containing vessel, this pressure will be *less* than the "ideal" pressure, due to intermolecular attraction. The "ideal" pressure (the pressure that the gas would exert if intermolecular attraction were absent) may be written $\left(p+\dfrac{a}{V^2}\right)$.

The van der Waals pressure term $\left(p+\dfrac{a}{V^2}\right)$ is not entirely satisfactory, and several other attempts have been made to improve on it. It now seems likely that no simple equation will ever adequately describe the real pressure exerted by a gas.

Inserting both correction terms, the van der Waals equation reads:

$$\left(p+\frac{a}{V^2}\right)(V-b) = RT$$

The term a will be relatively large for molecules with permanent dipole moments such as hydrogen chloride. What is surprising is that it is also quite large for some molecules which have no permanent dipole moment, such as chlorine. a increases with ease of liquefaction of the gas. Attractive forces between molecules are known as van der Waals forces.

Figure 8.3 shows a comparison between three sets of plots of pressure against volume. (a) is

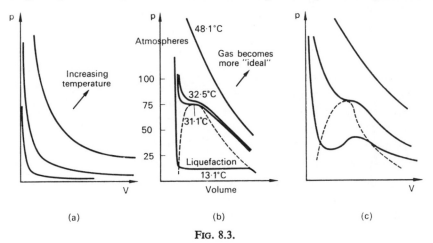

(a) (b) (c)

FIG. 8.3.

for an ideal gas: it will be seen that no effect such as liquefaction is predicted. (b) shows the actual plots for a real gas, carbon dioxide. It will be seen that below $31 \cdot 1°C$, the gas can be liquefied by applying pressure alone. $31 \cdot 1°C$ is called the **critical temperature** of carbon dioxide. (c) shows a set of plots using the van der Waals equation, with values of a and b chosen to give the closest possible fit to real conditions. Although graph (c) bears only superficial resemblance to (b) it will be seen that something akin to liquefaction and a critical temperature are in fact predicted.

It is instructive to re-examine Fig. 8.1 at this stage, and attempt to see which of the factors a and b are responsible for the deviations in the plot of pV against p for the gases considered.

8.9 Heat capacities of gases

The specific heat capacity of a substance is normally defined as the amount of heat required to raise the temperature of 1 kg of the substance through 1 K. A chemist is generally more interested in mole quantities than gram quantities, and so in this chapter we shall refer to *molar* heat capacity—the amount of heat to raise 1 *mole* of the substance through 1 K.

The molar heat capacity of a gas will differ, depending upon whether the measurements are made at constant volume or at constant pressure. If a gas is heated at constant pressure it will expand and will do mechanical work as it expands. Thus the molar heat capacity at constant pressure C_p, will be greater than the molar heat capacity at constant volume, C_v, by an amount equal to the mechanical work done.

Suppose the gas expands from V_1 to V_2 when the temperature is raised one degree, from T_1 to T_2, $(T_2 - T_1 = 1)$. We know that mechanical work done $=$ force $\times$ distance moved by force, and this may be rewritten as (force/area)$\times$

increase in volume. In other words:

$$\text{Work done} = p(V_2 - V_1).$$
$$\text{Since} \quad pV = RT$$
$$\therefore p(V_2 - V_1) = R(T_2 - T_1),$$

and since we made the rise in temperature 1 K,

the work done in expansion $= R$.

$$\therefore C_p - C_v = R \qquad (3)$$

The heat absorbed at constant volume on increasing the temperature of a gas from T_1 to T_2 is $C_v(T_2 - T_1)$. This quantity is equal to the increase in the kinetic energy of the molecules. In section 8.5 the total kinetic energy of the molecules was shown to be $\frac{3}{2}pV$, which for an ideal gas is $\frac{3}{2}RT$. The increase in the kinetic energy on going from T_1 to T_2 is therefore $\frac{3}{2}R(T_2 - T_1)$.

Thus $\quad C_v(T_2 - T_1) = \frac{3}{2}R\,(T_2 - T_1)$
$$Cv = \tfrac{3}{2}R \backsimeq 12 \cdot 6 \text{ J K}^{-1}$$

At first sight it is perhaps surprising that the gas constant, R, should turn out to be a difference of molar heat capacities. It is not really so odd, however, since pV has the dimensions of work or energy, and molar heat capacity has the dimensions of energy divided by temperature:

$$R = \frac{pV}{T} = \frac{\text{energy}}{\text{temperature}} \qquad (4)$$

For equation (4) to be meaningful to us, however, the energy term pV must be expressed in joules (J), and R expressed in the usual units for molar heat capacity, J mol^{-1} K^{-1}. This can be done if we remember that $0 \cdot 0224$ m^3 of an ideal gas at 273 K, exerts one atmosphere pressure (equal to $0 \cdot 76 \times 9 \cdot 81 \times 13\,600$ N m^{-2} where

$0 \cdot 76 =$ barometric height, in m;
$9 \cdot 81 =$ acceleration due to gravity, in m s^{-2};
$13\,600 =$ density of mercury, in Kg m^{-3}).

$$R = \frac{pV}{T} = \frac{0 \cdot 76 \times 9 \cdot 81 \times 13\,600 \times 0 \cdot 0224}{273} \backsimeq$$

$8 \cdot 4$ N m K^{-1} ($8 \cdot 4$ J K^{-1}).

TABLE 8.1

Gas	C_p, J K^{-1}	C_v, J K^{-1}	$C_p - C_v$, J K^{-1}	$C_p/C_v = \gamma$
Hydrogen	28·8	20·5	8·3	1·41
Helium	20·9	12·6	8·3	1·66
Oxygen	24·2	20·9	8·3	1·41
Carbon monoxide	24·3	21·0	8·3	1·41
Carbon dioxide	37·2	28·7	8·5	1·30
Ammonia	36·7	28·1	8·6	1·31
Ethylene	43·9	34·2	8·7	1·26

Since $R = 8\cdot4$ J K^{-1} for an ideal gas,

$\qquad C_v = 12\cdot6$ J K^{-1} for an ideal gas,

and $\qquad C_p = C_v + R$.

$\qquad \qquad \simeq 21\cdot0$ J K^{-1}.

Summarizing the results we have so far:

(1) The gas constant of an ideal gas $\simeq 8\cdot4$ J K^{-1}.

(2) C_p for an ideal gas $\simeq 21\cdot0$ J K^{-1}.

(3) C_v for an ideal gas $\simeq 12\cdot6$ J K^{-1}.

(4) The ratio C_p/C_v, often denoted by the symbol γ, for an ideal gas approximately equals 5/3.

Let us see how real gases compare with these predictions. Table 8.1 gives some values for well-known gases. It will be seen that only helium comes approximately near to the predictions, although all give roughly the right answer for $(C_p - C_v)$.

A closer study of the data in Table 8.1, and other figures, reveals that it is only *monatomic* gases which behave like ideal gases with respect to their molar heat capacities. The diatomic gases in the table have values rather close to $C_p = \frac{7}{2}R$, and $C_v = \frac{5}{2}R$; the higher the **atomicity** of the gas (the number of atoms per molecule), the higher the values of C_p and C_v become.

The reason for this discrepancy with polyatomic gases is as follows: monatomic gas molecules can move from place to place—translation—but they cannot undergo rotation or vibration. Monatomic gases can therefore only possess **translational energy,** while polyatomic gases can possess **rotational energy** and (sometimes) **vibrational energy.** Our arguments so far have been concerned with the kinetic energy of molecules, equally distributed between three translational **degrees of freedom.** A diatomic molecule can in addition possess *two* rotational degrees of freedom; it is not meaningful to talk about the rotation of a diatomic molecule about its axis, any more than we can talk about the rotation of an isolated atom. An isolated atom has an electron charge cloud which we cannot legitimately think of as being able to rotate in the usual sense of the word.

8.10 Principle of equipartition of energy

The observed molar heat capacities of gases can be explained by assuming that each available degree of freedom contributes an *equal amount* of energy—$\frac{1}{2}RT$ energy units—to the total energy. This is the principle of **equipartition of energy.** Hence a monatomic gas, having three degrees of translational freedom, has a molar heat capacity of $3R/2$. A diatomic gas, having five degrees of freedom (three translational and two rotational), has a molar heat capacity of $5R/2$.

TABLE 8.2

THE EFFECT ON C_v OF INCREASING TEMPERATURE (J K^{-1})

Gas	0°C	100°C	200°C	500°C	1200°C	2000°C
Argon	12·5	12·5	12·5	12·5	12·6	12·6
Nitrogen	20·9	21·2	21·6	22·0	24·1	26
Hydrogen chloride	20·9	21·7	22·1	22·8	25·7	29
Chlorine	24·9	26	28	29	30	30

specific heats C_v increase
(except argon)

A polyatomic molecule will have more degrees of freedom, the exact number depending upon its shape and the extent to which bending and vibration can occur. The higher values for carbon dioxide, ammonia and ethylene in Table 8.1 can be interpreted on this basis.

So far one possible type of movement of a diatomic molecule has been neglected, namely vibration. It appears from the above data that the diatomic molecules considered, $H_2(g)$, $O_2(g)$, and $N_2(g)$, do not vibrate at room temperature. If such vibration did occur, *two* extra degrees of freedom would have to be added because a vibrating bond possesses kinetic energy and potential energy. If all the molecules in a diatomic gas were freely vibrating, its predicted molar heat capacity, C_v, would therefore be $7R/2$, which is approximately 29 J K^{-1} mol^{-1}. Table 8.2 shows that diatomic molecules do indeed start to vibrate significantly as the temperature is raised.

If the temperature is raised indefinitely, there comes a point where the molecule will vibrate so violently that it will be disrupted completely. It will dissociate into atoms. It is instructive to examine the values of C_v at 2000°C for the molecules N_2, HCl, and Cl_2. Even at this high temperature it appears that the nitrogen molecules are not all vibrating. It would be misleading to expect any close correlation between "freedom of vibration" and energy of dissociation,

but the following figures show that a qualitative correlation does exist:

$$N_2(g) \longrightarrow 2\ N(atoms);$$
$$\Delta H = +\ 940\ kJ\ mol^{-1}.$$
$$HCl(g) \longrightarrow H(atoms) + Cl(atoms);$$
$$\Delta H = +\ 420\ kJ\ mol^{-1}.$$
$$Cl_2(g) \longrightarrow 2\ Cl(atoms);$$
$$\Delta H = +\ 242\ kJ\ mol^{-1}.$$

8.11 The heat capacity of solids

In the previous sections the heat capacities of gases were explained in terms of their molecular structure and their number of degrees of freedom. Similar correlations can be found amongst solids, in particular amongst elements.

Despite the wide diversity in structure which solid elements possess, the remarkable fact is that practically all of them have similar heat capacities per *gram-atom*. This quantity is often referred to as the **atomic heat capacity**. For most solid elements it is in the region 25 to 27 J g-atom^{-1}, as shown by the data in Table 8.3.

This fact was first discovered as early as 1819 by Dulong and Petit, and is commonly known as **Dulong and Petit's law**. It was later shown that certain elements, notably beryllium, boron, carbon and silicon, do not obey the law at room temperature, though they do as the temperature

TABLE 8.3

Element	Atomic Weight	Specific heat (J g^{-1} K^{-1})	Atomic heat J g-atom^{-1} K^{-1}
Li	7	3·8	27
Al	27	0·87	23
Ca	40	0·63	25
Fe	56	0·43	24
Ag	108	0·23	25
I	127	0·22	28
U	238	0·11	26

is raised. Further work has shown that the law breaks down for all elements at very low temperatures. Nevertheless, the fact that elements with such widely different atomic weights as lithium and uranium should have approximately the same atomic heat capacity cannot be mere coincidence. The principle of equipartition of energy, previously applied to gases, can also explain Dulong and Petit's law.

An ideal solid can be considered to be a lattice of *independently* vibrating atoms. Each atom can vibrate in three possible directions. Each mode of vibration is represented by two energy terms, a kinetic energy term and a potential energy term (cf. a vibrating diatomic molecule, section 8.9). Hence each atom has six degrees of freedom. Assuming the principle of equipartition of energy can be extended to solids, each degree of freedom should contribute $\frac{1}{2}RT$ to the total energy. Hence the atomic heat capacity, by an argument similar to that in section 8,8, will be $6 \times \frac{1}{2}R = 3R \simeq 25$ J g-atom^{-1} K^{-1}.

It is clear from the experimental data that the theory contains oversimplifications, although it gives reasonable agreement. It is reasonable to suppose that elements such as carbon deviate simply because their atoms do not vibrate independently at room temperature, and they therefore have fewer degrees of freedom. On raising the temperature, vibration becomes easier and the law is more closely obeyed.

8.12 The behaviour of liquids

The liquid state is the least understood of the three states of matter. The kinetic theory of matter must explain how liquids flow, and why they are almost incompressible. Liquids are often described as possessing *short-range order*, and long-range disorder. In other words if a very small region of a liquid, say 1 or 2 nm in diameter, were examined closely, it would appear to be relatively ordered—almost as well ordered as a crystal. If, on the other hand, a larger volume were observed it would appear to be highly disordered.

The disorder of a liquid is thought to be due to the movement of "holes" through the system. The molecules in a liquid can move rapidly relative to one another, provided there are vacancies into which they can move. The "holes" cannot contribute more than a small amount to the total volume of the liquid, otherwise liquids would be compressible.

As the liquid state is so complex, it is hardly surprising that no simple generalizations can be made about their heat capacities. Translation, rotation, and vibration all contribute to the degrees of freedom available to liquid molecules, to varying extents.

8.13 Kinetic theory and rate of chemical reactions

Most chemical reactions proceed faster as the temperature is raised; this is what would be expected from simple kinetic theory. Simple kinetic theory fails however to account for the very pronounced effect which a rise in temperature has on most chemical reactions. The rates of many chemical reactions are approximately doubled by raising the temperature about 10 K.

An understanding of simple kinetic theory —the various ways in which the particles of matter can move in different states—is essential to the broader understanding of chemical pro-cesses. **Chemical kinetics** (Chapter 17) is the study of rates of chemical reactions, and the deduction of the nature of collision processes and other effects from them.

Study Questions

1. (a) Which of the following would diffuse most quickly under comparable conditions?

    (i) Ar, Kr or Xe.
    (ii) H_2O, HDO or D_2O.
    (iii) $^{235}UF_6$ or $^{238}UF_6$.

   (b) In (iii), Calculate the relative rates of diffusion under comparable conditions.

2. (a) Under comparable conditions, 100 cm³ of nitrogen diffused in 21 s while 100 cm³ of a coloured vapour diffused in 63 s.

    (i) Calculate the approximate molecular weight of the vapour.
    (ii) Suggest what the colour of the vapour might be.

   (b) A sample of 50 cm³ of neon diffused in 26 s, while a sample of 50 cm³ of a compound of empirical formula BNH_2 diffused in 52 s under comparable conditions. Calculate the molecular formula of the compound.

3. (a) Predict the specific heat capacities, in J g⁻¹ K⁻¹, of cobalt (at. wt. 59), osmium (at. wt. 190), silver (at. wt. 108), and sodium (at. wt. 23).

   (b) At room temperature, boron has a specific heat capacity of 1·26 J g⁻¹ K⁻¹.
    (i) What is the atomic heat capacity of boron at this temperature?
    (ii) How does this compare with the values obtained for other solid elements?
    (iii) What values would you expect for the atomic heat capacity of boron at (a) about 100 K, and (b) about 1000 K?

4. (a) In Table 8·1, the value of γ is the same for hydrogen, oxygen and carbon monoxide. Why is this?

   (b) Predict values of γ for the following:
    (i) Argon.
    (ii) Fluorine.

    (iii) Nitrogen(I) oxide, N_2O.
    (iv) Nitrogen(II) oxide, NO.
    (v) A gas of high atomicity.

5. (a) How would you expect the molar heat capacity of bromine to vary with temperature?

   (b) What variations would you expect for the neighbouring element, krypton?

6. A gas, with $\gamma = 1·67$, weighed 284 mg at a pressure of 178 mm Hg in a 347 cm³ bulb at 19°C. What is the gas?

7. The following densities (all in g cm⁻³) were obtained for nitrogen and water:

	Solid	Liquid	Gas
Nitrogen	1·02	0·81	0·0012
Water	0·91	1·00	0·0008

   (a) Estimate the approximate volume that is taken up by a molecule in each phase.
   (b) Contrast the results obtained for each substance.
   (c) Is it nitrogen or water that is behaving in an unexpected manner?

8. Although Avogadro first formulated his hypothesis in 1811, it remained unknown for many years. Use your library to find out (a) why this was, and (b) when Avogadro's hypothesis was first accepted by chemists.

9. Explain the deviations in the plots of pV against p in Fig. 8.1 in terms of the van der Waals parameters a and b.

10. The rate of decomposition of HI at 400°C is almost three hundred times as fast as it is at 300°C (Bodenstein, 1899). Can this be accounted for by the increased number of collisions that occur at the higher temperature due to the increased molecular speeds?

CHAPTER 9

Enthalpy

9.1 The importance of energy changes in chemistry

Two important things which a chemist must understand about the behaviour of matter concern chemical reactions. He must ask:

(i) How *fast* does a chemical reaction go?
(ii) How *far* does it go?

The heat energy change of a chemical reaction may give a *partial* explanation of its kinetics (how fast), and its position of equilibrium (how far), although heat is not the *only* factor involved. It is a matter of common experience that chemical processes tend to reach a position of equilibrium, after which no further spontaneous change occurs. Quite often a spontaneous change takes place with the absorption of heat, indicating that the system has attained a state of higher energy. Cooling by evaporation is a well known example; another is the dissolution of sodium nitrate in water to form an aqueous solution, where a very noticeable cooling can be observed. Figure 9.1 represents the change

$$NaNO_3(c) + aq \longrightarrow Na^+(aq) + NO_3^-(aq);$$

$$\Delta H = + 21 \text{ kJ mol}^{-1}$$

by means of an energy diagram.

Other spontaneous changes take place with the evolution of heat energy, for instance the burning of hydrogen in oxygen. Although heat

energy is not the only factor influencing chemical change, it is very important.

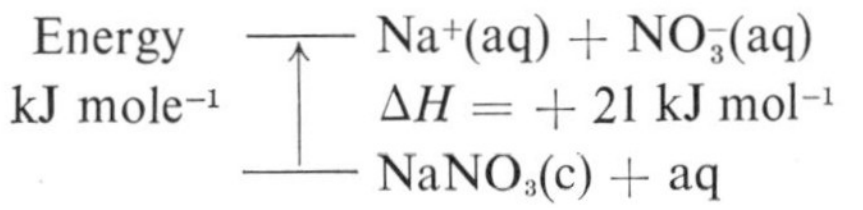

Energy kJ mole^{-1} Na$^+$(aq) + NO$_3^-$(aq)

$\Delta H = + 21$ kJ mol^{-1}

NaNO$_3$(c) + aq

FIG. 9.1.

9.2 Measurement of enthalpy change, ΔH

The symbol ΔH refers to the heat change of a process at constant pressure, often known as the **enthalpy** change.* Various ways are available for measuring ΔH, depending on the reaction conditions. Whatever quantities are actually used in the experiment, the enthalpy change is generally converted into molar units. For instance a calorimetric measurement may show that 33 J are required to convert 0·1 g of ice at 0°C into water at the same temperature. This is

---

* Many data books use composite symbols, in which a superscript zero denotes that standard conditions are observed, and a subscript numeral (usually 298) denotes the temperature to which the measurement refers. For instance ΔH^0_{298} denotes the standard enthalpy change at 298K (25°C). In this book all ΔH data are quoted for 25°C unless otherwise stated.

often expressed by saying that the specific latent heat of fusion of ice is 330 J g⁻¹. However the chemist expresses quantities per mole. ΔH for the reaction

$$H_2O(c) \longrightarrow H_2O(l) \text{ is}$$

$$+ \frac{(330 \times 18)}{1000} = + 5.94 \text{ kJ mol}^{-1}.$$

(a) CALORIMETRIC MEASUREMENTS

Since many chemical reactions occur in dilute aqueous solutions, the specific heat capacity of which can be assumed approximately to equal 4.2 J g⁻¹, a simple measurement of temperature change when two aqueous solutions are mixed gives a rapid estimate of ΔH. For instance, suppose 1 mole (98 g) of concentrated sulphuric acid (100%) was added to water to make 1 litre of solution, and a temperature rise of 17 K was noted. If we assume the specific heat capacity of a molar solution of sulphuric acid to be 4.2 J g⁻¹, we may write

$$H_2SO_4(l) + aq \longrightarrow H_2SO_4(aq, \text{M});$$

$$\Delta H = - (17 \times 4.2) \simeq -71 \text{ J mol}^{-1}.$$

Note that the rise in temperature in degrees is numerically equal to the heat evolved in kJ mol⁻¹ ÷ 4.2, in this case.

Similarly the neutralization of an acid with an alkali can be observed in this way. Suppose 100 cm³ of M NaOH(aq) were added to 100 cm³ of 0.5 M H₂SO₄(aq), and a rise in temperature of 6.8 K was noted. Assuming the specific heat capacities of the solutions to be 4.2 J g⁻¹,

$$\text{Heat evolved} = 6.8 \times 200 \times 4.2 \simeq 5600 \text{ J}$$

This is for mixing 0.1 mole of OH⁻(aq) with 0.1 mole of H⁺(aq)—note that every mole of **H₂SO₄** gives rise to 2 moles of H⁺(aq)—and so we may write:

$$H^+(aq) + OH^-(aq) \longrightarrow H_2O(l);$$

$$\Delta H = -54 \text{ kJ mol}^{-1}.$$

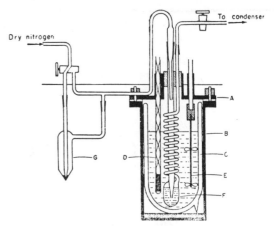

FIG. 9.2. Calorimeter for measuring heats of hydrolysis (with water vapour, at 25°C). A, lid; B, can; C Dewar-vessel; D, thermistor; E, stirrer; F, reaction chamber; G, water saturator.

For molar solutions the rise in temperature in degrees is numerically *one half* of the heat evolved in kJ mol⁻¹ ÷ 4.2, unlike the first case.

It is possible to find the end-point of an acid–alkali titration by taking various volumes of solutions and mixing them, noting the temperature rise in each case. This forms the basis of a **thermometric titration** (see Study Question 6).

The apparatus used for calorimetric measurements depends on the degree of accuracy required. For very rough measurements nothing more elaborate is needed than a thin plastic beaker or bottle, mounted in an insulating support, such as a block of foam polystyrene. For more accurate work a Dewar flask is essential. Figure 9.2 shows a typical experiment.

(b) ELECTRICAL MEASUREMENTS

An arrangement is sometimes employed, in which the heat rise produced by a chemical reaction is directly compared with the heat rise generated by an electric heating coil under identical conditions. In this way errors due to heat loss may be compensated. This type of measurement gives the result directly in joules:

Current(amperes) × time(seconds) ×
    P.D. (volts)
= charge(coulombs) × P.D.(volts)
= energy(joules).

(c) THE FLAME CALORIMETER

Simple measurements of heats of combustion require nothing more sophisticated than a burner with a wick, arranged to heat a known mass of water in a suitable heat exchanger. The weight of fuel burned is noted. This method is limited to inflammable liquids, and is in any case rather inaccurate, but is instructive for comparing such things as the molar heats of combustion of members of a homologous series, for instance aliphatic alcohols (see Study Questions 1, 2, 3). Figure 9.3 shows a more accurate arrangement,

used for measuring the heat of combustion of a gas like hydrogen.

(d) THE BOMB CALORIMETER

For more accurate work a bomb calorimeter is used, the original design being due to Berthelot (1867). The substance is placed in a platinum crucible in a cylindrical steel bomb lined with a resistant enamel, which is immersed in water. The bomb is filled with oxygen at a pressure greater than atmospheric, and the combustion is triggered off with a small platinum resistance wire. The heat evolved is measured by noting the rise in temperature of the water, and a small correction is made for the heating effect of the platinum resistance wire. The advantage of using pure oxygen is that the combustion is very rapid, and is certainly complete.

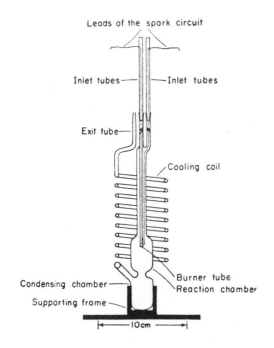

FIG. 9.3. A flame calorimeter.

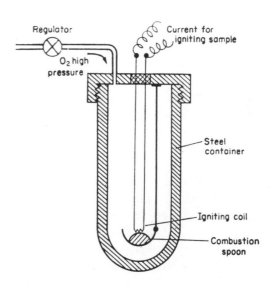

FIG. 1. Bomb calorimeter.

FIG. 9.4. A bomb calorimeter. The "bomb" is immersed in water (not shown)

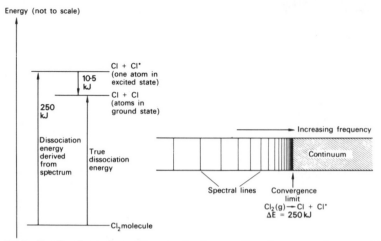

FIG. 9.5. Idealized vibrational spectrum of a gas; (real gases often show more than one series of lines).

†(e) SPECTROSCOPIC MEASUREMENTS

The heats of reaction of many processes involving ionization in the gas phase can be determined by measuring the wavelength and hence the frequency of the radiation absorbed in the change. In Chapter 3, the principles of the method used to determine the ionization energies of the alkali metals (e.g. $Na(g) \rightarrow Na^+(g)$) were outlined.

Spectroscopic measurements can also be used to determine the energy needed to dissociate molecules into atoms in the gas phase. As an example consider chlorine, Cl_2. When the chlorine molecule is made to vibrate, it absorbs a given quantity of energy and this gives rise to the absorption of radiation of definite frequency. The vibrational spectrum of chlorine consists of a number of lines, each at a definite frequency, which converge to a limit, beyond which energy is continuously absorbed (Fig. 9.5). The energy of this convergence limit corresponds to the dissociation energy, the energy necessary to split the molecule into atoms.

If this were the whole story, it would be a simple matter to compute dissociation energies from spectroscopic measurements: unfortunately some of the energy is also used to form an electronically excited chlorine atom, Cl^*, and this energy must be accounted for in the calculation of the dissociation energy of chlorine.

$$Cl_2(g) \longrightarrow Cl + Cl^*$$
$$\text{excited}$$
$$\text{atom}$$

The convergence limit for chlorine is 478·5 nm Using the relationship $\Delta E = h\nu$, we have:

$$\text{Wavelength, } \lambda, = 478 \cdot 5 \times 10^{-9} \text{ m}$$

$$\text{Frequency, } \nu, = \frac{3 \times 10^8}{478 \cdot 5 \times 10^{-9}}$$

$$= 6 \cdot 27 \times 10^{14} \text{ s}^{-1}.$$

$$\Delta E = h\nu$$

$$h = 6 \cdot 62 \times 10^{-34} \text{ J s}$$

$$\therefore \Delta E = 6 \cdot 62 \times 10^{-34} \times 6 \cdot 27 \times 10^{14}$$

$$= 4 \cdot 15 \times 10^{-19} \text{ J molecule}^{-1}$$

$$= 4 \cdot 15 \times 10^{-19} \times 6 \cdot 023 \times 10^{23} \text{ J mol}^{-1}$$

$$= 2 \cdot 50 \times 10^5 \text{ J mol}^{-1}$$

$$= 250 \text{ kJ mol}^{-1}$$

Atomic excitation energy is 10·5 kJ, so the dissociation energy of chlorine is 239·5 kJ mol⁻¹, in good agreement with the value of 238 obtained from thermal measurements.

The method, however, has severe limitations; it can only be applied to very simple molecules, and it is necessary to know the electronic states of the atoms after dissociation so that corrections can be applied. Within these limitations, the method is potentially extremely accurate.

(f) CALCULATION OF ΔH FROM EQUILIBRIUM CONSTANT MEASUREMENTS

In many cases an accurate value for the heat change of a reaction can be derived from measuring the way in which the equilibrium composition of a mixture in a reversible reaction varies with temperature.

(g) DERIVATION OF ΔH USING AN ENERGY CYCLE

It frequently happens that a measurement cannot be made directly on a chemical change, though separate measurements can be made on the values of ΔH for the process carried out in stages.

Hess's law of constant heat summation states that **the total heat evolved or absorbed in a given chemical reaction is the same whether the reaction proceeds directly or in a series of stages, and is dependent solely on the states of the initial reactants and final products.** Hess's law is really just another way of stating the law of conservation of energy: if it were not true it would be possible to reverse a given reaction using a different route with a different corresponding heat change. This is impossible since it would involve either the creation or the destruction of energy.

An important assumption which forms a part of Hess's law, and which is readily demonstrated experimentally in many cases, is that if a reaction is reversed then ΔH of the reverse reaction will be *minus* the value for the forward reaction. For instance, it is not at all easy to make direct measurements of the heat of reaction of $HCl(g) \rightarrow \frac{1}{2}H_2(g) + \frac{1}{2}Cl_2(g)$. The reverse

process is, however, readily carried out: when one mole of hydrogen chloride is formed by burning hydrogen in chlorine, 92 kJ are evolved. Hence ΔH for the decomposition of hydrogen chloride is $+92$ kJ mol^{-1}.

Hess's law may be illustrated by means of an energy level diagram, Fig. 9.6. ΔH is the algebraic sum of ΔH_1, ΔH_2, ΔH_3 and ΔH_4.

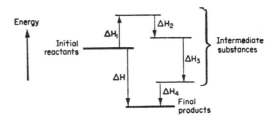

Hess's law: $\Delta H = \Delta H_1 + \Delta H_2 + \Delta H_3 + \Delta H_4$ (Algebraic sum)

FIG. 9.6.

9.3 Heat of formation of a substance

Since energy plays such an important part in our understanding of chemical substances, it is important that we should have some idea of the energy "stored up" in substances. For instance, if hydrogen gas is blown through an electric arc struck between carbon electrodes, some acetylene, C_2H_2, is formed. An extremely high temperature (above 2000°C) is needed in order to form appreciable quantities of acetylene and it is not possible to measure the heat of reaction directly:

$$2 \, C(c) + H_2(g) \rightleftharpoons C_2H_2(g);$$
$$\Delta H = + 240 \text{ kJ}$$

The heat of the above reaction, termed the **heat of formation** of acetylene, can be deduced from separate measurements of the **heats of combustion** of the three substances involved, carbon, hydrogen and acetylene. **Heat of combustion is defined as the heat evolved when one mole of a**

substance is completely burned in oxygen. (A minus sign denotes that heat is evolved.)

$$C(c) + O_2(g) \longrightarrow CO_2(g); \qquad (1)$$
$$\Delta H = -395 \text{ kJ}$$

$$H_2(g) + \tfrac{1}{2}O_2(g) \longrightarrow H_2O(l); \qquad (2)$$
$$\Delta H = -280 \text{ kJ}$$

$$C_2H_2(g) + \tfrac{5}{2}O_2(g) \longrightarrow 2\,CO_2(g) + H_2O(l); \quad (3)$$
$$\Delta H = -1310 \text{ kJ}$$

Remembering that reactions can be theoretically reversed, and the sign of their ΔH changed, the heat of formation of acetylene can be derived by taking $2 \times (1) + (2) - (3)$.

In order that data may be listed for reference purposes, heats of formation are quoted at a standard temperature, usually 25°C (298 K) which is conveniently near the temperature of the average laboratory. The substances must be in standard states and if there is any ambiguity, the state is given. For elements the standard used is that state of the element at 1 atm pressure and in its stable allotropic form.

Heat of formation is defined as the quantity of heat evolved or absorbed when one mole of a compound is formed from its elements in their standard states. It is often denoted by the symbol ΔH_f°.

A reaction which evolves heat is called an **exothermic reaction.** In an exothermic reaction the products will be shown on an energy level diagram *below* the reactants. A compound formed exothermically from its elements is called an *exothermic compound.* Carbon dioxide and water are examples of exothermic compounds.

A reaction which takes in heat is called an **endothermic reaction.** For instance the formation of acetylene from its elements is an endothermic reaction, and acetylene itself is called an *endothermic compound.* An endothermic compound can be regarded as having a larger amount of stored energy (it is higher up on the energy level diagram), and will give out a considerable amount of heat when burned in oxygen. The oxy-acetylene flame produces an extremely intense heat (about 3000°C) for this reason.

A reaction which neither evolves nor absorbs heat is sometimes termed a **thermoneutral** reaction.

9.4 Photochemical reactions

The endothermic formation of acetylene from its elements was caused to occur spontaneously by the choice of a very high temperature, in the example chosen above. Many endothermic compounds can in fact be formed in this way, and it is a fact of common experience that endothermic species tend to be stable at high temperatures. For instance if a diatomic gas is heated strongly enough it will dissociate into atoms, and these atoms are formed endothermically. Iodine vapour above 1000°C is present largely as atoms—an instance of the endothermic species being stable at the higher temperature. However, it is often not possible to form an endothermic compound simply by heating the appropriate elements together to a high enough temperature.

Some endothermic reactions proceed as a result of the absorption of electromagnetic radiation. Such reactions are known as **photochemical reactions.** Ultra-violet energy is the most useful for this purpose—X-rays have so much energy that they tend to disrupt molecules completely, while visible and infra-red light frequently contains insufficient energy. In the leaves of plants carbohydrates are formed endothermically from carbon dioxide and water. Light is essential for this reaction, it being absorbed in plant cells by coloured substances of which the green compounds, the chlorophylls, are examples. The detailed mechanism of the formation of carbohydrates is not yet understood.

Biochemistry provides many other examples of endothermic reactions, such as the formation of the metabolic substance, adenosine triphosphate, ATP, a substance which supplies energy which can be used in our muscles to do mechanical work. This is illustrated with an energy level diagram in Chapter 12, Fig. 12.3.

A mixture of hydrogen and chlorine will explode in the presence of sunlight. Even though the overall process is exothermic, the *mechanism* involves an endothermic stage, namely the dissociation of chlorine into atoms by the absorption of ultra-violet light. Chlorine atoms are highly reactive and give rise to a chain reaction with a sudden release of large amounts of energy. Chain reactions are dealt with in Chapter 17.

9.5 Energy changes when substances dissolve

The most useful generalization which can be made about the tendency of substances to dissolve in one another is *like dissolves like*. Even this is only a very rough rule-of-thumb, however, and we need to examine the energy changes a little more closely in an attempt to understand what is really happening.

(a) *Molecular substances*, such as iodine, white phosphorus, paraffin wax and so on, tend to dissolve in liquids which are themselves molecular, and non-polar. Highly polar liquids such as water are very poor solvents for essentially non-polar molecules.

(b) *Giant lattices* may or may not dissolve in solvents: those with extremely strong forces holding the atoms together have little tendency to dissolve in anything at all, as too much energy is required to break up the lattice; those which can break up to form ions may dissolve in polar solvents such as water or liquid ammonia. Metallic lattices do not dissolve except in

other metals, such as mercury: mercury will dissolve sodium to form an alloy known as an *amalgam*. When sodium is said to 'dissolve' in water the word 'dissolve' is being used rather loosely because a violent chemical reaction has occurred in which the sodium has formed $Na^+(aq)$ and the water has been changed into $OH^-(aq)$ and $H_2(g)$.

Lattices held together by covalent bonds, rather than by ionic forces, are generally insoluble in all solvents.

The dissolving of an ionic solid in a solvent can be visualized as taking place in two separate stages:

(i) The lattice must be broken into separate ions. The energy per mole required to effect this is known as the **lattice energy.**

(ii) The separate ions attach themselves electrostatically to the molecules of solvent. This is the attraction between a single charge and a dipole, known as an ion–dipole interaction. Energy will be released at this stage of the process. The amount of energy released when one mole of ions is combined with the solvent in this way is termed the **solvation energy.**

We can understand the process better with the aid of another energy level diagram, though it is important to bear in mind that stages (i) and (ii) do *not* take place separately. We can imagine, using Hess's law, that a mole of sodium chloride could first be converted *theoretically* into gaseous ions and that these ions could then be added to water. Hess's law states that the overall energy change is the same whatever the route chosen. Figure 9.7 is an attempt to show diagrammatically what really does happen, as far as we know, when sodium chloride dissolves in water. Figure 9.8 is an enthalpy diagram which analyses the process into stages (i) and (ii) above.

Experiments show that when sodium chloride is dissolved in water the rise or fall in temperature is negligible—we say that its heat of

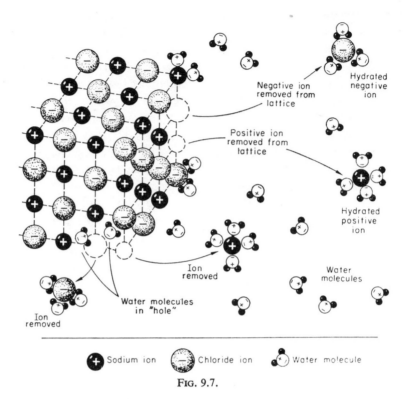

Negative ion
removed from
lattice

Hydrated
negative
ion

Positive ion
removed from
lattice

Hydrated
positive
ion

Ion
removed

Water
molecules

Ion
removed

Water molecules
in "hole"

Ion
removed

+ Sodium ion   − Chloride ion   Water molecule

FIG. 9.7.

solution in water is practically zero ($\Delta H = + 5\cdot3$ kJ mol^{-1}).

Heat of solution is defined as the heat change which takes place when one mole of a solute is added to a solvent to make a solution of stated concentration. If a given solution is diluted further there is usually a further small heat change, and this is referred to as heat of dilution. The more dilute a solution becomes the smaller will be the heat changes on further dilution, and for purposes of quoting data the heat of solution at infinite dilution may be referred to.

If the solvation energy of a substance is greater than its lattice energy, the substance will dissolve exothermically in the solvent. Anhydrous copper(II) sulphate and sodium hydroxide are examples. If the solvation energy is less than the lattice energy then the substance will dissolve endothermically. Sodium nitrate, ammo-

nium chloride, and many other highly soluble substances provide examples.

An ionic salt like sodium chloride fails to dissolve in a non-polar solvent, because there is no solvation energy. Water has a very large dipole moment, so that ion–dipole interactions will be high. A solvent like benzene or hexane is not attracted to ions in this way, and there is no gain in solvation energy to overcome the lattice energy term. Uncharged, non-polar solvent molecules such as benzene do not have the ability to insert themselves in the ionic lattice in order to break it up.

9.6 Other factors involved when substances dissolve

A simple consideration of ΔH does not provide us with all the explanations of why changes

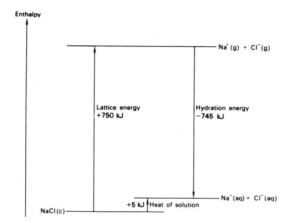

FIG. 9.8. Heat of solution of sodium chloride.

occur. Why for instance does iodine dissolve in benzene? Clearly the energy factor here is very small, because iodine is a molecular lattice and the molecules I_2 and C_6H_6 are both non-polar.

An even more extreme case is the question why gases dissolve in one another. Why does neon gas mix with argon gas completely in all proportions? There is no enthalpy loss or gain at all here, and the problem certainly cannot be treated in terms of energy levels. This aspect of the problem will be taken up in Chapter 12.

9.7 The Born–Haber cycle

Provided that large energy changes are involved, ΔH is generally the over-riding factor determining why chemical reactions take place. One chemical reaction of general interest is the combination of a metal with a non-metal to form an ionic solid, which releases considerable energy. For instance, the alkali metals (Group IA) and the alkaline earth metals (Group IIA) combine vigorously with the more electronegative non-metals such as fluorine, oxygen and chlorine. The question why such elements combine exothermically, can be answered by constructing an enthalpy diagram, in which the process is broken up into separate theoretical stages. Hess's law is assumed to apply. The whole process is referred to as an energy cycle, this particular sequence being named the **Born–Haber cycle** (1919).

The reaction between sodium and fluorine forming solid sodium fluoride is taken as a typical example, though a similar argument could be applied to similar pairs of elements. Direct measurement of ΔH gives

$$Na(c) + \tfrac{1}{2}F_2(g) \longrightarrow NaF(c);$$
$$\Delta H = -570 \text{ kJ mol}^{-1}.$$

The process can be imagined in various stages (Fig. 9.9).

(1) Sublimation: $Na(c) \longrightarrow Na(g)$;
$$\Delta H_1 = +109 \text{ kJ}$$

(2) Ionization: $Na(g) \longrightarrow Na^+(g) + e^-$;
$$\Delta H_2 = +495 \text{ kJ}$$

(3) Dissociation: $\tfrac{1}{2}F_2(g) \longrightarrow F(g)$;
$$\Delta H_3 = +75 \text{ kJ}$$

(4) Electron capture: $F(g) + e^- \longrightarrow F^-(g)$;
$$\Delta H_4 = -338 \text{ kJ}$$

(5) Lattice formation: $Na^+(g) + F^-(g) \longrightarrow NaF(c)$;
$$\Delta H_5 = -900 \text{ kJ}$$

$-\Delta H_5$ is the **lattice energy** of sodium fluoride. It cannot be measured directly but it may be calculated using the laws of electrostatics. Figure 9.9 shows that $\Delta H = \Delta H_1 + \Delta H_2 + \Delta H_3 + \Delta H_4 + \Delta H_5$. The same result could of course

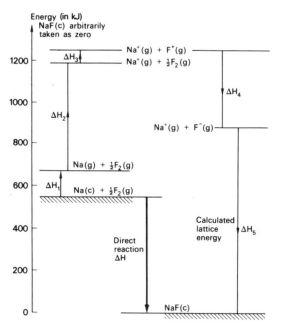

FIG. 9.9. The Born–Haber cycle for sodium fluoride, shown as an enthalpy diagram.

be obtained without actually constructing an energy level diagram simply by treating equations (1) to (5) algebraically.

The Born–Haber cycle provides a very good test of whether the ionic model which we use for sodium fluoride is valid. Direct measurement of ΔH gives -570 kJ mol^{-1}. Application of the Born-Haber cycle gives $\Delta H = 109 + 75 + 495 - 900 - 338 = -559$ kJ mol^{-1}. The assumption made in the Born-Haber cycle is that solid sodium fluoride can be regarded as being made up of singly charged ions. The very good agreement with experiment shows the assumption to be a good one in this case, though in many of the cases

met later in this book the agreement is less good, suggesting that the ionic model does not apply.

9.8 Bond energy

The **bond energy** of a diatomic molecule is easy to understand: it is simply the molar energy of dissociation of the molecule into atoms. Several values have already been quoted in this book.

The energy of a bond in a polyatomic molecule is not always something we can directly measure. For instance, if water is dissociated into atoms, the process can occur in two stages, each with a different intake of energy:

$$H_2O(g) \longrightarrow H(g) + OH(g);$$
$$\Delta H = +490 \text{ kJ} \quad (1)$$

$$OH(g) \longrightarrow H(g) + O(g);$$
$$\Delta H = +424 \text{ kJ} \quad (2)$$

This does not mean that the two H—O bonds in a water molecule differ from one another; it simply means that we are considering two different reactions in (1) and (2) above.

The bond energy of O—H in water is one-half of the energy required to atomize the entire molecule in one step, breaking both bonds simultaneously. The energy for this process cannot be directly measured, but it can be calculated by applying Hess's law, and adding equations (1) and (2).

$$H_2O(g) \longrightarrow 2 H(g) + O(g);$$
$$\Delta H = 914 \text{ kJ}$$

$$\therefore \text{ Bond energy of O—H in water} = \frac{914}{2}$$
$$= 457 \text{ kJ mol}^{-1}.$$

Similarly the bond energy of C—H in methane is one-quarter of the total energy needed to atomize a mole of methane:

$$CH_4(g) \longrightarrow C(g) + 4 H(g);$$

$\Delta H = 1660$ kJ (calculated using Hess's law).

∴ Bond energy of C—H in methane

$$= 1660/4 = 415 \text{ kJ mol}^{-1}.$$

The concept of bond energy is an important one to chemists even though, like the concept of atomic radius, it turns out to be only approximate in its applicability. The calculations done above for water and methane can be extended to other substances, and a complete set of bond energies derived. It is found that, to a first approximation, the bond energy of a bond between two given atoms is independent of other bonds in the molecule: For instance the same value of C—H bond energy applies approximately to all hydrocarbon molecules.

Where there is a pronounced discrepancy between the experimental heat of combustion and that predicted by bond energy, some structural peculiarity must be sought. For instance, if the bonds are under strain, some of the potential energy—**strain energy**—might be released as extra heat of combustion.

9.9 Le Chatelier's principle

The fact that chemical reactions can reach a position of equilibrium, instead of reaching completion in either direction, is a proof that ΔH cannot be the only factor which determines whether a reaction will go or not. An equilibrium can be approached from either direction, starting with the pure materials on either side of the equation, and this suggests that it represents a balance between opposing reactions. ΔH for the backward reaction must be numerically the same as, but opposite in sign to, the forward reaction.

If a substance is dissolved in water it does not matter whether the heat of solution is positive or negative—sooner or later a concentration will be reached at a given temperature, when no more solid will dissolve and equilibrium exists between the aqueous phase and the undissolved solid. The solution is said to be **saturated** with solute at that temperature. The mass of solute dissolved by a stated mass, or volume, of solvent at a given temperature, is defined as the solubility. ΔH cannot give any information about the solubility of a substance: many water-soluble substances dissolve exothermically in water, while many others of comparable solubility dissolve endothermically.

In other chemical processes where equilibrium is involved, it is also found that measurements of ΔH cannot give any direct information about the composition of the equilibrium mixture.

Despite all this, the magnitude of ΔH *does* have an effect on the way equilibrium composition varies with temperature. Take a very simple case, namely the dissociation of nitrogen (IV) oxide:

$$N_2O_4(g) \rightleftharpoons 2 NO_2(g);$$
pale yellow   brown

$$\Delta H = +61 \cdot 5 \text{ kJ}$$

This reaction is readily studied by taking the gas in a sealed tube and observing the darkening in colour when the tube is immersed in hot water. The result of the experiment can be summarized in an enthalpy diagram (Fig. 9.10). It shows that the species of higher energy, in this case $NO_2(g)$, is favoured at the higher temperature.

Many other examples can, be taken, for instance:

(a) $NH_4Cl(c) \rightleftharpoons NH_3(g) + HCl(g);$

$$\Delta H = +177 \text{ kJ}$$

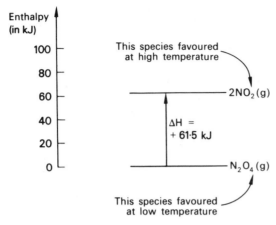

FIG. 9.10. Le Chatelier's principle.

(b) $PCl_5(g) \rightleftharpoons PCl_3(g) + Cl_2(g)$;

$$\Delta H = +120 \text{ kJ}$$

(c) $I_2(g) \rightleftharpoons 2 I(g)$;

$$\Delta H = +214 \text{ kJ}$$

These are all examples of **thermal dissociation.** The effect of temperature on such equilibria led Le Chatelier, in 1885, to propose the following general rule, known as **Le Chatelier's principle: when a constraint is applied to a system in equilibrium, the system will change in such a way as to try to remove the constraint.** The 'constraint' which concerns us here is that of increasing the temperature. If the temperature is increased, the system will behave as if it is trying to lower the temperature again by performing an endothermic reaction. Hence at higher temperatures the system will favour the species of higher energy. If ΔH for an equilibrium is very small, then changing the temperature will not have much effect on the equilibrium composition. For instance, the heat of solution of sodium chloride in water is extremely small, and as a consequence, the solubility of sodium chloride in water hardly alters with temperature.

Study Questions

1. The heats of combustion of C, H_2, CH_4, C_2H_6 and C_3H_8 are -395, -287, -880, -1545 and -2210 kJ mol^{-1} respectively.

    (a) Calculate the heat of formation of each hydrocarbon.

    (b) Predict the values of the heat of combustion and the heat of formation of n-butane, C_4H_{10}.

2. The heats of combustion of n-butylamine, n-propylamine and ethylamine are -2970, -2340 and -1710 kJ mol^{-1} respectively. What value would you expect for the heat of combustion of methylamine?

3. A number of alcohols were burned and the following results were obtained:

Alcohol	Weight loss	Heat given out
Methyl, CH_3OH	3·2 g	71 kJ
Ethyl, C_2H_5OH	9·2 g	268 kJ
Propyl, C_3H_7OH	7·5 g	252 kJ
n-Butyl, C_4H_9OH	7·4 g	263 kJ
iso-Butyl, C_4H_9OH	8·2 g	288 kJ

    (a) How many moles of each alcohol were burned?

    (b) How many kcal were given out per mole of each alcohol?

    (c) What trend do you observe in your results?

    (d) Comment on the values obtained for the two butyl alcohols.

4. When 8 g of $NH_4NO_3(c)$ was dissolved to give 100 cm^3 of solution, the temperature fell from 19° to 14·5°C. Calculate ΔH for the reaction

$$NH_4NO_3(c) \longrightarrow NH_4NO_3 \text{ (aq, molar)}$$

5. A series of neutralization experiments using molar solutions of sulphuric acid, nitric acid and sodium hydroxide, all initially at 20°C, gave the following results:

Acid	Base	Final temperature (°C)
100 cm^3 H_2SO_4	100 cm^3 NaOH	26·8
100 cm^3 H_2SO_4	200 cm^3 NaOH	29·1
100 cm^3 HNO_3	100 cm^3 NaOH	26·8
100 cm^3 HNO_3	200 cm^3 NaOH	24·6

(a) Account for these results.

(b) Calculate values for the heats of neutralization.

(c) What would the temperature rise have been in each case if the sodium hydroxide solution had been only M/2?

6. 25 cm³ of HCl of unknown concentration was titrated against M NaOH, both being initially at 20°C. The base was added at regular half-minute intervals, one cm³ at a time and the temperature was then taken. One mole of HCl exactly neutralizes one mole of NaOH.

cm³ base	temp. (°C)	cm³ base	temp. (°C)	cm³ base	temp. (°C)
1	20·5	6	22·4	11	23·9
2	20·9	7	22·7	12	24·1
3	21·3	8	23·0	13	24·2
4	21·7	9	23·3	14	24·1
5	22·1	10	23·6	15	24·0

(a) What is the approximate strength of the acid?

(b) Suggest two reasons why the temperature rise was less nearer the end-point than it was at the start of the reaction.

7. 1·16 g of acetone was burned in a bomb calorimeter. The same temperature rise was also caused when a current of 2 A at a P.D. of 10 V was passed through a resistance for 30 min.

(a) How many joules were needed to heat the water?

(b) How many moles of acetone were burned?

(c) What is the heat of combustion of acetone?

†**8.** (a) The vibrational spectrum of hydrogen has a convergence limit at 84·9 nm due to dissociation. 980 kJ are evolved when the dissociated atoms return to their ground state. Calculate the dissociation energy of hydrogen.

(b) The convergence limit in the spectrum of iodine occurs at 499·1 nm. The dissociation energy of iodine can be determined thermochemically as 148 kJ.

(i) Calculate the spectroscopic dissociation energy of iodine.

(ii) Comment on the value that you obtain.

9. Construct energy level diagrams to solve the following problems:

(a) The heats of formation of CO_2 from graphite and diamond under identical physical conditions are −393 and −396 kJ respectively. Calculate the heat of transition from graphite to diamond.

(b) The heat of formation of SnO_2 from white tin is −580 kJ. The heat of transition from white to grey tin is +2·5 kJ. Calculate the heat of formation of SnO_2 from grey tin.

(c) The heats of formation of Hg_2Cl_2 and $HgCl_2$ are −265 and −230 kJ mol⁻¹ respectively. Calculate the heat of the reaction $Hg_2Cl_2 + Cl_2 \rightarrow 2HgCl_2$.

(d) The heats of formation of HF, HCl, HBr and HI are −268, −92, −36 and +26 mol⁻¹ respectively.

(i) What is the heat of the reaction $2HBr + Cl_2 \rightarrow 2HCl + Br_2$?

(ii) Will chlorine displace bromine from HBr?

(iii) Arrange the halogens in a displacement order.

10. Sulphuric acid cannot be made directly from its elements. Using Chapter 22, state what data you would need to calculate the heat of formation of $H_2SO_4(l)$.

11. Suggest suitable solvents for (a) naphthalene, $C_{10}H_8$, (b) caesium iodide, CsI, (c) zinc, (d) carborundum, SiC.

12. (a) Which of the following will be more soluble in a given suitable solvent:

(i) White or red phosphorus?

(ii) Rhombic or plastic sulphur?

(b) White phosphorus is often stored under water while potassium is often stored under paraffin. Why are these particular liquids chosen?

(c) What would happen if the potassium were stored under water, and the phosphorus under paraffin?

13. Account for the following observations:

(a) Aniline, $C_6H_5NH_2$, is readily soluble in ether, but not in water.

(b) Aniline hydrochloride, $C_6H_5NH_3^+Cl^-$, is readily soluble in water, but not in ether.

(c) Silica, SiO_2, is insoluble in both water and ether.

(d) Sugar is a molecular substance, yet it dissolves in water.

(e) HCl(g), a molecular substance, dissolves in water to give a conducting solution.

(f) Water and benzene, C_6H_6, are immiscible.

†**14.** The lattice energy of potassium bromide is −656 kJ mol⁻¹, and the sum of the hydration energies of the gaseous ions is −640 kJ mol⁻¹.

(a) Draw an energy level diagram for this process.

(b) Calculate the heat of solution of potassium bromide in water.

†**15.** (a) Construct an energy level diagram for the heat of formation of HCl in terms of the

bond dissociation energies of H_2, Cl_2 and HCl.

(b) The bond dissociation energies of H_2, Cl_2 and HCl are 435, 242 and 431 kJ mol^{-1} respectively. Calculate the heat of formation of HCl.

†16. The heat of formation of AgCl $= -126$ kJ.
The first ionization energy of Ag $= +730$ kJ.
The bond dissociation energy of $Cl_2 =$
$+242$ kJ.
The electron affinity of chlorine $= +364$ kJ.
The heat of sublimation of Ag $= +272$ kJ.

(a) Calculate the lattice energy of AgCl from these figures.

(b) Assuming that AgCl is ionic, the lattice energy can be calculated to be -835 kJ. Comment on this.

17. (a) $N_2 + O_2 \rightleftharpoons 2NO$; $\Delta H = +90$ kJ

   (b) $N_2 + 3H_2 \rightleftharpoons 2NH_3$; $\Delta H = -92$ kJ.

What would be the effect of altering the temperature on each of these equilibria?

†18. The heats of combustion of the simple cyclic hydrocarbons of formula $(CH_2)_n$ are as follows (kJ mol^{-1}).

C_2H_4	-1420	C_3H_6	-2120
C_4H_8	-2780	C_5H_{10}	-3330
C_6H_{12}	-3930	C_7H_{14}	-4600

(a) Work out the heat of combustion "per CH_2 group" for each substance.

(b) How do these compare with the values given in question 1?

(c) What is the angle between adjacent C—C bonds in each compound?

(d) Can you discover a relationship between the bond angle and the heat of combustion "per CH_2-group"?

CHAPTER 10

Equilibrium

10.1 Steady state and equilibrium

A system is said to be in **equilibrium** when it has reached a "steady state" at a given temperature, with no matter passing into or out of the system. Some systems can be described as "steady states" even though strictly they are not in equilibrium: an example is the flame of a Bunsen burner with a constant flow of gas. Here the "macroscopic" properties of the flame, such as its temperature, colour, size and density, do not alter, but the flame is not a *closed system*, because matter is passing continually into and out of it.

When the properties of a closed system, such as a liquid and its vapour in a closed flask, reach a steady state at a given temperature, equilibrium is said to exist. Although individual molecules are continually passing between the liquid and vapour phases, measurement of large-scale properties (macroscopic properties) shows no evidence of overall change. In other words, microscopic changes are continuing but with no change in macroscopic properties.

The connection between the composition of a system at equilibrium and its energy has already been mentioned in Chapter 9. Here it was clearly shown that an equilibrium can exist between two sets of substances even when there is a large energy difference between them. The species of higher energy are favoured by high temperature, nevertheless equilibrium does exist at all temperatures. Equilibria which are *thermoneutral* do not have their compositions affected much by altering the temperature.

It can be shown theoretically that *all* chemical reactions are in reality equilibria. In this book we often write a single arrow → in a chemical equation, which implies that the backward reaction does not occur. Strictly we should always write the equilibrium sign ⇌ for any chemical process. Where the equilibrium sign is not written it may be assumed that the forward reaction is the one which predominates.

10.2 Representation of equilibrium by equations

The equilibrium between liquid water and its vapour is a balance between two opposing changes. On the one hand, molecules of water are leaving the liquid surface and changing into vapour, with consequent absorption of latent heat:

$$H_2O(l) \longrightarrow H_2O(g);$$
$$\Delta H = +44 \text{ kJ mol}^{-1} \text{ at } 25°C.$$

On the other hand, molecules of water vapour are returning to the liquid phase with consequent evolution of heat:

$$H_2O(g) \longrightarrow H_2O(l);$$
$$\Delta H = -44 \text{ kJ mol}^{-1} \text{ at } 25°C.$$

At equilibrium the two rates balance, and there is no overall heat change. Where the change in heat content has to be shown, it is expressed for the *forward* reaction, i.e. the reaction as written from left to right:

$$H_2O(l) \rightleftharpoons H_2O(g);$$
$$\Delta H = +44 \text{ kJ mol}^{-1} \text{ at } 25°C.$$

The chemical equation gives information about the number of moles of substances which react in either direction, but it does *not* give any information about the *composition* of the reaction mixture at equilibrium. The above equation, for example, does not tell us whether there is a large amount of liquid water in equilibrium with a small amount of steam, or vice versa.

Examples of other phase equilibria which can be represented in the form of chemical equations are:

$$CO_2(c) \rightleftharpoons CO_2(g);$$
$$\Delta H = +32·2 \text{ kJ mol}^{-1}$$

$$S(\text{rhombic}) \rightleftharpoons S(\text{monoclinic});$$
$$\Delta H = +0·4 \text{ kJ mol}^{-1} \text{ at } 25°C.$$

The first equation shows the equilibrium which can exist between solid carbon dioxide and the gas. At normal pressure carbon dioxide *sublimes* without intervention of the liquid phase. The second equation shows the equilibrium which can exist between two crystalline forms of sulphur.

All equilibria involve some form of chemical change, even though in the case of a simple phase change on a single chemical substance the chemical change may be simply the breaking and reforming of relatively weak hydrogen bonds or dipolar forces. Although changes are conventionally classified as "chemical" and "physical", the distinction is by no means a rigid one.

10.3 Rate of attainment of equilibrium

Many everyday systems are not true equilibria, even though they may not appear to change over a long time. Petrol in contact with air is one such system: equilibrium is not attained unless a spark is applied.

It is often necessary to examine a system very carefully in order to determine whether equilibrium truly exists, for in many cases the rate of reaction may be extremely slow. This is true for the transition between monoclinic and rhombic sulphur mentioned above. Monoclinic crystals of sulphur can be cooled to room temperature without any visible transformation to rhombic occurring, even though true equilibrium cannot exist at room temperature. The change from one solid phase to another is frequently slow since comparatively large forces have to be overcome in order to rearrange the crystal lattice. Similarly equilibrium cannot be attained at room temperature, in the reaction

$$N_2(g) + 3H_2(g) \rightleftharpoons 2NH_3(g);$$
$$\Delta H = -96 \text{ kJ}$$

because the rate of reaction between nitrogen and hydrogen at this temperature is very slow indeed.

10.4 Le Chatelier's principle

If a system in equilibrium is caused to change, for example by altering the conditions such as temperature or concentration of the reactants, then *processes occur which tend to nullify the change*, (le Chatelier's principle, Chapter 9). This enables us to predict what will happen to the equilibrium composition of a system, if an external constraint is applied to it. The following examples illustrate the principle.

(a) *The effect of concentration.* It is found experimentally that an equilibrium exists in aqueous solution between chromate ions $CrO_4^{2-}(aq)$ and dichromate ions $Cr_2O_7^{2-}(aq)$ as follows:

$$2 H^+(aq) + 2 CrO_4^{2-}(aq) \rightleftharpoons Cr_2O_7^{2-}(aq) + H_2O.$$
$$\text{yellow} \qquad\qquad \text{orange}$$

If the solution in equilibrium is acidified, i.e. $H^+(aq)$ ions are added, the change which occurs tends to remove the added hydrogen ions. The solution is seen to become deep orange. If sufficient hydrogen ions are added it is possible to effect an almost complete conversion of chromate into dichromate ions. If we now add an alkali to remove the hydrogen ions, the equilibrium composition promptly shifts back in an attempt to replenish some of the hydrogen ions removed.

$$H^+(aq) + OH^-(aq) \longrightarrow H_2O$$

Since at the same time this forms chromate ions at the expense of dichromate ions, the solution turns pale yellow. This process may be reversed any number of times by the addition of acid or alkali.

Another instance of reversible equilibrium is provided by the addition of acid or alkali to an indicator. A simple experiment shows that litmus, methyl orange, or any other acid-base indicator can undergo repeated colour changes to and fro, depending on whether the acid or alkali is present in excess. This is explained by the fact that an indicator is a weak acid, which can lose a proton to give an anion of different colour. Addition of $H^+(aq)$ will cause the acid form HX to predominate, while addition of $OH^-(aq)$ will shift the equilibrium back to the right, making the observed colour that of the anion $X^-(aq)$:

$$HX \rightleftharpoons X^-(aq) + H^+(aq).$$

(b) *The effect of pressure.* In the case of a gaseous system, which is compressible, the concentration of a reactant can be altered by varying the pressure. Consider again the equilibrium between nitrogen and hydrogen forming ammonia. The reaction is immeasurably slow at room temperature, but at about 500°C in the presence of a catalyst it proceeds quite rapidly.

$$N_2(g) + 3 H_2(g) \rightleftharpoons 2 NH_3(g);$$
$$\Delta H = -96 \text{ kJ at } 25°C.$$

The equation means that one mole of $N_2(g)$ would react with three moles of $H_2(g)$ to form two moles of ammonia, if the reaction were able to proceed completely from left to right. The equation does *not* tell us about the composition at equilibrium.

Since the molar volumes of gases are approximately equal under given conditions of temperature and pressure, the total volume of gas must decrease if nitrogen and hydrogen are made to react at a given constant pressure, because the number of moles of ammonia formed is less than the total number of moles of nitrogen plus hydrogen at the start.

Le Chatelier's principle predicts that increasing the pressure of the above equilibrium mixture will be to some extent nullified by the formation of more ammonia at the expense of nitrogen and hydrogen.

In general, an equilibrium involving a change in volume of reactants will be altered by increasing the pressure in such a direction as to favour the side of the equation with the smaller volume.

The same argument applies if the system is **heterogeneous,** that is, if there is more than one phase present. Consider the heterogeneous equilibrium between calcium carbonate and its decomposition products, calcium oxide and carbon dioxide:

$$CaCO_3(c) \rightleftharpoons CaO(c) + CO_2(g);$$
$$\Delta H = +178 \text{ kJ at } 25°C.$$

If calcium carbonate is heated in a closed container, carbon dioxide will be evolved until equilibrium is reached: thereafter the pressure of carbon dioxide will not alter at that temperature. If the volume of the container is now reduced, the equilibrium composition will shift in order to try to restore the original pressure. In other words carbon dioxide will react with calcium oxide to form calcium carbonate. Conversely if the volume of the container is increased the reaction will proceed from left to right, forming more gas.

(c) *The effect of temperature.* Le Chatelier's principle predicts that an endothermic change will be favoured by high temperature, while the reverse exothermic change will be favoured by low temperature. This is borne out by experiment. It is a familiar experimental fact, for example, that increasing the temperature of calcium carbonate causes it to decompose more readily to the oxide. If this experiment is done in a closed container, and the equilibrium pressure of carbon dioxide is measured with a manometer, the pressure is seen to rise with temperature. If heat is added to the system, a change will take place in such a way as to try to lower the temperature again. The endothermic change is therefore favoured, and more carbon dioxide is present in the equilibrium mixture.

10.5 Equilibrium as a balance between opposing changes

A chemical system in equilibrium has been described as a closed system in a steady state. It should not, however, be thought of as static.

There is a constant interchange of matter due to the existence of equal and opposite changes which balance out. For instance, when a liquid is in equilibrium with its vapour the vapour is said to be **saturated.** The vapour pressure of a given pure substance is determined by temperature, and when a liquid is in equilibrium with its saturated vapour we can say that the rate at which molecules leave the liquid surface = the rate at which other molecules return to the liquid from vapour.

A similar state of affairs occurs with a saturated solution in equilibrium with solute. For instance the dissolving of a simple ionic salt in water can be represented as a balance between two opposing reactions:

$$A^+ B^-(c) \rightleftharpoons A^+(aq) + B^-(aq).$$

It is impossible to state with any precision, by a mere examination of the above equation, the exact *mechanism* by which a salt dissolves in water. Beware of trying to guess the mechanism of a reaction by looking at the equilibrium equation. The familiar Haber equilibrium is indeed the result of opposing reactions balancing, but the forward reaction is *not* the result of three hydrogen molecules colliding simultaneously with one nitrogen molecule, nor does the backward reaction result from the immediate conversion of $2 NH_3(g)$ into $3 H_2(g) + N_2(g)$. Many intermediate stages are involved, and the overall equation does not show these.

Simple experiments can be done to show that the composition of an equilibrium mixture is the same regardless of the direction from which the composition is approached. These experiments do not *prove* that equilibrium is dynamic however. To be certain of this we need some way of "labelling" atoms in order to follow their progress during the exchange process. Use of a radioactive isotope is one way of doing this. Thorium salts contain radioactive lead, produced by disintegration of the thorium nucleus. If we precipitate the sparingly soluble salt lead(II) chloride, $PbCl_2$, by mixing a solution of $Cl^-(aq)$ with a solution of $Pb^{2+}(aq)$ which also contains some thorium ions, and hence radioactive lead ions, which we will denote as $^*Pb^{2+}(aq)$, the solid lead chloride now contains some *labelled* lead atoms—it is a mixture of $PbCl_2(c)$ with some $^*PbCl_2(c)$. Now prepare separately a saturated solution of unlabelled lead(II) chloride by allowing it to remain in contact with pure water for several days with occasional shaking. Test this solution in a radioactive counter and it will be found to be non-radioactive. Now decant some of the saturated solution and add some of the labelled solid lead(II) chloride to it. Since the solution was saturated the overall equilibrium composition cannot change. The question is whether an exchange of atoms can occur. To find out, test samples of the solution from time to time in the radioactive counter. It will be found to have become radioactive. Since the concentration cannot have altered (provided the temperature was kept constant) an exchange must have taken place. The radioactive lead(II) ions must have come from the labelled solid, by the forward reaction

$$^*PbCl_2(c) \longrightarrow \ ^*Pb^{2+}(aq) + 2\,Cl^-(aq)$$

The existence of dynamic equilibrium in more complex chemical processes can be similarly illustrated, provided that a suitable radioactive isotope is available. If all the available radioactive isotopes of a given element are too short-lived to be useful (as in the case of aluminium for instance) we can use a compound which has been enriched in one of its stable isotopes. Detection of the isotope is then more of a problem and the products have to be analysed in a mass spectrometer. The use of radioactive tracers and labelled compounds with enriched stable isotopes is very common in studies of reaction

mechanisms. It can be shown for instance that the rate of exchange in an equilibrium process is more rapid at a higher temperature.

10.6 The equilibrium law

A typical gaseous equilibrium, which was first thoroughly investigated by Bodenstein in 1897, is the following:

$$H_2(g) + I_2(g) \rightleftharpoons 2\,HI(g).$$

Bodenstein found that, at a given temperature, the concentrations in mol dm^{-3} of the three substances present, denoted by square brackets $[H_2]$, $[I_2]$ and $[HI]$, are always related by the following approximate expression:

$$\frac{[HI]^2}{[H_2]\,[I_2]} = K_c, \text{ a constant.}$$

The constant K_c is called the equilibrium constant. The subscript c is used to denote that molar concentration units are employed. Another equilibrium constant, K_p, can be written, in which the amounts of gases present are stated in terms of their partial pressures p_{H_2}, p_{I_2} and p_{HI}.

$$\frac{p_{HI}^2}{p_{H_2} \cdot p_{I_2}} = K_p.$$

In the case of the above equilibrium K_c and K_p are equal. This must be so, by the following argument: assuming the gases to be ideal, their molar concentrations will be directly proportional to their partial pressures. Partial pressure is therefore just another unit by which molar gas concentrations can be represented.

Worked Example 1. In an equilibrium mixture the following concentrations were observed at 0°C:

$$[H_2] = x \text{ mol dm}^{-3}$$
$$[I_2] = y \text{ mol dm}^{-3}$$
$$[HI] = z \text{ mol dm}^{-3}$$

Calculate (i) K_c, (ii) K_p.

(i) $K_c = \dfrac{[HI]^2}{[I_2][H_2]} = \dfrac{z^2}{xy}$

(ii) $[H_2] = 22 \cdot 4x$ mol per $22 \cdot 4$ dm^3.

$\therefore p_{H_2} = 22 \cdot 4x$ atmospheres.

Similarly,

$$p_{I_2} = 22 \cdot 4y \text{ atmospheres}$$
$$p_{HI} = 22 \cdot 4z \text{ atmospheres.}$$

$$K_p = \frac{p_{HI}^2}{p_{H_2} \cdot p_{I_2}} = \frac{z^2}{xy} = K_c$$

Bodenstein was able to show that altering the pressure did not have any appreciable effect on the equilibrium composition, and he also showed that the equilibrium composition for a given set of conditions was the same whether he started with pure hydrogen iodide or with pure hydrogen and pure iodine.

Accurate measurements by later workers have shown that the equilibrium law, where concentrations are used, is only approximate. The following figures obtained by Taylor and Crist (1941) illustrate this point (Table 10.1).

TABLE 10.1
TEMPERATURE = 457·6°C

$[H_2]$ mol dm^{-3}	$[I_2]$ mol dm^{-3}	$[HI]$ mol dm^{-3}	K_c
$5 \cdot 617 \times 10^{-3}$	$0 \cdot 5936 \times 10^{-3}$	$1 \cdot 270 \times 10^{-2}$	$48 \cdot 38$
$3 \cdot 841$	$1 \cdot 524$	$1 \cdot 687$	$48 \cdot 61$
$4 \cdot 580$	$0 \cdot 9733$	$1 \cdot 486$	$49 \cdot 54$
$1 \cdot 696$	$1 \cdot 696$	$1 \cdot 181$	$48 \cdot 48$
$1 \cdot 433$	$1 \cdot 433$	$1 \cdot 000$	$48 \cdot 71$
$4 \cdot 213$	$4 \cdot 213$	$2 \cdot 943$	$48 \cdot 81$

The first three rows of figures were obtained starting with pure hydrogen and pure iodine, and the remainder by starting with pure hydrogen iodide.

It will be noticed that in the expression for the equilibrium constant, the concentration of hydrogen iodide is raised to a power equal to its coefficient in the chemical equation. This is found to be generally true; e.g. for the Haber synthesis of ammonia the equilibrium constant is found to be

$$K = \frac{[NH_3]^2}{[H_2]^3[N_2]}.$$

Note that in this case K_c and K_p are *not* equal, as there are four concentration terms in the denominator and only two in the numerator.

Worked Example 2. In an equilibrium mixture the following concentrations were observed at 0°C:

$$[H_2] = a \text{ mol dm}^{-3}$$
$$[N_2] = b \text{ mol dm}^{-3}$$
$$[NH_3] = c \text{ mol dm}^{-3}$$

Calculate (i) K_c, (ii) K_p.

(i) $K_c = \dfrac{[NH_3]^2}{[H_2]^3[N_2]} = \dfrac{c^2}{a^3b}$ mol^{-2} dm^6

(ii) $p_{H_2} = 22 \cdot 4a$ atmospheres (see Worked Example 1).

Similarly,

$$p_{N_2} = 22 \cdot 4b$$
$$p_{NH_3} = 22 \cdot 4c$$

$\therefore K_p = \dfrac{p_{NH_3}^2}{p_{H_2}^3 \cdot p_{N_2}} = \dfrac{22 \cdot 4^2 \times c^2}{22 \cdot 4^3 \times a^3 \times 22 \cdot 4 \times b}$

$= \dfrac{1}{22 \cdot 4^2} \times \dfrac{c^2}{a^3b}$

$= \dfrac{K_c}{22 \cdot 4^2}$ atmospheres^{-2}

$\therefore K_c \neq K_p$.

Units. For convenience, the units of equilibrium constants are generally omitted. In the case of ammonia above, the dimensions of K_c are

$$K_c = \frac{(\text{mol dm}^{-3})^2}{(\text{mol dm}^{-3})^4}$$

$$= \text{mol}^{-2} \text{dm}^6.$$

The dimensions of K_p are (atmospheres)$^{-2}$, or N^{-2} m^4.

In the case of the hydrogen iodide equilibrium both K_p and K_c are dimensionless quantities since the concentrations and partial pressures cancel out.

The equilibrium law can be summarized by stating that for an equilibrium

$$aX + bY + cZ + \ldots \rightleftharpoons pL + qM + rN + \ldots$$

the expression $\dfrac{[L]^p[M]^q[N]^r \ldots}{[X]^a[Y]^b[Z]^c \ldots}$ will be constant at a given temperature. By convention, the substances on the right-hand side of the equation are written in the numerator.

10.7 Activity

The equilibrium law, as stated above, is only approximately true. It can be shown theoretically that, quite apart from experimental error, deviations are to be expected in the case of real gases on account of their failure to obey the gas laws. If we were always dealing with ideal gases the law would always hold exactly, but real gases are far from ideal in their behaviour, especially at high pressures and low temperatures.

It is also found that similar deviations occur in equilibria involving solutions, though for very dilute solutions the agreement is sufficient for the law to be of practical value.

In the accurate statement of the equilibrium law, a quantity known as the **activity** of each reactant is used in place of concentration. The activity is a quantity obtained by making allowance for non-ideal behaviour of the gas or solution, and is thus an "effective concentration" of a component. The measurement of activities is dealt with in section 11.6.

The activity of a pure liquid or solid is a constant at a given temperature and is generally taken as unity; (the concentration of a pure solid might be represented by its density).

10.8 The effect of a catalyst on equilibrium composition

A catalyst is a substance which can increase the rate of a chemical reaction without itself undergoing any overall chemical change. The effect of a catalyst on an equilibrium system will be to accelerate the attainment of equilibrium: it is thus acting as a kind of chemical lubricant. A catalyst cannot affect the equilibrium composition of a system since it does not alter its energy content.

When the equilibrium composition of a system is altered, there will be a change in energy: heat will either be evolved or absorbed depending on whether the forward reaction is exothermic or endothermic. Conversely if there is an energy change in an equilibrium system the equilibrium composition will alter. Since the catalyst is itself not undergoing any permanent chemical change it cannot contribute to the energy of the system, and hence the equilibrium composition must stay the same.

An everyday example of a catalyst is the platinum catalyst used in some automatic gas lighters. Living systems also use biological catalysts, called **enzymes,** for instance to enable "fuels" such as sugar to undergo "combustion" at body temperature. Catalysts are used extensively in manufacturing processes. The theory of catalysis is discussed in Chapter 17.

10.9 Solution equilibria

The equilibrium law can also be applied to solutions. Consider a solid which dissolves as molecules. Examples of this are naphthalene in benzene, and sucrose, $C_{12}H_{22}O_{11}$, in water:

$$C_{12}H_{22}O_{11}(c) \rightleftharpoons C_{12}H_{22}O_{11}(aq)$$

The activity of solid sucrose is a constant, and the equilibrium law predicts that the concentration of solute in the saturated solution will be a constant at a given temperature. In other words the solubility of a substance in a solvent at a given temperature is independent of the mass of undissolved solute present, or upon its surface area. This is borne out by experiment. Increasing the surface area of undissolved solute will increase the rate at which molecules leave the solid, but it will also increase the rate at which they return.

10.10 Henry's law

The solubility of a gas in a liquid, in mol dm^{-3}, is found to be directly proportional to the partial pressure (section 8.7) of the gas at a given temperature. This relationship, known as **Henry's law,** is also a consequence of the equilibrium law. For example, with nitrogen dissolving in water we have

$$N_2(g) \rightleftharpoons N_2(aq)$$

and therefore

$$\frac{[N_2(aq)]}{[N_2(g)]} = K$$

$$\therefore \frac{[N_2(aq)]}{p_{N_2}} = \text{constant. (Henry's law.)}$$

where p_{N_2} = partial pressure of nitrogen. Soda water (carbon dioxide dissolved in water) provides a good illustration of Henry's law.

10.11 The distribution law (partition law)

If a solute is added to two immiscible liquids, being soluble in both, it will distribute itself between them in such a way that its concentration in one solvent is directly proportional to the concentration in the other solvent at a given temperature. This relationship is known as the distribution law or partition law, and again it

is a form of the equilibrium law. Iodine, for example, dissolves both in benzene (readily) and in water (sparingly).

$$I_2(aq) \rightleftharpoons I_2(benzene)$$

$$\therefore \frac{[I_2(benzene)]}{[I_2(aq)]} = K.$$

For such systems, the constant ratio K between the molar concentrations is known as the **distribution coefficient** (or **partition coefficient**).

The distribution law is only obeyed accurately at low concentrations. This is because strictly we should write activities rather than concentrations.

10.12 Solubility product

Many solids dissolve in water to form ions. In the case of substances which are sparingly soluble in water the equilibrium law can again be applied. With silver chloride, for example, the solubility at room temperature is only about 10^{-5} mol dm^{-3} and we can write:

$$AgCl(c) \rightleftharpoons Ag^+(aq) + Cl^-(aq)$$

$$\therefore \frac{[Ag^+][Cl^-]}{[AgCl]} = K_c.$$

Since AgCl is a solid, [AgCl] is a constant.
$\therefore [Ag^+][Cl^-]$ = another constant, K_s, known as the **solubility product.** (The activity of a pure solid is a constant, which can be put equal to unity.)

Similarly for lead(II) chloride, we have

$$PbCl_2(c) \rightleftharpoons Pb^{2+}(aq) + 2Cl^-(aq)$$

and therefore $K_s = [Pb^{2+}][Cl^-]^2$. Similarly, with aluminium hydroxide Al(OH)$_3$, the relationship is $K_s = [Al^{3+}][OH^-]^3$.

The solubility product enables the effect on solubility of adding a *common ion* to be predicted. Returning once more to silver chloride, we have seen that its solubility at room temperature is 10^{-5} mol dm^{-3}. From this we can deduce the solubility product:

$$AgCl(c) \rightleftharpoons Ag^+(aq) + Cl^-(aq)$$

| 10^{-5} mol dissolves per dm^3 | 10^{-5} mol dm^{-3} | 10^{-5} mol dm^{-3} |

$$\therefore K_s = 10^{-5} \times 10^{-5} = 10^{-10}.$$

Now let us see what happens when silver chloride is added to a molar solution of sodium chloride. Since sodium chloride is an electrolyte it will exist in solution in the form of ions, so that $[Cl^-] = 1$ mol dm^{-3}. Now silver chloride will dissolve in this solution to a sufficient extent to satisfy the solubility product principle. Suppose that 10^{-10} mol of silver chloride dissolved. $[Ag^+]$ will now equal 10^{-10}, and Cl^- will equal $1 + 10^{-10}$, which is for practical purposes equal to 1. In fact this is exactly the amount of silver chloride which must dissolve in order that the solubility product principle can be satisfied:

$$[Ag^+][Cl^-] = 10^{-10} \times 1 = K_s.$$

By adding a common ion therefore, the solubility of silver chloride has been lowered from 10^{-5} mol dm^{-3} to 10^{-10} mol dm^{-3}, i.e. by a factor of 100 000.

Similar results would be observed if a solution of silver ions was taken originally. The solubility of silver chloride in a dilute solution of other ions, e.g. a dilute solution of potassium nitrate will be the same as the solubility in pure water, since a non-common ion does not enter into the solubility product relationship.

If two solutions containing ions are mixed, and the solubility product is exceeded for one pair of oppositely charged ions in the mixed solutions, then that substance will have to *precipitate* in order to restore equilibrium. In the above case, if a molar solution of silver ions was mixed with a molar solution of chloride ions the solubility product would momentarily be exceeded by a factor of 10^{10}, and very rapid precipitation would occur.

10.13 Limitations of the solubility product principle

The solubility product principle should only be used as an approximate guide when predicting whether substances will precipitate from solution. At least three factors cause deviations in practice:

(i) The use of ionic *concentrations* in the expression for K_s is only strictly permissible when the concentrations are low. We would not expect a substance such as potassium nitrate to obey the principle since it is too highly soluble, and indeed the deviations are very considerable. Similarly the calculation given above for the solubility of silver chloride in molar sodium chloride is only approximate, because even at a concentration of $[Cl^-] = 1$ mol dm^{-3} the difference between concentration and activity is very noticeable. Nevertheless the law holds qualitatively.*

(ii) In many cases, even though the solubility product of a substance may be exceeded, a precipitate may not form due to the slow *rate* of precipitation. This applies particularly to substances with very low solubility products.

An interesting experiment to illustrate this point is to take solutions of $Ba^{2+}(aq)$ (barium chloride or nitrate) and SO_4^{2-} (sodium sulphate) at concentrations varying from 10^{-1} to 10^{-4} mol dm^{-3}. It will be found that as the solutions

---

* An interesting quantitative study of the "common ion" effect has been carried out as a pupils' research project at Marlborough College. The results showed convincingly that the deviations from ideal behaviour of solutions can be very considerable (*Educ. in Chem.*, vol. 3, No. 4, p. 164).

are made more dilute the precipitate appears more slowly, and that with very dilute solutions it will not appear at all even though the solubility product is exceeded (K_s for barium sulphate $\simeq 10^{-10}$ mol^2 dm^{-6}).

Remember again therefore that the equilibrium law tells us nothing about the rate of a chemical process: it tells us *how far* a given chemical reaction will go, but *not how fast*.

(iii) In other cases the formation of a *complex ion* may be a complicating factor. With lead(II) chloride, for example, a rather curious situation results: the substance is less soluble in dilute hydrochloric acid than in water, as one would expect, but on the addition of concentrated hydrochloric acid, the lead chloride becomes *more* soluble. We have in fact two equilibria to consider:

(a) The dissolving of the solid to form aqueous ions:

$$PbCl_2(c) \rightleftharpoons Pb^{2+}(aq) + 2\,Cl^-(aq)$$

(b) The formation of a complex ion:

$$Pb^{2+}(aq) + 4\,Cl^-(aq) \rightleftharpoons PbCl_4^{2-}(aq)$$

$$\frac{[PbCl_4^{2-}]}{[Pb^{2+}][Cl^-]^4} = K.$$

If [Cl$^-$] is made very large (e.g. 10 mol dm^{-3}) then the effect will be most marked in the equilibrium involving the complex ion, for now [Cl$^-$]^4 will equal 10^4. A larger amount of the complex ion PbCl$_4^{2-}$ has to form, and this has the effect of lowering the concentration of Pb^{2+}(aq). More solid lead chloride therefore dissolves to restore equilibrium between itself and the simple aqueous ions.

Silver chloride, AgCl(c), although only sparingly soluble in water, dissolves readily in aqueous ammonia (ammonium hydroxide). This is due to the formation of the complex diammine silver(I), which is quite stable:

$$Ag^+(aq) + 2\,NH_3(aq) \rightleftharpoons Ag(NH_3)_2^+(aq)$$

simple hydrated silver(I) ion   diammine silver(I) ion

$$K = \frac{[Ag(NH_3)_2^+]}{[Ag^+][NH_3]^2}$$

Application of le Chatelier's principle to this reaction shows that the equilibrium composition will move to the right, favouring the complex ion, if the concentration of ammonia, [NH$_3$(aq)], is increased. For the same reason, a high concentration of ammonia will reduce the concentration of Ag$^+$(aq). Since the ammine complex of silver is quite stable ($K \simeq 10^7$), the solubility product of silver chloride, [Ag$^+$][Cl$^-$], is not attained, and silver chloride therefore dissolves. The constant K is called the **stability constant** of the complex ion.

The dissolution of an insoluble precipitate in a reagent can often be attributed to the formation of a complex ion. Other examples will be found in the text, and by reading the exercises at the end of this chapter.

10.14 Auto-ionization and ionic product

Some liquids, although they consist almost entirely of molecules, are found to have a very small electrical conductivity due to the presence of traces of ions. Water is a very good example: no matter how carefully it is purified it will always show a small residual conductivity, and a high enough applied potential will cause electrolysis, giving hydrogen at the cathode and oxygen at the anode. This phenomenon is due to an equilibrium between water molecules and the ions H$^+$(aq) and OH$^-$(aq).

$$H_2O(l) \rightleftharpoons H^+(aq) + OH^-(aq);$$
$$\Delta H = +58\ kJ\ mol^{-1}$$
$$\therefore K = \frac{[H^+][OH^-]}{[H_2O]}$$

[H_2O] is constant, since the extent of ionization is small.

$\therefore$ [H^+] [OH^-] = a constant, K_w, termed the **ionic product** of water.

The reaction from left to right is an endothermic process with an enthalpy change of $+58$ kJ mol^{-1}. Application of le Chatelier's principle to this shows that the formation of ions will be favoured by high temperature, and this is borne out by experimental measurements on the variation of conductivity of pure water with temperature. Table 10.2 gives values for K_w at temperatures between 0°C and 100°C.

It will be noticed that at 25°C the value for K_w is very near to 10^{-14}. Since this is close to normal room temperature it is a useful figure to remember, and for approximate calculations involving K_w the figure 10^{-14} is often taken.

Some other liquids auto-ionize like water: in every case the concentration of ions is very low. Liquid ammonia (obtained by cooling ammonia to -33°C, not to be confused with an aqueous solution of ammonia) is a good example, and so is anhydrous acetic acid. Table 10.3 summarizes some of the liquids which auto-ionize.

The phenomenon of auto-ionization plays a very important part in the behaviour of acids and bases, and is discussed further in Chapter 19.

TABLE 10.2

Temperature (°C)	$K_w \times 10^{14}$ mol^2 dm^{-6}
0	0·11
25	1·01
50	5·47
100	51·3

TABLE 10.3

Liquid	Equation	Ionic product (mol^2 dm^{-6})
Water	$H_2O \rightleftharpoons H^+(aq) + OH^-(aq)$	10^{-14} at 25°C
Ammonia	$NH_3 \rightleftharpoons NH_4^+ + NH_2^-$	10^{-22} at -33°C
Acetic acid	$CH_3COOH \rightleftharpoons CH_3COOH_2^+ + CH_3COO^-$	10^{-17} at 25°C

Study Questions

1. The following equilibrium exists between iron, steam, hydrogen and iron(II,III,III) oxide:

$$Fe_3O_4(c) + 4 H_2(g) \rightleftharpoons 3 Fe(c) + 4 H_2O(g);$$
$$\Delta H = +138 \text{ kJ}.$$

(a) What would be the effect of (i) increasing the amount of steam present in a closed vessel, (ii) passing a continuous flow of hydrogen over the heated iron oxide, removing the steam as it formed, (iii) passing a continuous flow of steam over heated iron and removing the hydrogen as it formed?

(b) What information about the reaction does the chemical equation *not* contain?

2. Bismuth(III) chloride is soluble in water, but aqueous Bi^{3+} ions are hydrolysed according to the equation:

$$Bi^{3+}(aq) + Cl^-(aq) + H_2O \rightleftharpoons BiOCl(c) + 2H^+(aq)$$

(a) What would happen if a solution of bismuth(III) chloride were poured into an excess of water?

(b) What would happen if solid bismuth oxychloride (BiOCl) were shaken with a solution of nitric acid?

(c) What is meant by the statement that this equilibrium is "dynamic"?

(d) Suggest an experiment which would show that this equilibrium was dynamic.

3. Dinitrogen tetroxide, N_2O_4, is colourless when pure. In the vapour phase it dissociates into brown nitrogen dioxide, NO_2.

$$N_2O_4(g) \rightleftharpoons 2\,NO_2(g)$$

(a) The vapour darkened on warming to 100°C, but when cooled, it became pale again. What can you say about ΔH for this reaction?
(b) Would the vapour darken or grow paler if the pressure was decreased?

4. Magnesium carbonate, $MgCO_3$, dissociates as follows:

$$MgCO_3(c) \rightleftharpoons MgO(c) + CO_2(g)$$
$$\Delta H = +117\ kJ$$

In a closed vessel, how will the pressure of carbon dioxide alter:

(a) If the temperature is raised?
(b) If an inert gas, such as argon, is pumped into the container, the temperature remaining constant?
(c) If more magnesium carbonate is added?

5. What effect would the following changes have on the equilibrium concentration of ammonia in the reaction:

$$4\,NH_3(g) + 5\,O_2(g) \rightleftharpoons 4\,NO(g) + 6\,H_2O(g);$$
$$\Delta H = -900\ kJ.$$

(a) Increasing the temperature?
(b) Decreasing the total pressure?
(c) Increasing the volume of the reaction vessel?
(d) Decreasing the oxygen concentration?
(e) Adding a catalyst?

†**6.** Log K_p for the reaction $N_2 + 3H_2 \rightleftharpoons 2\,NH_3$ is 5·8 at 300°C, −1·0 at 500°C and −4·1 at 700°C. What will the effect of the equilibrium composition be if
(a) More nitrogen is pumped into the system?
(b) The pressure is increased?
(c) The temperature is increased?

7. 531 g of ammonia will dissolve in 1000 cm³ of water at 20°C, giving an alkaline solution.
(a) Suggest an experiment to investigate the conditions under which ammonia obeys Henry's law.
(b) How closely would you expect ammonia to obey the law?

8. Acetic acid, CH_3COOH, is soluble in both water and carbon tetrachloride, CCl_4, which are themselves immiscible. If solutions of different concentrations are shaken together, it is found that

$$\frac{(\text{concentration in } CCl_4)}{(\text{concentration in water})^2}\ \text{is a constant.}$$

(a) What can be deduced from this result?
(b) Suggest how you could determine the value of the constant.

9. Write down expressions for the solubility products of
(a) Silver(I) bromide.
(b) Barium sulphate.
(c) Lead(II) iodide.
(d) Bismuth(III) sulphide.
(e) Aluminium hydroxide.
(f) Calcium oxalate.

10. The solubility products of CuS and MnS are 10^{-36} and 10^{-15} mol² dm⁻⁶ respectively.
(a) Calculate the solubilities of the two compounds in g dm⁻³.
(b) What would happen if a 10^{-10} M solution of copper ions was mixed with a 10^{-10} M solution of sulphide ions?
(c) What would happen if a 10^{-10} M solution of manganese ions was mixed with a 10^{-10} M solution of sulphide ions?

11. The solubility products of barium and calcium sulphate are 10^{-10} and 2×10^{-5} respectively.
(a) What is the concentration of a saturated solution of each salt in g dm⁻³?
(b) Both salts can be used to determine sulphate gravimetrically. Which is preferable?

12. In the manufacture of soaps, such as sodium stearate, common salt is often added to precipitate the solid. Why is this?

13. At 25°C the solubility product for AgCl is 2×10^{-10}, while at 100°C it is $1·4 \times 10^{-4}$.
(a) Calculate the solubility of silver chloride at 100°C, in g dm⁻³.
(b) What can you say about the heat of solution of AgCl in water?

14. The solubility of iron(II) hydroxide is 0·1 g dm⁻³ at 20°C. When a solution of Fe^{2+} ions is boiled with potassium cyanide solution, the complex ion, $Fe(CN)_6^{4-}$, hexacyanoferrate(II), is formed. If NaOH (aq, M) is added to an aqueous solution of hexacyanoferrate(II) ions, no precipitate forms.
(a) Write down an expression for the solubility product of iron(II) hydroxide.
(b) Calculate the solubility of iron(II) hydroxide in mol dm⁻³.
(c) Calculate the solubility product of iron(II) hydroxide.
(d) Write down an expression for the stability constant of the hexacyanoferrate(II) ion.
(e) What can you say about the magnitude of this stability constant?

15. (a) What happens if gaseous hydrogen chloride is added to a saturated solution of sodium chloride?
(b) How would you try to determine the solubility product of sodium chloride in the presence of varying concentrations of hydrochloric acid?
(c) Why would the values obtained in (b) not be constant?

CHAPTER 11

Cells

11.1 The importance of electrical measurements in chemistry

This chapter is concerned with the measurement of the e.m.f.s of cells as a means of obtaining information about chemical processes. It is not always easy in chemistry to make accurate quantitative measurements, but voltage measurements can be made with a high degree of precision, leading to precise knowledge of energy changes.

Chapter 9 was concerned mainly with measurements of heat energy, and it was shown that exothermicity was not a criterion of whether a reaction would proceed spontaneously or not. It will be shown in this chapter that measurement of e.m.f. *does* provide a criterion: if a cell gives an e.m.f. then the cell reaction will proceed spontaneously. The larger the e.m.f. the greater the "driving force" of the chemical process.

The use and design of cells is of great technological importance today. Apart from the familiar examples such as the dry battery and the lead accumulator used in motor cars, recent years have seen the development of miniature cells (such as those for hearing aids) and fuel cells capable of delivering considerable electrical power.

11.2 The cell

You will probably have done experiments with a Daniell cell consisting of copper and zinc plates (electrodes) each dipping into an electroiyte (Fig. 11.1). An electromotive force

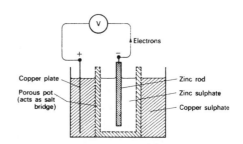

FIG. 11.1. The Daniell cell.

(e.m.f.) is developed and the current generated is sufficient to light a torch bulb. If a voltmeter is inserted in the circuit, a potential difference of about 1·1 V is observed, the zinc plate being negative with respect to the copper.

Varying the size of the plates and their distance apart does not affect the e.m.f. produced, though it does affect the capacity of the cell for giving current. Varying the concentration and composition of the electrolytes does have an effect, however, and it is necessary to devise some standard condition if meaningful measurements are to be made.

113

The processes which occur at the electrodes may be summarized by equations:

zinc electrode: $Zn(c) \longrightarrow Zn^{2+}(aq) + 2e^-$;

copper electrode: $Cu^{2+}(aq) + 2e^- \longrightarrow Cu(c)$;

The zinc electrode is acting as a source of electrons and these flow around the external circuit, and the electrode dissolves forming ions, $Zn^{2+}(aq)$. At the copper electrode the electrons combine with $Cu^{2+}(aq)$ ions in the solution and copper is deposited.

The above equations represent **half-cell reactions.** It is impossible for a half-cell reaction to occur on its own—there must always be two half-reactions together to make up a complete chemical process. In the zinc–copper case above the overall chemical process which occurs is that which results from adding the two half-reaction equations algebraically, in such a way that the electrons "cancel out":

$$Zn(c) + Cu^{2+}(aq) \rightleftharpoons Zn^{2+}(aq) + Cu(c)$$

In this reaction the zinc has lost electrons, and is said to have been **oxidized.** The copper ions have gained electrons, and are said to have been **reduced.**

The loss of electrons by a species is termed oxidation. The gain of electrons by a species is termed reduction.

The reaction between zinc metal and copper ions occurs when zinc is added to copper sulphate solution. It is a typical example of a **redox** (*red*uction-*ox*idation) reaction:

loses 2e⁻   ∴ oxidized

$$Zn(e) + Cu^{2+}(aq) \longrightarrow Zn^{2+}(aq) + Cu(c)$$

gains 2e⁻   ∴ reduced

This reaction, when carried out in a test tube, gives out heat. When put to use in a simple cell it gives rise to electrical energy.

If two different metals are made to touch under the surface of an electrolyte they really constitute a "short-circuited" cell. A current flows between the two metals and a reaction occurs.

Everyday life contains many examples of cell action. Chromium plated articles corrode very rapidly once the chromium layer has been penetrated, and so do tin-plated metal cans, the corrosion taking place where the two metals are simultaneously in contact with air and moisture. Corrosion of car fittings is rapid in winter when salt has been put down on the roads—chromium and iron are the two metals, and sodium chloride is the electrolyte in a cell. Corrosion of iron is oxidation forming rust—hydrated iron oxide—assisted by the electrolytic process.

11.3 Displacement reactions of metals

Metals may be classified according to their ability to *displace* the ions of other metals from solution. Adding zinc powder to copper(II) sulphate solution, which contains $Cu^{2+}(aq)$, causes metallic copper to be deposited and the zinc to dissolve:

$$Zn(c) + Cu^{2+}(aq) \longrightarrow Cu(c) + Zn^{2+}(aq);$$
$$\Delta H = -210 \text{ kJ.}$$

There will also be a certain amount of hydrogen evolved, due to the fact that a solution of $Cu^{2+}(aq)$ contains some $H^+(aq)$ as well, but this does not detract from the main argument.

If zinc powder is added to a series of aqueous solutions of metal ions, it will be found to displace certain ones with ease, such as silver(I) $Ag^+(aq)$, mercury(II) $Hg^{2+}(aq)$, lead(II) $Pb^{2+}(aq)$, and tin(II) $Sn^{2+}(aq)$. Others it will not displace at all, such as sodium $Na^+(aq)$, magnesium $Mg^{2+}(aq)$, and aluminium $Al^{3+}(aq)$.

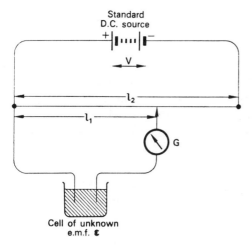

FIG. 11.2. A potentiometer: $\varepsilon = V(l_1/l_2)$.

The metals can in fact be arranged in a sequence in such a way that a given metal will displace the ions of all metals below it in the series. This series is known as the **electrochemical series,** which will be considered in section 11.9.

The "driving force" of a displacement reaction is obtained by constructing a cell in which the reaction can occur, and finding the magnitude of an external opposing e.m.f. which has to be applied in order *just* to prevent the reaction happening. A potentiometer circuit is ideal for this, provided that it has an accurate source of d.c. supply as a standard (Fig. 11.2). The chief problem lies not in measuring the e.m.f., but in constructing a cell which performs the desired reaction.

Each electrode is placed in a solution of its own ions at a standard concentration (molar will do, but strictly we should use solutions which are at "unit activity"—see section 11.6.) A piece of wire is not suitable for connecting these two half-cells together as it would introduce its own voltages at the points where it contacts the different electrolytes. A much better circuit is made by employing a **salt bridge.** This is generally an inverted U-tube containing a concentrated, non-reacting elec-

trolyte such as potassium chloride solution KCl (aq). Bulk diffusion between the liquids is eliminated either by cotton wool plugs or, better, sintered glass discs at the two ends of the U-tube. Even a salt bridge produces small liquid-junction potentials where it is in contact with the other electrolytes, but it is more reliable than a metal wire.

Figure 11.3 shows the cell which would be required in order to measure the e.m.f. of the displacement reaction between zinc and copper(II). The slide wire of the potentiometer is adjusted until the opposing e.m.f. just prevents current flowing through the galvanometer. The reactants are then being *artificially held at equilibrium* by the applied e.m.f. We can therefore say that the applied e.m.f. is a measure of the "driving force" of the displacement reaction.

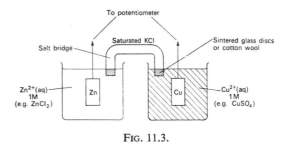

FIG. 11.3.

11.4 Conventions used in writing cell reactions

In order to save space, and the labour of drawing a separate diagram for each cell we set up, a shorthand notation is used in describing cells. The cell used in Fig. 11.3 is written down thus:

$$\text{Zn} \mid \text{Zn}^{2+}(\text{aq, 1 M}) \mid \text{Cu}^{2+}(\text{aq, 1 M}) \mid \text{Cu};$$
$$\epsilon = +1\cdot10 \text{ V}.$$

For brevity, symbols showing the states of reactants are omitted when these are obvious, but concentrations or activities of the electrolytes are written in. The symbol ε is used to denote e.m.f. in this book, and should not be confused with E for energy.

Sign convention. The sign of the e.m.f. is the sign of the right-hand electrode. In the above case the copper terminal is positive with respect to the zinc.

Many displacement reactions can be investigated in a cell, as well as in the test tube. For instance the reaction

$$Cu(c) + 2\,Ag^+(aq) \rightleftharpoons Cu^{2+}(aq) + 2\,Ag(c),$$

which gives rise to very beautiful crystals of metallic silver when done in a test tube, can be investigated quantitatively in the cell:

$$Cu\,|\,Cu^{2+}(aq)\ \vdots\ Ag^+(aq)\,|\,Ag;$$
$$\epsilon = +0.46\ V.$$

---

†11.5 Concentration cells

We have seen that the e.m.f. of a cell is a measure of the driving force of the cell reaction. If this is reliable, then *any* spontaneous change ought to be capable of being harnessed to give an e.m.f. Take, for instance, the mixing of

two solutions of an electrolyte at different concentrations. This will occur spontaneously, quite regardless of whether heat energy is evolved or absorbed in the process. It is possible to set up a cell in which the driving force of the mixing process is harnessed: such a cell is called a **concentration cell.**

If 0·01 M copper(II) sulphate solution is added to 0·1 M copper(II) sulphate, mixing will occur until the concentration is uniform; thereafter spontaneous change will cease. Such a mixing process is represented by the following cell (Fig. 11.4).

$$Cu\,|\,Cu^{2+}(0.01\ \text{M})\ \vdots\ Cu^{2+}(0.1\ \text{M})\,|\,Cu;$$
$$\epsilon = +0.029\ V.$$

The sense of the e.m.f. is such as to cause some copper from the left-hand electrode to dissolve, thereby increasing the concentration of $Cu^{2+}(aq)$ in the electrolyte, and to cause some of the copper ions in the solution of greater concentration to deposit on the right-hand electrode. The driving force of this process can be measured by applying an external e.m.f. from a potentiometer in order *just* to prevent current flowing. A sensitive potentiometer is needed, showing

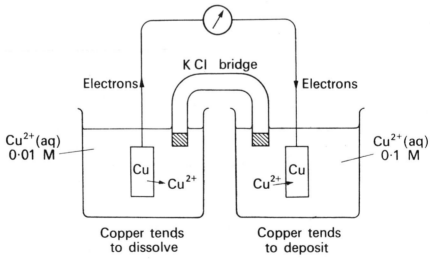

FIG. 11.4. A concentration cell.

that the driving force in a concentration cell is very much smaller than when two different metals are used in a displacement reaction.

Further experiments show that, provided the solutions are dilute, it is the *ratio* of ion concentrations which determines the e.m.f. and hence the driving force. A mathematical expression can be derived relating the concentrations to the e.m.f., and this is found to be logarithmic in form.

$$\varepsilon = \frac{RT}{nF} \log_e \frac{c_1}{c_2}$$ (provided the solutions are *dilute*)

where ε = e.m.f. of cell,

T = temperature in K,

F = the faraday constant, 96 490 coulombs,

n = the number of faradays transferred per gram–equation,

R = the gas constant, expressed in J K^{-1} mol^{-1},

c_1 = concentration, in mol dm^{-3}, of the solution in the left-hand cell,

c_2 = concentration, in mol dm^{-3}, of the solution in the right-hand cell.

Concentration cells are of little practical importance, as the e.m.f.s they give are very feeble, though they play an important role in living systems, for instance in muscles. There are two important things to be learned from studying them however:

(i) *Concentration of electrolyte does have an effect on the e.m.f. of a cell. If meaningful data are to be quoted, the ion concentration must be known.*

(ii) *Even in the simple case of 'physical' mixing of two electrolytes, it is possible to measure the 'driving force' of the process quantitatively. Some solutions mix endothermically, some exothermically; others do not give a detectable temperature change. Thus whereas heat change is not a criterion of spontaneous change, e.m.f. is.*

†11.6 The concept of activity

In the argument above it was stressed that the solutions were dilute. Concentration cells can certainly be made to work at higher concentrations, but their e.m.f.s no longer fit the mathematical relationship given above. This is one of the reasons why chemists often use the term **activity.** At very low concentrations the terms activity and concentration may be taken as synonymous—a 0·001 M solution of Cu^{2+}(aq) can be taken as having an activity of $a = 0·001$— but at higher concentrations this may not be so.

One way of *defining* activity would be "that quantity which fits the equation

$$\varepsilon = \frac{RT}{nF} \log_e \frac{a_1}{a_2}$$

in a concentration cell". A concentration cell can be used to *measure* activities, once we have defined activity in this way.

11.7 Standard electrode potentials

We shall in future frequently refer to solutions of unit activity. To a very rough approximation, activity = molar concentration. For instance it is calculated that a molar solution of sodium chloride has an activity of 0·68. For practical purposes activity may be regarded as a sort of "effective concentration".

In order to quote meaningful data for electrodes in cells, the concentration of ions has to be stated. In theory any concentration could be taken, but in practice we choose that concentration which corresponds to unit activity.

It is not possible to quote the potential of a single electrode: the only way of measuring it would be to insert a second electrode, and we would then be measuring the difference between

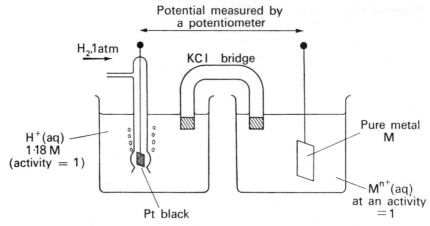

FIG. 11.5. Standard electrode potential on the hydrogen scale.

two electrode potentials! This being so, we must choose one electrode arbitrarily as a standard, and measure the potential difference between this standard electrode and any other standard electrode we require.

By general agreement the **normal hydrogen electrode** is taken as the reference point from which data on other electrodes are quoted. Since hydrogen is a gas at room temperature we cannot use it as an electrode in the ordinary way, and instead we use a piece of platinum coated with finely divided 'platinum black' containing adsorbed hydrogen, which catalyses the half-cell reaction:

$$2 H^+(aq) + 2e^- \rightleftharpoons H_2(g).$$

Pure hydrogen at 1 atm pressure must be used, and this is bubbled continuously over the platinum black electrode to maintain equilibrium (except momentarily when taking the actual e.m.f. reading). A solution of dilute hydrochloric acid is generally used, having a concentration of hydrogen ions of 1.18 mol dm^{-3}, which corresponds to unit activity.

Figure 11.5 shows the arrangement which would be used to measure the standard electrode potential of a metal M on the hydrogen scale. The data in Table 11.1, and in Appendix

1, are quoted for these conditions. The symbol ε° is used when giving standard potentials.

The cell arrangement may be written down thus:

$$H_2(g), Pt \mid H^+(aq, a = 1 \; \vdots \; M^{n+}(aq, a = 1) \mid M,$$
$$(e.m.f. = \epsilon^\circ)$$

where $n =$ number of positive charges on the metal ion.

TABLE 11.1

SOME STANDARD HALF-CELL POTENTIALS

Metal	Half-reaction	ϵ° (in V)
Sodium	$Na^+ \quad + e^- \rightleftharpoons Na$	-2.714
Magnesium	$Mg^{2+} +2e^- \rightleftharpoons Mg$	-2.370
Zinc	$Zn^{2+} +2e^- \rightleftharpoons Zn$	-0.763
Iron	$Fe^{2+} +2e^- \rightleftharpoons Fe$	-0.440
Hydrogen	$2 H^+ \quad +2e^- \rightleftharpoons H_2$	0.000
Copper	$Cu^{2+} +2e^- \rightleftharpoons Cu$	$+0.337$
Silver	$Ag^+ \quad + e^- \rightleftharpoons Ag$	$+0.799$

The half-cell potentials quoted in Table 11.1 are *standard* potentials, which we denote by using the superscript zero in the symbol ε°. Section 11.5 showed that e.m.f.s are affected by concentration according to a logarithmic relationship. The expression for a half-cell potential, ε, of any couple in which *oxidized form*

+n *electrons* ⇌ *reduced form* is similar to that quoted for concentration cells:

$$\varepsilon = \varepsilon^\circ + \frac{RT}{nF} \log_e \frac{[\text{oxidized form}]}{[\text{reduced form}]}$$

where square brackets denote activities, or for approximate purposes, concentrations. In the case of a metal in equilibrium with its ion, this expression reduces to

$$\varepsilon = \varepsilon^\circ + \frac{RT}{nF} \log_e [\text{ion}]$$

For practical purposes, it is more convenient to use logarithms to base 10, and since data are usually quoted at 25°C, we may write

$$\varepsilon = \varepsilon^\circ + \frac{2 \cdot 3RT}{nF} \log_{10} \frac{[\text{oxidized form}]}{[\text{reduced form}]}$$

$$= \varepsilon^\circ + \frac{0 \cdot 059}{n} \log_{10} \frac{[\text{oxidized form}]}{[\text{reduced form}]} \, .$$

These equations are all forms of the **Nernst equation.**[*]

11.8 Using half-cell potentials to predict reactions

Although ε° values are quoted relative to the normal hydrogen electrode they can be also used to predict the direction of reactions other than those involving hydrogen as well. First, however, we will show how the ε° values predict whether or not metals will reduce hydrogen ions.

(a) DISPLACEMENT OF HYDROGEN BY METALS

Consider the reaction which occurs when magnesium is added to dilute acid:

$$Mg(c) + 2H^+(aq) \rightleftharpoons Mg^{2+}(aq) + H_2(g).$$

---

[*] By international agreement the sign given to ε° values is the sign of the charge on the *electrode*, relative to the hydrogen electrode. Alternatively, it is the e.m.f. of the cell written down with the standard electrode on the left, and the electrode being measured on the right.

This reaction may be investigated in various ways, for instance:

(i) In a test tube, rapid and vigorous evolution of hydrogen is observed. Heat is evolved, that is, ΔH is negative.

(ii) A standard cell might be set up, and an external e.m.f. applied, in order *just* to prevent reaction taking place. Table 11.1 tells us that the e.m.f. required will be 2·37 V. If the opposing e.m.f. is removed, spontaneous reaction will occur in the direction left to right in the equation above. We may term this e.m.f. the standard e.m.f. of reaction. In this book it is denoted by the symbol $\Delta\epsilon^\circ$.

Observation (ii) is more valuable, because it tells us accurately how great is the "driving force" of the reaction. The procedure for predicting a reaction from half-cell potentials is as follows:

(1) Write down the half-reactions, together with their ε° values from a book of data:

$$2\,H^+ + 2e^- \rightleftharpoons H_2;$$
$$\epsilon^\circ = 0 \cdot 00 \text{ V, by definition}$$
$$Mg^{2+} + 2e^- \rightleftharpoons Mg;$$
$$\epsilon^\circ = -2 \cdot 37 \text{ V.}$$

(2) Obtain the overall equation by subtraction of the second half-reaction from the first, and obtain the reaction e.m.f. $\Delta\varepsilon^\circ$, by subtracting the second ε° from the first:

$$2H^+ + Mg \rightleftharpoons Mg^{2+} + H_2;$$
$$\Delta\epsilon^\circ = 0 \cdot 00 - (-2 \cdot 37) = +2 \cdot 37 \text{ V.}$$

(3) If $\Delta\varepsilon^\circ$ is positive, the reaction will proceed from left to right as written. If it is negative the equilibrium composition will lie to the left. If the e.m.f. is almost zero, then the reactants will come to equilibrium with all species present in appreciable amounts.

Consider another case: predict from ε° data whether metallic silver will reduce the hydrogen ion in dilute acid. Following the procedure above:

(1) Write down the half-reactions:

$$2H^+ + 2e^- \rightleftharpoons H_2; \quad \epsilon^\circ = 0.00 \text{ V.} \qquad \text{(i)}$$

$$Ag^+ + e^- \rightleftharpoons Ag; \quad \epsilon^\circ = +0.799 \text{ V.} \qquad \text{(ii)}$$

(2) Subtract as before. In this case equation (ii) has to be doubled in order that the electrons "cancel out" in the overall equation. This has no effect on its ϵ° value.

$$2H^+ + 2Ag \rightleftharpoons 2Ag^+ + H_2; \quad \Delta\epsilon^\circ = -0.799 \text{ V.}$$

(3) Since the resultant e.m.f. is negative the reaction will *not* proceed from left to right. Instead the equilibrium composition will lie heavily to the left. The ϵ° data predict that if hydrogen gas is bubbled into a solution of, say, silver nitrate, metallic silver should be precipitated.

This prediction is however not borne out by experiment: in the absence of a catalyst there is no observable change when hydrogen is bubbled into $Ag^+(aq)$. This illustrates one of the limitations of using half-cell potentials, namely that they cannot predict reaction *rates*.

(b) DISPLACEMENT OF ONE METAL BY ANOTHER

Although ϵ° values are quoted relative to the hydrogen electrode the same data can be used to predict the outcome of reactions which do not involve hydrogen. Suppose for instance we wish to predict the outcome of adding zinc powder to a solution of lead(II) ions, $Pb^{2+}(aq)$. A data book gives two half-reactions which are relevant, namely:

$$Pb^{2+} + 2e^- \rightleftharpoons Pb; \quad \epsilon^\circ = -0.126 \text{ V.}$$

$$Zn^{2+} + 2e^- \rightleftharpoons Zn; \quad \epsilon^\circ = -0.763 \text{ V.}$$

We require the e.m.f. of the reaction

$$Pb^{2+} + Zn \rightleftharpoons Zn^{2+} + Pb;$$

$$\Delta\epsilon^\circ = -0.126 - (-0.763) = +0.537 \text{ V.}$$

This equation is obtained by subtracting the second equation from the first, and hence the reaction e.m.f. is obtained by subtracting -0.763 from -0.126 V. Since the e.m.f. is positive for the equation as written, the data predict that adding zinc powder to $Pb^{2+}(aq)$ should cause metallic lead to be precipitated. In fact this reaction is a very effective way of growing crystals of lead—a so called "lead tree" is produced.

11.9 The electrochemical series of metals

The metals in Table 11.1 are arranged in the order of their ϵ° values. The order is seen to be the same as the electrochemical series produced by conducting simple displacement reactions. This is not surprising, for we have seen that a comparison of ϵ° values enables us to predict the direction of displacement reactions. A given metal in the series will *reduce* the ions of all metals below it, and a given ion will *oxidize* all the metals above it. The advantage of using e.m.f. data is that it enables the electrochemical series to be placed on a quantitative basis. For instance, simple displacement reactions in a test-tube are not accurate enough to distinguish between tin and lead. ϵ° measurements, made to ± 0.001 V or better, show that tin comes just above lead in the series.

In a case like tin and lead, the reaction e.m.f. is extremely low, and we should expect to be able to observe an equilibrium rather than a practically complete displacement of one metal by the other:

$$Sn^{2+} + 2e^- \rightleftharpoons Sn; \quad \epsilon^\circ = -0.136 \text{ V}$$

$$Pb^{2+} + 2e^- \rightleftharpoons Pb; \quad \epsilon^\circ = -0.126 \text{ V}$$

Subtracting,

$$Sn^{2+} + Pb \rightleftharpoons Sn + Pb^{2+};$$

$$\Delta\epsilon^\circ = -0.136 - (-0.126) = -0.01 \text{ V.}$$

Hence only 0·01 V would need to be applied to a cell to prevent reaction occurring. We therefore expect an equilibrium, slightly favouring Sn^{2+} and Pb. The equilibrium constant can be determined experimentally for such a system:

$$K = \frac{[Sn][Pb^{2+}]}{[Sn^{2+}][Pb]} = \frac{[Pb^{2+}]}{[Sn^{2+}]} = 0·46.$$

†11.10 Prediction of equilibrium constants from e.m.f. measurements

It can be shown theoretically, and verified by experiment in many cases, that the equilibrium constant, K, of a reaction is related to the reaction e.m.f., $\Delta\varepsilon°$, by an expression which is similar to that given earlier for concentration cells:

$$\Delta\varepsilon° = \frac{RT}{nF}\log_e K = \frac{2·303\,RT}{nF}\log_{10} K$$

$$= \frac{0·059}{n}\log_{10} K \text{ at } 25°C.$$

Figure 11·6 shows this relationship graphically and may be used for the problems on equilibrium constant at the end of this chapter, as well as elsewhere in the book.

Substituting in this equation, using the data for tin and lead,

$$\log_{10} K = \frac{n\Delta\varepsilon°}{0·059} = \frac{2\times0·010}{0·059} = 0·339$$

$$\therefore\ K = 2·18 = \frac{[Sn^{2+}]}{[Pb^{2+}]}.$$

Worked Example 1. Calculate the equilibrium constant of the reaction between cobalt and nickel(II).

Half-reactions: $\quad Ni^{2+}+2e^- \rightleftharpoons Ni;$

$$\epsilon° = -0·250 \text{ V}$$

$$Co^{2+}+2e^- \rightleftharpoons Co;$$

$$\epsilon° = -0·277 \text{ V}$$

Subtract for complete reaction:

$$Ni^{2+}+Co \rightleftharpoons Ni+Co^{2+};$$

$$\Delta\epsilon° = -0·250-(-0·277) = +0·027 \text{ V}.$$

$$K = \frac{[Co^{2+}][Ni]}{[Co][Ni^{2+}]} = \frac{[Co^{2+}]}{[Ni^{2+}]}$$

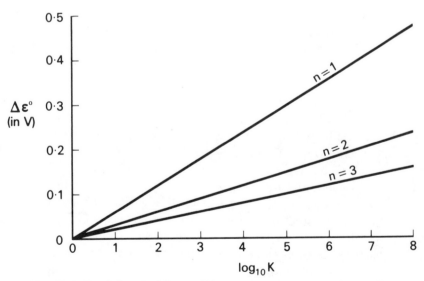

FIG. 11.6. Graph for obtaining equilibrium constant from reaction e.m.f.

$$\frac{2 \cdot 303\ RT}{nF} \log_{10} K = \frac{0 \cdot 059}{n} \log_{10} K = 0 \cdot 027$$

$$\log_{10} K = \frac{0 \cdot 027 \times 2}{0 \cdot 059} = 0 \cdot 915$$

$$K = 8 \cdot 23.$$

The positive value of the reaction e.m.f. $\varDelta \varepsilon^{\circ}$ indicates that cobalt(II) ions should reduce metallic nickel. The fact that it is a low e.m.f. suggests that a measurable equilibrium ought to exist. The calculation shows us that $K = 8 \cdot 23$, and hence the reaction does not proceed completely to the right. A reaction is virtually complete when $K > 10^6$. This corresponds to a $\varDelta \varepsilon^{\circ}$ value of $0 \cdot 36$ V where $n = 1$. ∴ 1.

A similar calculation can be applied to any reaction for which $\varDelta \varepsilon^{\circ}$ is known. Remember that a favourable equilibrium does not necessarily *prove* that the reaction is feasible, for it might be too slow, or there might be side reactions which interfere. However, the calculations do tell us with complete certainty those reactions which are *impossible* to carry out; this in itself is very important. For instance, a chemical engineer may save his company many thousands of pounds by carrying out a five-minute calculation to show that a proposed reaction is impossible!

11.11 Redox reactions among non-metals

So far this chapter has been concerned with redox reactions among metals, apart from hydrogen (which is "metallic" insofar as it forms positive ions in aqueous solution). Displacement reactions among metals are easily carried out in a test-tube and are readily studied quantitatively by means of cells.

It is quite possible to set up electrodes for the study of non-metals, as has already been done in the case of hydrogen, and redox reactions similar to those among metals occur here also. It is quite natural therefore to extend the redox series to include non-metallic half-reactions. A standard electrode can be constructed for say, chlorine, representing the following:

$$Cl_2(g) + 2e^- \rightleftharpoons 2\,Cl^-(aq);$$
$$\epsilon^{\circ} = +1 \cdot 36 \text{ V}.$$

Chlorine has a strong tendency to gain electrons and form $Cl^-(aq)$ in solution, as its highly positive half-cell potential indicates. It is therefore a *strong oxidizing agent*.

Chlorine is an element of Group VIIB of the periodic table, one of the halogens. The halogens show the similarity expected of members of the same group or chemical family, and also a gradation in properties. The most powerful oxidizing agent of all is fluorine, and the oxidizing power decreases down the series.

$$F_2(g) + 2e^- \rightleftharpoons 2\,F^-(aq);$$
$$\epsilon^{\circ} = +2 \cdot 87 \text{ V}$$
$$Cl_2(g) + 2e^- \rightleftharpoons 2\,Cl^-(aq);$$
$$\epsilon^{\circ} = +1 \cdot 36 \text{ V}$$
$$Br_2(l) + 2e^- \rightleftharpoons 2\,Br^-(aq);$$
$$\epsilon^{\circ} = +1 \cdot 07 \text{ V}$$
$$I_2(c) + 2e^- \rightleftharpoons 2\,I^-(aq);$$
$$\epsilon^{\circ} = +0 \cdot 54 \text{ V}$$

The halogens are an example of a redox series exactly analogous to the electrochemical series of metals. Examination of the above series indicates that fluorine gas ought to be capable of oxidizing all the other halogen ions $X^-(aq)$. Bromine ought to be capable of oxidizing $I^-(aq)$ to iodine, but will have no effect on $Cl^-(aq)$ or $F^-(aq)$.

An examination of the reaction e.m.f. $\varDelta \varepsilon^{\circ}$ gives us a closer insight into these reactions. Let us see for example what might happen when chlorine gas is bubbled into a solution of potassium bromide, which contains $Br^-(aq)$:

$$Cl_2 + 2\,Br^- \rightleftharpoons Br_2 + 2\,Cl^-$$
$$\varDelta \epsilon^{\circ} = 1 \cdot 36 - 1 \cdot 07 = +0 \cdot 29 \text{ V}.$$

A positive value for $\Delta\varepsilon°$ indicates that the reaction will proceed from left to right as written. Although strictly an equilibrium it will be complete for practical purposes ($K \simeq 10^5$).

The displacement reactions of the halogens (except fluorine gas) are readily carried out on a test-tube scale in the laboratory.

11.12 The general redox series

Appendix 1 gives a selection of standard half-cell potentials, $\varepsilon°$, for a wide variety of elements under various conditions. It contains both metals and non-metals, and there is considerable overlap although metals like the alkali metals occur at the "negative" end of the scale, and the most active non-metals like fluorine and oxygen occur at the "positive" end.

These data give an enormous amount of information about redox reactions when correctly used. If we wish to determine whether a given substance will oxidize or reduce another substance, then it is simply necessary to imagine setting up a cell. If the $\varepsilon°$ values of the relevant half-reactions differ by a large amount, indicating that a large e.m.f. would have to be applied in opposition to the cell by means of a potentiometer, to prevent reaction occurring, then we may predict a strong tendency for the reaction to occur. If the reaction e.m.f. is very small then we predict incomplete reaction.

The data in Appendix 1 have been obtained by a variety of methods, but in most cases it is possible to set up some sort of cell. For instance an electrode might be set up to investigate the reducing power of iron(II) ions, which are readily oxidized to iron(III):

Fe^{3+}(aq)+e$^-$ $\rightleftharpoons$ Fe^{2+}(aq); $\epsilon° = +0.771$ V
iron(III)   iron(II)
ions   ions

The standard electrode for this would be a piece of platinum dipping into a solution containing

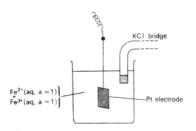

FIG. 11.7. Standard half-cell for the reaction
Fe^{3+}+e$^-$ $\rightleftharpoons$ Fe^{2+}.

iron(II) and iron(III) ions mixed together, each at a concentration corresponding to unit activity (Fig. 11.7).

Similarly, an electrode can be set up to measure the oxidizing power of permanganate ions, MnO_4^-(aq), in acid solution. Potassium permanganate is a powerful oxidizing agent, and is itself reduced to manganese(II) ions, Mn^{2+}(aq). The half-reaction is as follows:

$$MnO_4^-(aq)+8\,H^+(aq)+5e^-$$
$$\rightleftharpoons Mn^{2+}(aq)+4\,H_2O;$$
$$\epsilon° = +1.51 \text{ V}$$

The standard electrode for this would be a platinum plate dipping into a solution of MnO_4^-(aq), Mn^{2+}(aq) and H^+(aq), all at unit activity.

Worked Example 2. Predict from redox data whether permanganate ions should oxidize iron(II) ions in acid solution, and obtain a balanced equation.

To obtain the overall equation, the electrons must "cancel out", and we must therefore take five moles of Fe^{2+}(aq) for every mole of MnO_4^-(aq):

$$MnO_4^- + 8\,H^+ + 5e^- \rightleftharpoons Mn^{2+} + 4\,H_2O;$$
$$\epsilon° = +1.51 \text{ V}$$

$$5\,Fe^{3+} + 5e^- \rightleftharpoons 5Fe^{2+};\qquad \epsilon° = +0.77 \text{ V}$$

Subtract:

$$MnO_4^- + 8\,H^+ + 5\,Fe^{2+} \rightleftharpoons Mn^{2+} + 5\,Fe^{3+} + 4\,H_2O.$$

$$\Delta\epsilon° = 1.51 - 0.77 = +0.74 \text{ V}.$$

The equilibrium will therefore lie completely to the right for practical purposes, and iron(II) is quantitatively oxidized by permanganate ions.

This reaction is frequently performed as a titration, for estimating the concentration of iron(II) ions in a solution.

Worked Example 3. 20 cm³ of a solution of Fe²⁺(aq) were titrated with 0·01 M KMnO₄(aq) until reaction was complete. 21·6 cm³ of permanganate solution were required. Calculate the molarity of the iron(II) solution.

21·6 cm³ of 0·01 M KMnO₄ contain

$$\left(\frac{21\cdot6}{1000} \times 0\cdot01\right) \text{ moles } MnO_4^-.$$

∴ the 20 cm³ of iron(II) solution contained

$$\left(5 \times \frac{21\cdot6}{1000} \times 0\cdot01\right) \text{ moles of } Fe^{2+}$$

∴ 1 dm³ (1000 cm³) of iron(II) solution contains

$$5 \times \frac{21\cdot6}{1000} \times 0\cdot01 \times \frac{1000}{20} \text{ moles}$$

$$= 0\cdot054 \text{ moles}.$$

∴ the concentration of Fe²⁺(aq) is 0·054 M.

It is sometimes a little complicated to set up standard cells in the laboratory for determining the redox potentials given in Appendix 1, though it is relatively easy to set up a cell to demonstrate qualitatively that a redox reaction can generate an e.m.f. Figure 11.8 shows how this can be done for the reaction in which MnO_4^- oxidizes Fe²⁺(aq). Such a cell reaction is sometimes described as *oxidation at a distance*. If a trace of ammonium thiocyanate is added (NH₄CNS, which provides thiocyanate ions, CNS⁻) a blood-red colour is seen to develop around the platinum plate in the iron(II) solution. Although iron(II) ions do not react with CNS⁻ ions, iron(III) react to form deep blood-red complexes, such as $Fe(CNS)_6^{3-}$:

$$Fe^{3+}(aq) + 6\,CNS^-(aq) \rightleftharpoons Fe(CNS)_6^{3-}$$
iron(III)   blood-red
                                                 complex ion

Note that the blood-red colour forms at the platinum electrode, not at the porous plug salt bridge. This is a clear indication that the oxidation is taking place at a distance, due to electron transfer round the external wire.

11.13 Oxidation number

The concept of oxidation number is an important one, especially as a "book-keeping" device for balancing equations and for doing calculations involving oxidation and reduction. In the previous section we referred to the existence of two aqueous ions for iron, namely Fe²⁺(aq) and Fe³⁺(aq). These are named iron(II) and iron(III), and in these cases iron has an **oxidation number** of +2 or +3 respectively. Similarly the oxidation number of chlorine is −1 in Cl⁻(aq).

The concept of oxidation number is not necessarily limited to ionic compounds. Silver chloride, for instance, shows few, if any, of the properties of an ionic substance. Nevertheless, the oxidation number of silver is still +1 in silver chloride as it is in the aqueous ion; the oxidation number of chlorine is again −1.

$$Ag^+(aq) + Cl^-(aq) \longrightarrow AgCl(c)$$
O.N.:   +1   −1   +1 −1

Magnesium oxide, MgO, has properties which suggest that it is made up of ions Mg²⁺ and

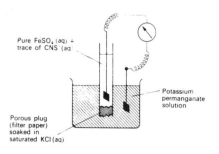

Pure FeSO₄ (aq) +
trace of CNS⁻(aq)

Potassium
permanganate
solution

Porous plug
(filter paper)
soaked in
saturated KCl(aq)

FIG. 11.8.

O^{2-}, and they are assigned oxidation numbers of $+2$ and -2. By analogy, other oxides (with one or two exceptions, see below) are assigned oxidation numbers on the basis that oxygen always has an oxidation number of -2. For instance, FeO. O.N. of iron $= +2$, therefore the compound is named iron(II) oxide. The oxidation numbers must add up algebraically to zero. Fe_2O_3 is iron(III) oxide.

$$\begin{array}{cc} Fe_2 & O_3 \end{array}$$
$$\text{O.N.:} \quad (2\times +3) + (3\times -2) = 0.$$

The compound Fe_3O_4, magnetic iron oxide, presents problems: the oxidation number of iron works out to $+2\frac{2}{3}$. Alternatively it can be regarded as a mixture of two oxidation states, as if it were $FeO + Fe_2O_3$. This is the method adopted in naming this compound systematically, and it may be called iron(II,III,III) oxide. Similarly lead forms two simple oxides, lead(II) oxide PbO, and lead(IV) oxide PbO_2. Red lead oxide, Pb_3O_4, can be regarded as a compound of these two simple oxides, $2PbO + PbO_2$, and the name becomes lead(II,II,IV) oxide.

Compounds which are markedly non-stoichiometric have to be assigned fractional oxidation numbers, and here the system consequently finds only limited application. Silver oxide for instance is only approximately Ag_2O, and silver is only approximately $+1$.

The system of oxidation numbers also applies to elements, which are assigned the number zero. To some extent the assignment of oxidation numbers is arbitrary, but for convenience, where an electron-pair bond exists, the oxidation number is assigned by transferring the electrons completely to the more electronegative element. In hydrogen chloride:

$$\begin{array}{cc} \text{less} & \text{more} \\ \text{electronegative} & \text{electronegative} \end{array}$$
$$H \text{---} Cl$$
$$\delta + \quad \delta -$$

Transferring the electron pair completely to the chlorine atom, we have:

$$H^+ \qquad :Cl^-$$

Therefore the oxidation numbers are: hydrogen $= +1$, chlorine $= -1$. Where two identical atoms are covalently linked, the electron-pair is assumed equally shared. This point is brought out by comparing water and hydrogen peroxide:

$$\begin{array}{cccc} +1 & +1 & +1 & +1 \\ H & H & H & H \end{array}$$

Compounds which contain the peroxo-link, $-O-O-$, are an exception to the general rule that oxygen atoms are assigned -2. Barium peroxide, BaO_2, is correctly regarded as barium $= +2$, oxygen $= -1$, since its properties suggest that it is structurally built up of ions, Ba^{2+} and $[O-O]^{2-}$.

The oxidation numbers in a neutral molecule must add up algebraically to zero; in a complex ion they must add up to the overall charge on the ion. The complex ion $CuCl_4^{2-}$ has copper in an oxidation state of $+2$:

$$[CuCl_4]^{2-} + 2 + (4\times -1) = -2.$$

This ion is named tetrachlorocuprate(II), the ending -ate implying an anion. The ion $[Cu(NH_3)_4]^{2+}$ is named tetra-ammine copper (II). The four ammonia molecules are neutral with an overall oxidation number of zero, so the oxidation number of the copper atom is the overall charge on the ion.

The oxidation number of the central atom in an oxo-ion is calculated on the assumption that the oxygen atoms are -2. Most oxo-ions have common names which are rather unsystematic, but they can also be named systematically on the basis of their oxidation states. The oxo-ions of chlorine illustrate this point (Table 11.2).

TABLE 11.2

Formula of acid	Formula of ion	O.N. of chlorine	Common name of ion	Systematic name of ion
HClO	ClO^-	+1	Hypochlorite	Chlorate(I)
$HClO_2$	ClO_2^-	+3	Chlorite	Chlorate(III)
$HClO_3$	ClO_3^-	+5	Chlorate	Chlorate(V)
$HClO_4$	ClO_4^-	+7	Perchlorate	Chlorate(VII)

This system of nomenclature was invented by Stock, to avoid the confusion which common names often cause. It is often referred to as Stock nomenclature. The salts of vanadium and manganese, for instance, occur in a wide variety of oxidation states which often make the older names inadequate.

It is customary to define oxidation of a substance as the removal of electrons, and reduction as the addition of electrons. A more general definition is:

oxidation = increase in oxidation number
reduction = decrease in oxidation number

Thus the definition of oxidation and reduction is not restricted to ionic substances.

Worked Example 4. For each of the following changes, write "oxidized", "reduced" or "no change".

(a) $SnCl_2 \rightarrow SnCl_4$

Oxidation number of Sn increases from +2 to +4.
∴ oxidized.

(b) $MnO_4^-(aq) \rightarrow Mn^{2+}(aq)$

Oxidation number of Mn decreases from +7 to +2.
∴ reduced.

(c) $MnO_4^{2-}(aq) \rightarrow MnO_4^-(aq)$

Oxidation number of Mn increases from +6 to +7.
∴ oxidized.

(d) $2\,CrO_4^{2-}(aq) \rightarrow Cr_2O_7^{2-}(aq)$

Oxidation number of Cr is +6 on both sides.
∴ no change.

Study Questions

1. (a) Show by means of labelled sketches how you would set up the following cells. Assign the polarity of each electrode by reference to Appendix 1.
   (i) $Zn|Zn^{2+}(M)$ $Sn^{2+}(M)|Sn$
   (ii) $Pb|Pb^{2+}(2\,M)$ $Pb^{2+}(0.1\,M)|Pb$
   (iii) Pt, $H_2(1$ atm.$)|H^+(M)$ $H^+(10^{-7}\,M)|H_2$ (1 atm.), Pt

   (b) Write a complete equation for the cell reaction that you would expect to occur in (i).

2. Account for the following observations using electrochemical theory.

(a) If a tin-can (iron dipped in molten tin) is scratched, rusting occurs where the tin surface has been penetrated: if a galvanized bucket (iron dipped in zinc) is scratched, no rusting occurs, but the zinc is slowly eaten away.

(b) Chromium plated articles, such as steel motor car fittings, only remain rustproof as long as the chromium surface is unbroken.

(c) Corrosion of motor cars is a greater problem in the winter, when salt is spread on the roads to melt ice and snow.

(d) Corrosion of the iron in the hull of a ship can

be prevented by attaching pieces of zinc to the hull.

(e) Copper and silver will not react at all with dilute sulphuric acid. On the other hand, zinc and magnesium react rapidly.

(f) Aqueous solutions of potassium permanganate, $KMnO_4$, and of chlorine will decompose if kept for a long time.

(g) If chlorine is bubbled through a solution of Br^-(aq), the colour of bromine is immediately observed. The addition of a crystal of iodine to Br^-(aq) does not liberate bromine.

3. Suggest a simple way of preparing metallic lead from a solution of lead nitrate, $Pb(NO_3)_2$(aq).

4. Suppose that the half-cell Cu^{2+}(aq)$+2e^- \rightleftharpoons Cu$(c) had been selected as the "standard electrode". Calculate the standard potentials, ε^*, for the following half cells, ε^* for Cu^{2+}/Cu being zero.

(a) Mn^{2+}(aq)$+2e^- \rightleftharpoons Mn$(c).

(b) Cl_2(g)$+2e^- \rightleftharpoons 2Cl^-$(aq).

(c) $2H^+$(aq)$+2e^- \rightleftharpoons H_2$(g).

5. Explain why it is not possible to measure a single electrode potential, such as the potential difference that is supposed to exist between a copper rod and a solution of Cu^{2+} ions.

6. (a) What would happen if a nickel spatula were used to stir a solution of copper(II) sulphate, $CuSO_4$(aq)?

(b) Could silver(I) nitrate, $AgNO_3$(aq), solution be stored in a copper container?

†7. Use the data in Appendix 1 to calculate the equilibrium constants for the reactions:

(a) $Ni+Sn^{2+} \rightleftharpoons Ni^{2+}+Sn$.

(b) $Cr^{3+}+Al \rightleftharpoons Cr+Al^{3+}$

8. Can you discover any relationship between a metal's position in the electrochemical series and its position in the periodic table?

9. How would ε° for Na^+/Na be affected (qualitatively) if the sodium was present as a 1% amalgam in mercury?

10. 20 cm³ of 0·1 M iron(II) ammonium sulphate (containing Fe^{2+}) was titrated with a solution of potassium dichromate (containing $Cr_2O_7^{2-}$) of unknown concentration. 25·36 cm³ of the latter were needed.

(a) Six Fe^{2+} ions react with one $Cr_2O_7^{2-}$ ion. Write down an ionic equation for this reaction.

(b) What is the molarity of the dichromate solution?

11. Use your library to list the different kinds of cell that are in everyday use. Try to discover the chemical reactions that each depends on.

12. Calculate the oxidation numbers in the following:

(a) N in NH_3, N_2H_4, NH_2OH, N_2, N_2O, NO, NO_2^+, NO_2, NO_2^-, NF_3, NCl_3 and NOCl.

(b) Mn in $MnCl_2$, MnO_2, Mn_2O_7, $KMnO_4$, MnO_4^{2-}, MnF_6^{2-}, $Mn(CN)_6^{4-}$, $MnSO_4$, $Mn_2(CO)_{10}$ and Mn?

13. (a) Label the atoms in the following equations with oxidation numbers:

(i) $2CO \longrightarrow C+CO_2$.

(ii) $2HI+H_2O_2 \longrightarrow I_2+2H_2O$.

(iii) $2H_2O_2 \longrightarrow 2H_2O+O_2$

(iv) $SiH_4+HI \longrightarrow SiH_3I+H_2$

(v) $NH_3+HCl \longrightarrow NH_4Cl$

(vi) $Ca+H_2 \longrightarrow CaH_2$

(b) Deduce from your numbers which species are oxidized and which are reduced.

CHAPTER 12

Free Energy

12.1 The limitations of ΔH measurements

A chemical engineer frequently needs to know the heat evolved or absorbed in a chemical process. He may require information on the efficiency of a new fuel, or the energy input required to maintain a reaction in a chemical plant. A simple measurement of ΔH will however only give a limited amount of *theoretical* information about a reaction. Calculation (by Hess's law) of the ΔH of a new, uninvestigated reaction will *not* tell him whether or not the reaction will go. Spontaneous chemical changes can be either exothermic, endothermic or thermoneutral, hence ΔH is no criterion.

ΔH measurements will not tell us how far a chemical change will proceed towards equilibrium. For instance, data on the heat of solution of a salt in water give no information on whether that salt is highly soluble or sparingly soluble.

If then ΔH measurements are apparently so limited in their application to theoretical problems, why are we so concerned with energy measurements in chemistry? It was shown in the last chapter that we *can* measure the "driving force" of a reaction, in terms of the e.m.f. which has to be applied to a cell in order to prevent the reaction taking place. The e.m.f. of a cell *is* a criterion of whether or not

the reaction will proceed. It is now time to investigate the nature of this 'driving force' in more detail.

12.2 The nature of driving force of a reaction—free energy

When we use the term *driving force* we are using the word force rather loosely. The term e.m.f. (electromotive *force*) is similarly confusing. It is necessary to think of the driving force in terms of *energy* units.

Imagine once again a cell in which standard electrodes of silver and zinc are coupled together. Let us calculate the energy *per mole* of reactants which will be liberated, if a small amount of reaction (that is, insufficient to run the cell down) is allowed to occur.

$$Zn(c) + 2\,Ag^+(aq) \rightleftharpoons 2\,Ag(c) + Zn^{2+}(aq);$$

$$\Delta\epsilon° = +1{\cdot}54 \text{ V}.$$

Charge which flows per mole $= 2\,F$ coulombs where $F =$ the faraday constant, 96 490 coulombs.

(Note that *two* faradays of charge are transferred per mole, because the half-reactions involve two electrons per ion.)

$\therefore$ Energy change per mole $= -2\,F\Delta\epsilon°$ J

128

(Remember: energy change in joules = charge × potential through which that charge is transferred. The minus sign is inserted to indicate that there is a decrease in energy as the reaction proceeds, by analogy with the convention used for ΔH.)

∴ energy change under equilibrium conditions

$$= -297 \text{ kJ mol}^{-1}$$

This energy change at *equilibrium* is called the **free energy** of reaction. It is given the symbol ΔG. In the above case a cell containing *standard* electrodes was considered, and the free energy change is then termed the **standard free energy of reaction,** and is denoted by adding a superscript zero, $\Delta G°$. In general,

$$\Delta G° = -nF\Delta\varepsilon°$$

where $n =$ number of faradays of charge transferred per gram-equation.

ΔG normally has a different numerical value from ΔH. It may even be of opposite sign.

12.3 ΔG as a criterion of spontaneous change

It has already been shown that a reaction e.m.f. is a measure of the tendency of a reaction to proceed in a particular direction. It follows from the relationship $\Delta G° = -nF\Delta\varepsilon°$ that free energy change is an alternative criterion. A cell reaction proceeds spontaneously in the direction determined by its positive e.m.f., which is in the direction of *negative ΔG*.

Spontaneous changes occur with a decrease in free energy (ΔG negative).

It appears then that ΔG, not ΔH, is the chemical analogue of "potential energy" of mechanical systems. In mechanical systems, bodies tend to move downhill under the action of gravity—a decrease in potential energy. In chemistry, substances move 'downhill' in terms of free energy.

12.4 Factors which determine the magnitude of ΔG

The enthalpy change, ΔH, of a reaction appears nevertheless to have some importance in deciding the outcome of chemical reactions. It is a matter of common experience that the *majority* of vigorous chemical changes at room temperature seem to give out heat, and it has already been observed when considering le Chatelier's principle that endothermic reactions seem to become more important as the temperature is increased. ΔH is one factor involved, but not the only one.

Another factor which influences spontaneous change is the *tendency of matter to reach a state of maximum disorder* (Fig. 12.1). A few examples may be quoted to illustrate this:

(i) If oxygen and nitrogen are mixed together in a vessel, the random motion of molecules determines the mixing process. There is no energy change. We cannot say for certain that the gases will never spontaneously "unmix". We can only say that it would be exceedingly improbable. Note that there is no energy released when oxygen and nitrogen are mixed. Note also that theoretically we should not require to use up any energy in separating oxygen and nitrogen. Yet in practice, the separation of oxygen and nitrogen from liquid air consumes quite a considerable expenditure of energy.

(ii) If ink is dissolved in water there is practically no change in temperature (ΔH is

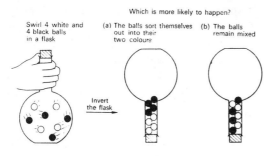

Which is more likely to happen?

Swirl 4 white and 4 black balls in a flask

(a) The balls sort themselves out into their two colours

(b) The balls remain mixed

Invert the flask

FIG. 12.1.

approximately zero). Ink dissolves because the system reaches a state of greater randomness or disorder when the molecules disperse in the water.

(iii) When a liquid boils there is an increase in enthalpy. For instance:

$$H_2O(l) \rightleftharpoons H_2O(g);$$

$$\Delta H = +41 \text{ kJ mol}^{-1} \text{ at } 100°C.$$

Nevertheless, the vapour state is more disordered, more random, than the liquid state since the molecules have freedom to move. This effect becomes more pronounced at a higher temperature.

(iv) If iodine vapour is heated it dissociates. This again is endothermic:

$$I_2(g) \rightleftharpoons 2 I(g);$$

$$\Delta H = +150 \text{ kJ mol}^{-1} \text{ at } 25°C.$$

Note that iodine atoms represent a state of greater disorder than the diatomic molecule.

We may summarize these observations by saying

$$\begin{pmatrix} \text{tendency for spontaneous} \\ \text{change} \end{pmatrix} = \begin{pmatrix} \text{tendency for system} \\ \text{to give out heat} \end{pmatrix}$$
$$+ \text{ (tendency for system to become disordered)}$$

12.5 Entropy

The degree of disorder of a system, or its randomness, is measured by a quantity called the **entropy,** S. It is in fact easier to talk about

changes in entropy, and these are shown by the symbol ΔS.

ΔS **positive** means *system becomes more disordered.*

ΔS **negative** means *system becomes less disordered.*

A few examples of physical and chemical change ought to make this clear:

(i) $H_2O(l) \longrightarrow H_2O(g)$; ΔS positive.

(ii) $H_2O(g) \longrightarrow H_2O(l)$; ΔS negative.

(iii) $Cl_2(g) \longrightarrow 2 Cl(atoms)$; ΔS positive.

(iv) $O_2(pure) + N_2(pure) \longrightarrow$ mixture;

ΔS positive.

(v) $C(c) + CO_2(g) \longrightarrow 2 CO(g)$; ΔS positive.

The relationship summarized at the end of section 12.4 may now be restated:

$$\begin{pmatrix} \text{tendency for} \\ \text{spontaneous change} \end{pmatrix} = \begin{pmatrix} \text{tendency for system} \\ \text{to give out heat} \end{pmatrix}$$
$$+ \text{ (tendency for system to increase in entropy)}$$

The entropy factor becomes more important at higher temperatures. Tendency to spontaneous change is free energy, and free energy consists of a "heat factor" and an "entropy factor". The mathematical relationship between them is as follows:

$$\Delta G = \Delta H - T\Delta S$$

When a system is in equilibrium $\Delta G = 0$. This must be the case, since ΔG determines whether spontaneous change will take place. To summarize, not all reactions proceed spontaneously with the evolution of heat, but *all* spontaneous reactions involve *either* the evolution of heat *or* increase in disorder.

†12.6 Entropy of vaporization

Consider again the example of boiling water in equilibrium with steam at 1 atm pressure:

$$H_2O(l) \rightleftharpoons H_2O(g); \qquad \Delta G = 0.$$

$$\Delta H = +41 \text{ kJ mol}^{-1}$$
$$\therefore T\Delta S = \Delta H - \Delta G,$$
$$= +41 \text{ kJ mol}^{-1}.$$

$\therefore$ under equilibrium conditions,

$$\Delta S = \frac{\Delta H}{T} = \frac{41}{373} = 0 \cdot 11 \text{ kJ mol}^{-1} \text{ K}^{-1}.$$

This quantity is the entropy change which occurs when water is vaporized. It may seem strange that entropy change is measured in the same units as specific heat, but remember that ΔS is only a *measure* of the disorder of a system. We now have a way of expressing ΔS numerically, and calculating it in simple cases.

How does the entropy of vaporization of water at 1 atm pressure compare with other liquids? If measurements are made at the same pressure it is simply necessary to take the molar latent heat of vaporization at the boiling point and divide it by the boiling point in degrees Kelvin. Figure 12.2 shows a graph of ΔH against T for a number of common liquids. They mostly lie on the same straight line, indicating that the entropy of vaporization of all liquids is approximately constant. This con-

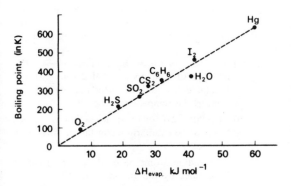

FIG. 12.2. The entropy of vaporization of all liquids is approximately constant (see text).

IPC—K

stant is known as **Trouton's constant.** It is approximately 88 J mol⁻¹ K⁻¹.

Water is an exception, but in general it is true to say that all liquids when they become vapour become disordered to approximately the same extent. Water is rather more highly ordered in the liquid state than most other liquids due to the formation of hydrogen bonds which have to be broken when water vaporizes, and its entropy of vaporization is abnormally high.

†12.7 Free energy and equilibrium constant

In section 11.10 the relationship between K and the standard reaction e.m.f. ε was stated:

$$\varepsilon = \frac{RT}{nF} \log_e K$$

Combining this with the relation $\Delta G^\circ = nF\varepsilon$, we have
$$\Delta G^\circ = -RT \log_e K$$

This equation is known as the **van't Hoff isotherm.** To derive it rigorously would take us even further into the realm of chemical thermodynamics, but it is a simple relationship and very useful. If we can calculate the standard free energy change, ΔG°, of a reaction, then it is easy to derive its equilibrium constant. This sort of calculation is frequently done by chemists and chemical engineers. For instance, a metallurgist may need to find out the optimum temperature at which to carry out a reaction. To do this he must first calculate ΔG° and then work out the equilibrium constant. Chapter 13 shows how this reasoning is put into effect in the extraction of metals from their ores.

Worked Example 1. In the reaction Cu(c) + 2 Ag⁺(aq) ⇌ Cu²⁺(aq) + 2 Ag(c), the following values were obtained at 25°C:

(*a*) $\Delta\epsilon°$ *for the cell* = *0·46 V*

(*b*) ΔH = −121 *kJ g-equation*$^{-1}$

Calculate (i) $\Delta G°$, (ii) ΔS.

(i) $\Delta G°$ = −$nF\Delta\epsilon°$

    = $2 \times 96\ 490 \times 0·46$ J g-equation^{-1}

    $\simeq$ −89 kJ g-equation^{-1}.

(ii) $\Delta S = \dfrac{\Delta H - \Delta G°}{T}$

    = $\dfrac{-121 + 89}{298}$

    $\simeq$ −0·108 kJ K^{-1} mol^{-1}

    = −108 J K^{-1} mol^{-1}

In this example there is a decrease in entropy, due to the decrease in randomness on the right-hand side of the equation. There are less copper ions than silver ions.

ions than silver ions.

12.8 Free energy and living systems

Free energy is stored chemically in muscles in the form of adenosine triphosphate (ATP). This can be illustrated by the use of a free energy diagram (Fig. 12.3) analogous to the enthalpy diagrams used in Chapter 9. Figure 12.3 shows that there is 34 kJ mol^{-1} of *available* energy in ATP.

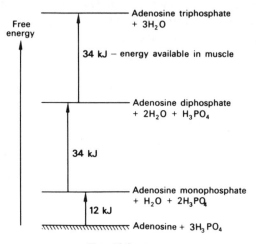

FIG. 12.3.

Study Questions

1. Predict whether the following changes involve an increase or a decrease in entropy:

(a) $Br_2(g) \longrightarrow Br_2(l)$.

(b) $Br_2(g) \longrightarrow 2\ Br(g)$.

(c) $O_2(g) + 2\ CO(g) \longrightarrow 2\ CO_2(g)$.

(d) $NaCl(c) \longrightarrow NaCl(aq)$.

(e) $NH_4Cl(c) \longrightarrow NH_3(g) + HCl(g)$.

(f) $C(c) + ZnO(c) \longrightarrow Zn(g) + CO(g)$.

(g) $2\ Al(c) + 3\ Fe^{2+}(aq) \longrightarrow 2\ Al^{3+}(aq) + 3\ Fe(c)$.

(h) Naphthalene dissolving in benzene.

(i) The crystallization of salt from brine.

2. When ammonium nitrate dissolves in water, there is an appreciable drop in temperature. Explain why the salt dissolves, when the process is absorbing so much heat.

†**3.** (a) At 25°C, ΔH for the reaction $I_2(g) \longrightarrow 2\ I(g)$ is +151 kJ, and ΔS is +101 J mol^{-1} deg^{-1}. Calculate ΔG, and hence the equilibrium constant at this temperature.

(b) At 2000K, ΔG for the same reaction is −50 kJ. What is the equilibrium constant at this temperature?

(c) **Can iodine atoms be produced from iodine molecules by the action of heat?**

†**4.** (a) Use the data given in the Appendix 1 to calculate ΔG for the following reactions.

(i) $Zn(c)+Cu^{2+}(aq) \longrightarrow Cu(c)+Zn^{2+}(aq).$

(ii) $Cl_2(g)+2 Br^-(aq) \longrightarrow$
$Br_2(g)+2 Cl^-(aq).$

(iii) $Au^{3+}(aq)+3 Ag(c) \longrightarrow$
$3 Ag^+(aq)+Au(c).$

(b) What can you say about ΔH in each case?

†5. Methanol can be synthesized according to the equation: $CO(g)+2 H_2(g) \rightleftharpoons CH_3OH(g)$: $\Delta H = -90$ kJ and $\Delta S = -220$ J K^{-1} at 25°C.

(a) Calculate $\Delta G°$, and hence the equilibrium constant, at 25°C.

(b) Since reactions proceed faster at higher temperatures, a chemical firm is thinking of making methanol by this method at 700°C, at which temperature ΔG for the reaction is $+125$ kJ. What advice would you give to the firm?

6. $C(c)+CO_2(g) \rightleftharpoons 2 CO(g)$: $\Delta H = +173$ kJ and $\Delta S = +176$ J K^{-1} at 25°C.

(a) How will ΔG and the equilibrium constant be affected by increasing the temperature?

(b) Are your conclusions consistent with le Chatelier's principle?

†7. (a) Work out the entropies of vaporization of the following substances:

(b) Comment on the values that you obtain.

(c) Can you suggest a reason for any exceptional results that you have obtained?

Substance	Boiling point (K)	Latent heat of evaporation (kJ mol^{-1})
Hydrogen	20	0·84
Methane	108	8·15
Ether	307	27·0
Bromine	332	30·5
Ethanol	351	39·5
Zinc	1180	116

CHAPTER 13

Extraction of metals from their ores

Sections 13.1 to 13.7 deal with the underlying principles of the extraction of metals, in terms of free energy. They may be omitted by readers whose syllabuses do not require this treatment.

13.1 The free energy of formation of metal oxides

It may be taken as a general guide that the most electropositive metals in nature occur in combination with the most electronegative anions such as chloride and nitrate, while the "weaker" metals (lead, zinc, mercury for instance) are found in combination with "weaker" anions such as sulphide, S^{2-}. The weakest metals of all, gold and the platinum metals, generally occur *native*, that is to say, as the free elements.

The Earth's crust is a very complex system, and the study of the occurrence and distribution of the elements is part of the subject of **geochemistry.** The factor which enables the whole problem to be approached rationally is, perhaps not surprisingly, that of free energy. In order fully to understand what is happening, we should have to consider the relative free energies of formation of a great many compounds of metals, such as chlorides, sulphides, silicates, sulphates and so on, as well as oxides. However, the problem of extracting a metal

from its ore is very often essentially concerned with decomposing the *oxide* of the metal, since (apart from simple binary compounds such as metal sulphides and chlorides which occur in nature) most metal ores consist essentially of a metal oxide in association with one or more non-metal oxides.

Hence one of the important factors in determining whether a metal can be extracted from, say a carbonate or silicate, is the standard free energy of formation of the oxide. Free energy of formation, ΔG_f, is the standard free energy of the reaction:

$$x M + \frac{y}{2} O_2 \longrightarrow M_x O_y; \quad (1)$$

ΔG_f = free energy of formation.

or

$$\frac{2x}{y} M + O_2 \longrightarrow \frac{2}{y} M_x O_y; \quad (2)$$

ΔG_f = free energy of formation *per mole of* $O_2(g)$

If the standard free energy of formation, ΔG_f, has a minus sign at a given temperature then the oxide can be expected to form spontaneously from the metal plus oxygen. If ΔG_f has a positive sign, the oxide will be expected to decompose spontaneously into its elements.

13.2 Factors determining free energy of formation of oxides

Unfortunately we cannot devise a cell capable of obtaining an e.m.f. for equation (2) in the majority of cases. $\varepsilon°$ measurements were referred to in Chapter 12, for reactions occurring in aqueous solutions, but here we are dealing with a metal oxide which may or may not be molten, and with a metal which may be solid, liquid, or gaseous, depending on the temperature.

However, ΔH_f is very easily measured by burning the metal in oxygen. If the overall heat change is obtained with the reactants starting at, say 25°C, and with the products finally cooled to 25°C, the heat of reaction will be ΔH_f for 25°C.*

The heats of formation of a number of oxides, *expressed per mole of $O_2(g)$*, are listed in Table 13.1. Notice that their order roughly parallels the $\varepsilon°$ values for the appropriate aqueous ions, which are given for comparison.

TABLE 13.1
THE HEATS OF FORMATION OF SOME METAL OXIDES

Oxide	ΔH_f (per mole of $O_2(g)$)	$\varepsilon°$ (M^{n+}/M) for comparison
MgO	-1200 kJ	$-2·37$ V
Al_2O_3	-1110 kJ	$-1·66$ V
ZnO	-695 kJ	$-0·76$ V
NiO	-490 kJ	$-0·25$ V
CuO	-310 kJ	$+0·34$ V
Ag_2O	-63 kJ	$+0·78$ V

The free energy of formation of an oxide can now be determined, provided we know its entropy of formation, ΔS_f.

$$\Delta G_f = \Delta H_f - T\Delta S_f$$

can be meas-ured   can be calculated

---

* Since the specific heats of reactants and products are not equal, ΔH_f will change slightly with temperature but we can disregard this relatively small effect.

The entropy change which takes place is essentially that which occurs when one mole of gas phase $O_2(g)$ is removed from the system, provided that neither the metal nor its oxide M_xO_y are vaporized:

$$\frac{2x}{y} M + O_2(g) \longrightarrow \frac{2}{y} M_xO_y; \qquad (2)$$

solid or molten metal   one mole of gas   solid or molten oxide

ΔS_f negative

There will be small changes in disorder, and hence entropy, associated with the change from the metal phase to the oxide phase, but the *main* factor influencing ΔS_f is the using up of the highly disordered gas phase.

For this reason, the entropy change of reaction (2) is roughly the same for all metal oxide systems, provided that the boiling point of neither metal nor oxide is exceeded. It is approximately 200 J mol⁻¹ K⁻¹.

This information enables us to plot the variation of ΔG_f with temperature for metal oxides (Fig. 13.1). Notice that, below the boiling point

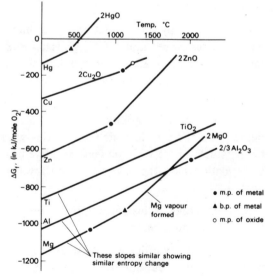

FIG. 13.1.

of the metal, the slopes of all the graphs are *roughly* the same, since the $T\Delta S$ factor is the same whatever the metal. Where the boiling point of the metal is exceeded however, the slope increases since the reaction is now involving a bigger entropy change. For instance, above 1110°C, three moles of gas phase are converted into solid phase, in the reaction:

$$2\,Mg(g) + O_2(g) \longrightarrow 2\,MgO(c).$$

magnesium	gaseous	solid
vapour	oxygen	oxide

Above a certain temperature, ΔG_f becomes positive for some of the oxides. This explains why mercury(II) oxide, for instance, decomposes spontaneously into its elements when heated. The diagram predicts that tin(IV) oxide and zinc oxide ought to decompose if heated strongly enough, but it does not hold out much hope for obtaining, say, pure magnesium by straightforward heating of the oxide to a high temperature.

The main points to be learned from Fig. 13.1 are therefore as follows:

(a) The entropy of formation of a metal oxide, expressed per mole of $O_2(g)$, is approximately the same whatever the metal, as shown by the similar slopes of the plots of ΔG_f against temperature.

(b) The thermal stability of metal oxides depends on the way in which ΔG_f varies with temperature. Those with the least negative heat of formation have the lowest stability towards heat.

(c) Heat alone is insufficient to decompose the oxides of most metals.

(d) The stability order of the oxides of metals parallels roughly, but not exactly, their order in the redox series.

†13.3 The "Thermit" process

Figure 13.1 also enables us to predict whether a given metal will reduce the oxide of another metal. Consider the reaction between chromium(III) oxide, and aluminium.

$$\tfrac{4}{3}\,Al + O_2 \longrightarrow \tfrac{2}{3}\,Al_2O_3;$$
$$\Delta H = -1110 \text{ kJ at } 25°C \quad (3)$$

$$\tfrac{4}{3}\,Cr + O_2 \longrightarrow \tfrac{2}{3}\,Cr_2O_3;$$
$$\Delta H = -755 \text{ kJ at } 25°C \quad (4)$$

$$\tfrac{4}{3}\,Al + \tfrac{2}{3}\,Cr_2O_3 \longrightarrow \tfrac{2}{3}\,Al_2O_3 + \tfrac{4}{3}\,Cr;$$

$$\Delta H = (-1110+755) = -355 \text{ kJ at } 25°C \quad (5)$$

The standard free energy, $\Delta G°$, of this reaction can similarly be derived by subtracting the two relevant ΔG_f values at a given temperature. Figure 13.2 shows the two curves plotted alone. At all accessible temperatures, ΔG is markedly

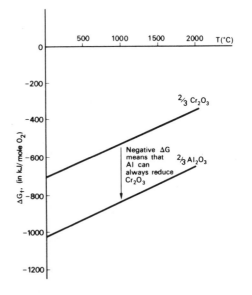

FIG. 13.2.

negative, and we should expect the reaction to proceed.

In fact the entropy change, ΔS, of this reaction is quite small, for the simple reason that no gaseous products or reactants are involved. The quite large entropy terms in the oxide-formation reactions cancel out, and we are left with comparatively small effects due to the different structures of the various phases.

ΔG is approximately the same at room temperature as at higher temperatures, but the reaction needs to be raised to a high temperature to "trigger it off". This can be done in the laboratory by priming it with magnesium ribbon and barium peroxide. Once started the reaction is highly exothermic, and very intense temperatures are reached. Care must be taken with this reaction—it is essential either to do it out of doors or to provide a good area of asbestos board and a sand tray.

The "Thermit" reaction, as reduction with aluminium is called, finds relatively little application on an industrial scale, because cheaper reducing agents than aluminium are available (see next section). However, some manganese and chromium are produced in this way. Earlier in this century, before it became more economical to use portable welding equipment, the reaction between iron(III) oxide and aluminium was used for making welded joints, for instance in tramway track.

Another name for the "thermit" reaction is the Goldschmidt process. It could, in principle, be applied to all but the most electropositive metals, but in practice other methods are employed on a large scale.

The conversion of a metal oxide to the metal is reduction, and the reverse process, addition of oxygen, is oxidation. Oxidation is the removal of electrons, and reduction is the addition of electrons. This covers the reaction of metal oxides if we assume them to be ionic, or at any rate partially ionic.

electrons added

$$Fe^{3+} + Al \rightleftharpoons Fe + Al^{3+} \qquad (6)$$

electrons removed

The concept is made more general by using the term oxidation number. The rules for assigning oxidation numbers were given in section 11.13. The system can be expressed as a redox process as follows:

decrease in oxidation number $\therefore$ reduction

$$Fe(III) + Al(0) \longrightarrow Fe(0) + Al(III)$$

increase in oxidation number $\therefore$ oxidation

†13.4 Carbon as a reducing agent

Since ancient times, carbon has been used as a reducing agent in the extraction of metals like iron, lead and copper, for carbon is the only reducing element which occurs native in large enough quantities. Below about 700°C carbon burns in oxygen to form carbon dioxide, CO_2, with little entropy change because one mole of gaseous O_2 forms one mole of gaseous CO_2. The disappearance of the well-ordered solid phase lattice of carbon has little effect on the entropy change ΔS.

$$C(c) + O_2(g) \rightleftharpoons CO_2(g);$$
$$\Delta G° \simeq \Delta H = -395 \text{ kJ at } 25°C$$
$$\Delta S = +3 \text{ J K}^{-1} \quad (7)$$

However, there is a possible reaction between carbon and carbon dioxide to form carbon monoxide, which is favoured at high temperatures because it involves an increase in the disorder. ΔS is positive because *two* moles of $CO(g)$ are formed from only one of $CO_2(g)$:

$$C(c) + CO_2(g) \rightleftharpoons 2 CO(g);$$
$$\Delta H = +173 \text{ kJ at } 25°C$$
$$\Delta S = +176 \text{ J K}^{-1} \quad (8)$$

Figure 13.4 explains what happens in practice. Adding (7) and (8) we see that equation (9) also has a favourable entropy change which will make the entropy term $T\Delta S$ more significant at higher temperatures.

$$2C(c) + O_2(g) \rightleftharpoons 2\,CO(g);$$
$$\Delta H = (-395 + 173) = -222 \text{ kJ mol}^{-1}$$
$$\Delta S = +179 \text{ J K}^{-1} \text{ (9)}$$

Figure 13.3 shows that, of the two possible reactions between carbon and oxygen, the one which actually occurs at a given temperature is the one which has the more negative ΔG. The bold line on the free energy graph indicates this. It is the positive ΔS for equation (9) which causes the free energy plot to slope downwards, above about 800°C.

In the region around 700°C, where the two reactions (7) and (9) both have approximately equal ΔG values (where the two lines cross), the products of combustion will be a mixture of $CO(g) + CO_2(g)$.

The downward slope of the carbon–oxygen graph shows carbon to reduce the oxides of most metals, provided the temperature is high enough. Consider the possible reaction.

$$MgO + C \rightleftharpoons Mg + CO; \quad\quad \text{(10)}$$
$$\Delta H = +492 \text{ kJ at } 25°C$$

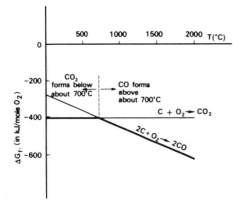

FIG. 13.3.

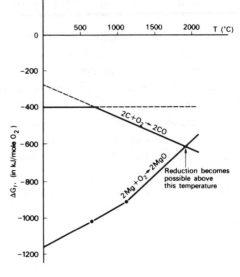

FIG. 13.4.

The free energy of reaction can be derived by considering the separate ΔG values for the reactions:

$$2C + O_2 \rightleftharpoons 2\,CO \quad\quad \text{(9, above)}$$
$$\text{subtract} \quad 2Mg + O_2 \rightleftharpoons 2\,MgO \quad\quad \text{(10)}$$

Figure 13.4 shows that this reaction has $\Delta G = 0$ at the point where the two lines representing (9) and (10) cross over. At 1900°C, from the graphs, the equilibrium constant equals unity. Above this temperature ΔG for the reaction becomes negative and the reaction is favoured. Note that the formation of magnesium vapour steepens the "MgO" curve, and makes the temperature for reduction lower than it would otherwise be.

A qualitative argument using entropy can be applied to this reaction: above the boiling point of magnesium metal, there is an increase of two moles of gas and therefore the ΔS of reaction is strongly positive. Provided that T is high enough, the $T\Delta S$ term will be sufficient to offset the adverse positive ΔH value. This is another example of an endothermic reaction being favoured by high temperature.

$$\Delta G \ = \ \Delta H \ - \ T\Delta S$$

reaction	this is	ΔS is
feasible	positive	positive, so
only if	opposing	$T\Delta S$ overcomes
this is	the	ΔH if T high
negative	reaction	enough

This qualitative argument leads to the same conclusion as does le Chatelier's principle. This would predict that the position of equilibrium would shift in such a direction as to tend to lower the temperature. The species of higher energy content [r.h.s. of equation (10)] will be favoured by the higher temperature.

Figure 13.5 shows a more complete diagram of the free energies of some metal oxide systems, together with carbon (and also hydrogen). The temperature at which the "carbon" curve intersects the "metal" curve is the temperature at which the equilibrium constant of the reduction becomes unity. The diagram shows, for instance, that $TiO_2(c)$ requires about 1700 °C for reduction to titanium, and $ZnO(c)$ only about 1000°C. In the case of the reduction of copper(II) oxide the diagram predicts that carbon dioxide, rather than the monoxide, will be the main product of the reaction:

$$2CuO(c) + C(c) \rightleftharpoons 2\,Cu(1) + CO_2(g).$$

Table 13.2 gives a summary of the methods used in practice for the extraction of metals.

†13.5 Hydrogen as a reducing agent

Hydrogen is not a very effective reducing agent for obtaining metals from their oxides, as Fig. 13.5 shows. The reason is that ΔS is negative for the reaction:

$$2\,H_2(g) + O_2(g) \rightleftharpoons 2\,H_2O(g);$$

$\underbrace{\qquad\qquad\qquad}_{\text{3 moles of gas}} \quad \underbrace{\qquad\qquad}_{\text{2 moles of gas}}$

ΔS negative; r.h.s. less disordered.

The plot of ΔG against T therefore rises with temperature, meaning that not many metal oxide plots are intersected. Hydrogen will therefore reduce oxides such as copper(I) oxide, and copper(II) oxide, but not the oxides of aluminium, magnesium and calcium. Oxides of iron are reduced only with difficulty. In the case of magnetic iron oxide, Fe_3O_4, an equilibrium composition is readily established.

†13.6 Reduction of sulphides and chlorides of metals

Although carbon reduces oxides quite effectively, it is not so effective when used directly on sulphides or chlorides of metals. In the case of chlorides, reaction is prevented by the fact that the entropy of reaction is not sufficiently favourable. To take a hypothetical case, there is not sufficient increase in entropy to make the

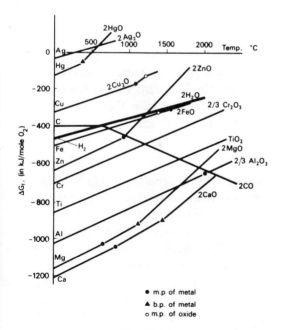

FIG. 13.5.

reaction

$$2\,MgCl_2 + C \rightleftharpoons CCl_4(g) + 2\,Mg$$

go, even at high temperatures, despite the fact that ΔS is positive.

A similar case occurs with sulphides. Carbon forms only one stable sulphide, CS_2 (b.p. 46°C). The situation can be likened to the analogous case which would occur if carbon was incapable of forming CO. The downward-sloping part of the curve in Fig. 13.3 would not exist. Hence carbon will reduce sulphides directly but only with difficulty. The situation is overcome by **roasting** the sulphide ore by heating it in a stream of air. This reaction converts it to the oxide, with the formation of sulphur dioxide, $SO_2(g)$. Lead(II) sulphide, found in the ore *galena*, is converted into the oxide in this way:

$$PbS + \tfrac{3}{2}\,O_2 \longrightarrow PbO + SO_2.$$

Metallic lead may then be obtained by incomplete roasting, to produce a mixture of PbS and PbO, followed by heating in the absence of air:

$$PbS(c) + 2\,PbO(c) \longrightarrow 3\,Pb(l) + SO_2(g);$$
$$\Delta S \text{ positive.}$$

Such a method is possible for lead since the formation of PbO is not very exothermic. It would not be possible for metals where ΔH_f of the oxide is strongly negative—in this case a reducing agent such as carbon must be used.

Mercury(II) oxide, which occurs as the ore *cinnabar*, converts directly to the metal when roasted:

$$HgS + O_2 \longrightarrow Hg + SO_2.$$

This happens in preference to the reaction observed for lead above, because at the roasting temperature the free energy of formation of HgO is positive. HgO therefore decomposes into $Hg + \tfrac{1}{2}\,O_2$.

†13.7 Slag formation

In many metal extraction processes, an oxide is added deliberately to combine with other impurities and form a stable molten phase immiscible with the molten metal, called a **slag**. The principle of slag formation is essentially

non-metal oxide $+$ metal oxide $\longrightarrow$
(acidic oxide)   (basic oxide)
                        fusible (easily melted) slag.

Two instances will be given: in the first case an acidic oxide is added to remove basic oxide impurities, and in the second case a basic oxide is added to remove acidic oxide impurities.

(1) *Removal of unwanted basic oxides.* If metal oxides are present in an ore that would otherwise interfere with the main extraction process, an acidic oxide can sometimes be used to remove them as slag. Sand, which is silicon(IV) oxide, SiO_2, (silicon dioxide, silica) is chosen because it is cheap, involatile, and leads to silicates which are themselves stable though fusible.

For example, a common source of copper is *copper pyrites*, $CuFeS_2$. On roasting, this mixed sulphide produces a mixture of oxides of copper and iron. The iron oxides are not wanted, and since they are more basic than those of copper they combine preferentially with added sand:

$$3\,SiO_2(c) + Fe_2O_3(c) \longrightarrow Fe_2(SiO_3)_3(l)$$
iron(III) silicate slag

Alternatively the iron(III) oxide is allowed to combine with the surface of the converter in which the ore is roasted.

Once an oxide is "tied up" as silicate it is exceedingly difficult to recover it since the free energy of formation of silicates is very negative. Aluminium occurs very widely in nature as clays and other minerals which are silicates, but these cannot be used economically for the extraction of aluminium.

(2) *Removal of unwanted acidic oxides*. Interfering acidic oxide impurities which can occur in metal ores, such as the oxides of phosphorus and silica (in combination) can be removed by the addition of a strongly basic oxide. For this purpose calcium oxide is used, obtained from limestone, $CaCO_3$.

In the blast furnace for extracting iron there are many complex reactions occurring simultaneously, but essentially iron(III) oxide is reduced to iron by carbon monoxide:

$$Fe_2O_3(c) + 3\,CO(g) \rightleftharpoons 2\,Fe(l) + 3\,CO_2(g) \quad (11)$$

This reaction is achieved by charging the blast furnace with a mixture of coke, iron(III) oxide

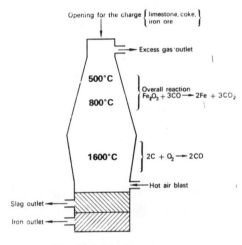

FIG. 13.6. The blast furnace.

and limestone, and heating it with a blast of air going through. The heat liberated is sufficient to maintain the temperature once the reactions have started (Fig. 13.6). At the bottom of the furnace the combustion of coke generates carbon monoxide, with intense heat. This intense heat decomposes the limestone into $CaO + CO_2$. The carbon dioxide passes from the top of the furnace, and the calcium oxide combines with unwanted acidic impurities, for instance:

$3\,CaO$	$+$	P_2O_5	$\longrightarrow$	$Ca_3(PO_4)_2(l)$
from lime stone		phosphorus(V) oxide in combination with metal oxides		fusible slag ("basic slag")

Note that carbon monoxide, rather than carbon itself, is the reducing agent at the top of the furnace. Nevertheless the *overall* reaction is the reduction of oxides of iron by carbon. The free energy diagrams such as Fig. 13.5 can be applied to the overall process, even if the reaction is thought to proceed in stages. Considerations of free energy have no bearing on the *mechanism* of the reaction. It is indeed highly improbable that two solid phases will react directly together as implied by the overall equation,

$$Fe_2O_3(c) + 3\,C(c) \rightleftharpoons 2\,Fe(l) + 3\,CO(g).$$

The same consideration applies to the reduction of other oxides with carbon considered in section 13.4.

13.8 Factors affecting choice of extraction method

(1) *Economic considerations*. The method of reduction must be as cheap as possible consistent with producing a product of the required degree of purity. In some cases the purity required is the main factor determining the choice of method. Carbon (in the form of coke) is undoubtedly the cheapest reducing agent, but it does not give a very pure product in all cases owing to the formation of carbides. Iron from a blast furnace contains a few per cent of carbon, and this has to be removed in a later process because it makes the metal brittle. An extreme case is calcium, which fails to form the metal at all, no matter how high the temperature. Instead,

$$CaO(c) + 3\,C(c) \longrightarrow CaC_2(c) + CO(g).$$

In many cases electrolytic extraction is used in preference to chemical reduction, particularly when a pure product is needed. In other cases, an impure product is obtained by carbon reduction, and is refined where it is required pure. Impure copper, although good enough for many purposes, is not good enough for electrical conduction where high conductivity is required; electrolytic refining is therefore used (section 13.9).

Aluminium is only chosen as reducing agent in one or two special cases, where the quantity of metal is not critical, and where the cost of raw materials is high anyway (Thermit process). One or two interesting cases may be instanced:

(i) a Fe–Ti alloy, *ferrotitanium*, is made by the Thermit reaction between aluminium and the ore *ilmenite*, $FeTiO_3$; this alloy is more useful for adding to steel than pure titanium itself, since the overall cost is less and iron is present anyway.

(ii) a Cr–Fe alloy, *ferrochrome*, is made similarly by the reduction of the ore *chromite*, $FeCr_2O_4$, with aluminium.

(2) *Practical considerations.* The choice of a reducing method for extracting a metal depends upon the availability of suitable ores. Many metals occur in unsuitable ores—the case of aluminium has already been mentioned, and silicates in general do not afford promising material despite their widespread abundance.

The metals at the top of the electrochemical series (which also have large negative values for ΔH_f of oxide, see Table 13.1) tend to occur mainly with electronegative anions. The alkali metal cations associate with singly charged anions on the whole, and are found largely as chlorides and nitrates. The alkaline earth metals (Group IIA of the periodic table) occur commonly with doubly charged anions, such as SO_4^{2-} and CO_3^{2-}. Further down the electro-chemical series the preference appears to be for silicates, sulphides, S^{2-} and oxides O^{2-}, while the metals at the bottom of the series may occur native. In order to understand the factors at work in deciding the method used for extracting the metal, a few cases will be given.

13.9 Specific methods for extracting metals

Potassium. Highly reactive, occurring in the ore *carnallite*, $KCl.MgCl_2.6H_2O$, and in living matter. Chemical reduction is insufficiently powerful. The chloride can be obtained from carnallite but it is unsuitable for electrolysis when fused, largely because potassium is volatile at the temperature involved. Potassium hydroxide is first made, and this is made anhydrous, fused (m.p. 360°C), and electrolysed. Demand for metallic potassium is not great—it is used in certain specialized reductions—and so the cost of the process is not an overriding factor.

Sodium. Large quantities of sodium are required for a diversity of purposes—as a heat exchanger in nuclear reactors, as a chemical reducing agent, and even as an electrical conductor—and it is essential that the process be as cheap as possible. Originally the electrolysis of fused sodium chloride (Downs process) failed in competition with the electrolysis of fused sodium hydroxide. The higher cost of producing sodium hydroxide was offset by the technical problems concerned in producing sodium from the fused chloride. Sodium is appreciably volatile unless the melt temperature is kept low, and it tends to disperse in the melt in the form of tiny droplets instead of collecting as a liquid at the cathode. Nowadays these technical difficulties have been overcome, and a modern Downs cell uses added calcium chloride to lower the melting point of the electrolyte. The discharge potential of calcium is very close to that of sodium in such a melt, and both metals are discharged simultaneously. They are however immiscible, the less dense sodium rising to the top, and they can be readily separated (Fig. 13.7). Reduction with an element like carbon might be just feasible at a high enough temperature, but could not compete economically with the electrolysis of the fused chloride.

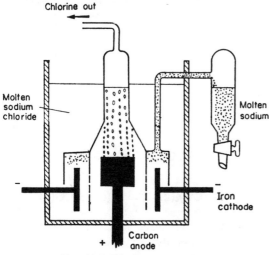

FIG. 13.7. The Downs process.

Magnesium. A variety of processes have been used for making magnesium, which is an important metal today for making light alloys, for use as aircraft materials. Electrolytic processes generally depend on the use of molten anhydrous magnesium chloride (obtained from dolomite or sea water) with added substances to lower the melting point.

Aluminium. Aluminium oxide, Al_2O_3, has a very high negative (exothermic) heat of formation. Furthermore the binding energy of aluminium metal, with three bonding electrons per atom, is rather higher than that of magnesium metal with only two bonding electrons per atom. This means that aluminium is not easily vaporized, unlike magnesium, and the reduction of the oxide with carbon is therefore not feasible. (Fig. 13.5.)

In default of a chemical reducing agent, we are forced to use electrolytic reduction, where there is no limit to the reducing cathode potential which can be applied. However, there are problems here too: aluminium chloride is structurally unsuitable, being a molecular solid Al_2Cl_6 (b.p. 180°C, sublimes), while aluminium oxide has an extremely high lattice energy and hence cannot be easily melted. Other salts of aluminium present similar problems—the most abundant of all, the silicates in clay, are far too stable and cannot be electrolysed—and we are forced to electrolyse the oxide under rather special conditions.

It is found that the oxide dissolves to form ions when added to molten cryolite, Na_3AlF_6. Ions Al^{3+} and O^{2-} are produced. (The ions Na^+ and the complex $[AlF_6^{3-}]$ are also present but they are not discharged in electrolysis.) The source of oxide is the ore *bauxite*, $Al_2O_3 \cdot xH_2O$, which is refined by:

(i) Dissolving in concentrated NaOH(aq) to separate from Fe_2O_3 impurity (see section 20.16 for an explanation of why this occurs).

(ii) Reprecipitating the aluminium as $Al(OH)_3 \cdot xH_2O$(c), by passing carbon dioxide in to remove the excess OH^- ions.

(iii) Heating the hydroxide to give pure oxide. The pure oxide, dissolved in cryolite, is electrolysed at 800–900°C at high current density. Molten aluminium *sinks* to the bottom (contrast magnesium which floats). Oxygen is liberated at the carbon anodes, and slowly burns these away.

Cathode half-reaction: $Al^{3+} + 3 e^- \longrightarrow Al(l)$

Anode half-reaction: $O^{2-} \longrightarrow \frac{1}{2}O_2(g) + 2e^-$;

followed by $\frac{1}{2}O_2(g) + C \longrightarrow CO(g)$.

Copper. Copper pyrites, $CuFeS_2$, is a commonly used ore. Incomplete roasting is carried out, forming sulphides and oxides of copper, mainly copper(I). Iron sulphide, forming iron oxide, is removed as iron silicate slag by adding sand. On heating these together in the absence of air, copper is produced (the reaction is feasible because ΔG_f of Cu_2O is only slightly negative, Fig. 13.5):

$$2\,Cu_2O + Cu_2S \longrightarrow 6\,Cu + SO_2(g).$$

Alternatively, carbon may be used to assist the reduction of oxide.

Copper of high purity for electrical conductors is made by electrolytic refining. This is an application of the principle of preferential discharge: the ions with half-cell potentials more positive than Cu^{2+}/Cu do not dissolve at the anode, since they require too much free energy. Impurities such as silver and gold sink to the bottom of the cell as a sludge (anode slime). Of the ions which do dissolve, only Cu^{2+} (aq) deposits on the cathode, because again the discharge potential for the remainder is too negative. Electrolytic refining is nowadays becoming much more common for metals which are relatively low in the electrochemical series, (see also zinc, and chromium below).

Zinc. The common ore of zinc is the sulphide, *zinc blende*, ZnS. Zinc oxide is obtained by roasting the ore, which is then reduced with anthracite (impure carbon) at 1100°C. At this temperature the zinc vapour is distilled.

Chromium. Chromium occurs as *chromite*, $FeCr_2O_4$. Direct reduction with carbon or aluminium gives the alloy *ferrochrome*. To obtain the pure metal, the ore is converted to Cr_2O_3, which is then reduced with aluminium.

Nickel. Nickel is a useful metal on account of its remarkable resistance to attack by corrosive substances. Its alloy with copper, *cupronickel*, is used for making "silver" coinage. It is extracted from its sulphide by roasting followed by reduction with carbon, but the process is complicated by the fact that nickel is found in association with other metals. The refining is rather unusual, for nickel forms a compound with carbon monoxide, tetracarbonylnickel(0) $Ni(CO)_4$. This substance is molecular in structure, and readily volatilized, (b.p. 43°C). It is made by heating nickel powder to 50°C in a stream of carbon monoxide, and is then decomposed at 200°C. The sequence of reactions is

$$\underset{\text{steam}}{H_2O(g)} + \underset{\text{coke}}{C(c)} \longrightarrow \underset{\text{"water gas"}}{\underline{CO(g) + H_2(g)}};$$

$$\Delta H = +131 \text{ kJ}$$
$$\Delta S = -420 \text{ J K}^{-1}$$
$$\Delta G = -38 \text{ kJ}$$

$$Ni(c) + 4\,CO(g) \xrightarrow{50°} Ni(CO)_4\,(g);$$
$$\Delta H = -164 \text{ kJ}$$

$$Ni(CO)_4(g) \xrightarrow{200°} Ni(c) + 4\,CO(g)$$

Many other transition metals form carbonyls, but none so readily and rapidly as nickel. Carbonyls are analogous to complex ions, but are uncharged and have the metal in zero oxidation state. The decomposition of nickel carbonyl is accompanied by a large entropy increase (four moles of gas formed from one mole) and hence it takes place spontaneously at higher temperatures, when the $T\Delta S$ term becomes significant.

Silver. Silver occurs native in Norway, Chile and Peru. In combination it is found chiefly as the sulphide, Ag_2S, and the chloride, AgCl. It is extracted by adding aqueous sodium cyanide, NaCN(aq): in whatever form silver occurs, cyanide ions will dissolve it, due to the formation of the very stable ion $Ag(CN)_2^-$(aq), $(K = 10^{19})$:

$$AgCl(c) + 2CN^-(aq)$$
$$\longrightarrow Ag(CN)_2^-(aq) + Cl^-(aq)$$

Silver is then precipitated by reducing this ion with aluminium or zinc, in alkaline solution:

$$2\ Ag(CN)_2^-(aq) + Zn(c) + 4\ OH^-(aq) \rightarrow 2\ Ag(c) + 4\ CN^-(aq) + Zn(OH)_4^{2-}(aq)$$

(oxidized ... reduced)

13.10 Summary of methods used for extraction of metals

Table 13.2 summarizes the methods used for extracting metals from their ores. It is only to be used as a general guide, and the details will be found either in this chapter, or in the relevant chapter dealing with the metal in question.

TABLE 13.2

Metal	Common ores	Source used for extraction	Method of extraction	Metal	Common ores	Source used for extraction	Method of extraction
K	$KCl.MgCl_2.$ $6\ H_6O$	KOH	Electrolysis of the fused salt using a carbon anode and usually an iron cathode. Another salt is added to lower the melting point.	Zn	$ZnCO_3$ ZnS	ZnO	Sulphide roasted to oxide, and the oxide is then heated with coke. Slag formation is important.
Na	NaCl NaNO$_3$	NaCl		Fe	FeS_2, Fe_2O_3	Fe_2O_3	
				Ni	$NiFeS_2$	NiO	
Ca	$CaCO_3$ $CaSO_4$	$CaCl_2$		Sn	SnO_2	SnO_2	
				Pb	Pbs	PbO	
Mg	$MgCO_3$ $MgCl_2$	$MgCl_2$		Cu	$CuFeS_2$	$CuFeS_2$	
Al	$Al_2O_3.xH_2O$	Al_2O_3		Hg	HgS	HgS	Roasting, and thermal decomposition of HgO.
Mn	MnO_2	MnO_2	Thermit process or reduction with carbon.				
Cr	$FeCr_2O_4$	Cr_2O_3 or $FeCr_2O_4$		Ag	Ag, AgCl Ag_2S	AgCl Ag_2S	Cyanide process.

Study Questions

1. Use Fig. 13.5 to determine approximately what temperatures are needed for (i) carbon (ii) hydrogen to reduce the following oxides to the metals:

(a) CaO.
(b) FeO.
(c) TiO_2.

2. (a) In what form would you expect the following metals to occur in nature?

(i) Bismuth, (ii) cadmium, (iii) cobalt, (iv) germanium, (v) platinum, (vi) rubidium, (vii) strontium, (viii) tungsten.

(b) How is each metal likely to be extracted?

3. What factors must be taken into consideration when deciding on a suitable reducing agent for (a) metal oxides, (b) metal chlorides?

4. What relationships can you discover between the method used to extract a metal and

(a) its position in the periodic table?

(b) its position in the electrochemical series?

5. (a) Use a data book to arrange the metals in order of decreasing heats of formation of (i) chlorides, (ii) oxides.

(b) What other series do these orders resemble?

†**6.** (a) Use the following information to sketch the Ellingham diagram for sulphides.

Com-pound	ΔG_f at 0°C	ΔG_f at 1000°C	ΔG_f at 2000°C
CS$_2$	− 42	− 42	− 42
HgS	− 188	−63 (500°C)	
Cu$_2$S	− 230	−188	−167
FeS	− 272	−176	− 84
ZnS	− 418	−230	
MnS	− 480	−335	−188
SO$_2$	− 690	−542	−398
H$_2$S	− 146	− 42	+ 63
CaS	−1020	−815	−522

(All values in kJ per mole of S$_2$(g).)

(b) Which of these sulphides can be reduced directly to the element using carbon?

(c) Why is the line for calcium sulphide not straight?

(d) Ignoring changes in ΔH, what is ΔS_f for (i) CS$_2$, (ii) FeS?

(e) Why are these figures not valid below the boiling point of sulphur?

(f) Use Fig. 13.7 and your sketch to work out which sulphides can be converted to the oxides by roasting in air.

(g) Comment on the methods used for the extraction of (i) copper (ii) mercury (iii) zinc, in the light of these results.

7. Library project. Discover what you can about the extraction of (a) Be (b) Ti and (c) U. How do these elements fit into Table 13.2?

CHAPTER 14

Equilibria between phases

14.1 Definitions

This chapter is concerned with systems in equilibrium in which there is more than one **phase** present.

A phase is a homogeneous region of a system, separated from other phases by a boundary surface. For instance, a system containing pure water in equilibrium with its vapour constitutes two phases. A solution of alcohol in water is one phase, while an immiscible "mixture" of carbon tetrachloride and water represents two phases. A saturated solution of sodium chloride in equilibrium with undissolved solid constitutes two phases: the liquid phase (solution) is homogeneous even though it contains two substances, and the solid phase is regarded as one phase even though it may consist of many separate crystals.

The term **homogeneous** requires comment. Homogeneous means *uniform* in properties, but since matter is composed of atoms and molecules which are definitely non-uniform when viewed under sufficiently high magnification, we must be clear what this means. A solution is regarded as homogeneous because the particles are fully dispersed as separate molecules or ions. It is possible to buy milk which is said to be "homogenized", but it is not truly homogeneous since it consists of tiny fat globules, whose dimensions greatly exceed those of the molecules which they contain, suspended in water. Milk is therefore an **emulsion**, and consists of two phases, fat and aqueous. Both these phases are homogeneous, but milk itself is not.

A system contains one or more **components. The number of components in a system is defined as the least number of substances whose quantity must be determined in order to fix the composition of all the phases present at equilibrium, at a given temperature.** Some examples are given to make this clear:

(a) *Water $\rightleftharpoons$ steam.* Two phases but only one component, namely H_2O.

(b) *A solution of alcohol in water, in equilibrium with vapour.* Two phases (liquid and vapour) and two components (C_2H_5OH and H_2O). If the amounts of alcohol and water in the liquid phase are defined, the composition of the vapour phase is fixed also. Conversely if the composition of the vapour is fixed, then only one composition of liquid can be in equilibrium with it.

(c) *Liquid nitrogen(IV) oxide in equilibrium with its vapour.* In both phases there is an equilibrium present:

$$N_2O_4 \rightleftharpoons 2\,NO_2.$$

However, there is only one component. The concentration of NO_2 molecules in a given

IPC—L

phase is fixed once the concentration of N_2O_4 has been fixed, and once the composition of one phase is fixed, the composition of the other phase is fixed too. It does not matter whether we regard the component of the system as being NO_2 or N_2O_4. The important thing is that the *number* of components is one.

14.2 Vapour pressure

A pure liquid in equilibrium with its vapour constitutes a two-phase, one-component system. Water can remain in equilibrium with steam, and if the temperature is specified then only one vapour pressure is possible. Similarly ice can be equilibrated with vapour, even when there is no liquid water present—snow will evaporate slowly in a dry wind even when the temperature is well below freezing—and ice too has a vapour pressure curve.

At the melting point of ice, its vapour pressure must *equal* that of water. The two vapour pressure curves must meet at that point (Fig. 14.1). This can be proved by the following argument (Fig. 14.2). Suppose we have some ice in equilibrium with its vapour, at a

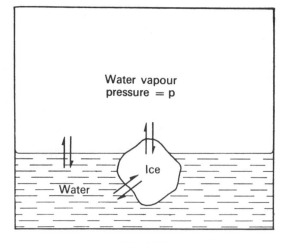

FIG. 14.2.

vapour pressure p. Let the ice be in equilibrium with water at the same time. Let the vapour pressure of the water be p', and suppose that $p' > p$, H_2O molecules will evaporate from the liquid phase to establish a pressure of p' in the vapour phase. The vapour pressure of ice is now exceeded (because $p' > p$) and so H_2O molecules will condense on to the ice phase. However, this cannot happen, because we stated that the water and ice phases were in equilibrium. Therefore p' cannot be greater than p. Conversely, if $p' < p$, H_2O molecules would have to evaporate from the ice phase and condense in the water phase, and again this would contradict the fact that the water and ice are in equilibrium. It follows therefore that, for equilibrium $p = p'$.

When a solid and liquid phase are in equilibrium with one another, they both exert the same vapour pressure.

14.3 Phase diagrams

Figure 14.1 was a plot of temperature against pressure, showing the vapour pressure curves of ice and water. It is possible to add a third line

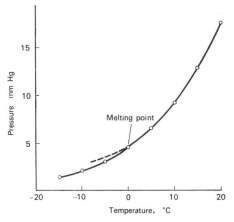

FIG. 14.1. Vapour pressure of water and ice.

to Fig. 14.1, namely the set of conditions where ice and water are in equilibrium with one another. The resultant plot is now a complete **phase diagram** for the H_2O system (Fig. 14.3).

The temperature at which ice melts to form water depends upon the pressure. If the pressure is increased, melting point is lowered. The creeping movement of glaciers is attributed to this effect. Application of le Chatelier's principle suggests that the phase of higher density will be favoured by increasing the pressure, and this is exactly what happens, as the slope of the line *TY* in Fig. 14.3 indicates.

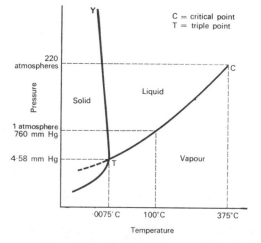

FIG. 14.3. A phase diagram for water (not to scale) (cf. Fig. 14.1).

The point *T* is termed the **triple point** of the H_2O system. It is the *unique* set of physical conditions (4 mm Hg, $+0.0075°C$) where all three phases can co-exist in equilibrium.

A phase diagram is more meaningful than a simple vapour pressure curve. In Fig. 14.3, every point on the diagram has a significance. Each *area* represents the conditions of temperature and pressure under which a given phase is stable. Each *line* represents the conditions under which two given phases may be in equilibrium. The *point* at which three areas are in contact is the condition for all three phases being in equilibrium.

The H_2O system has been chosen because it illustrates a simple phase diagram, but the same idea will be applied in this chapter to other systems, including two-component systems.

14.4 Allotropy of elements

If the density of sulphur is investigated at different temperatures, by heating it in a dilatometer such as that shown schematically in Fig. 14.4 then, provided the heating is carried

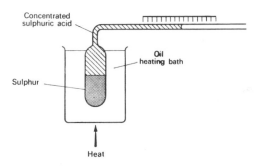

FIG. 14.4. Dilatometer.

out *slowly*, an abrupt decrease in density is observed around 96°C. Sulphur above 96° is less dense than that below, once equilibrium is established. Furthermore the crystals which form above 96°C when a solution of sulphur is cooled are monoclinic (needle-shaped) while those which form below 96°C are rhombic. Careful measurements of vapour pressure of sulphur indicate a discontinuity in the curve at 96°C, as shown at *B* in Fig. 14.5.

Here is an example of an element capable of existing in two different structural forms, **allotropes,** each of which forms a different phase. 96°C is the **transition temperature** between the two phases: it is the temperature at which both phases can be in equilibrium because they exert the same vapour pressure.

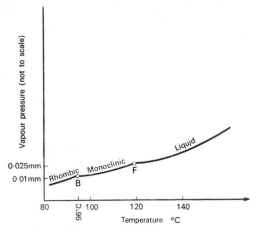

FIG. 14.5. Vapour pressure of sulphur.

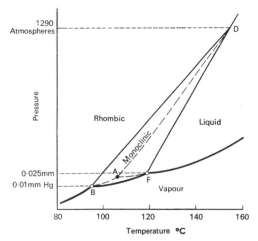

FIG. 14.6. Complete phase diagram for sulphur.

Figure 14.6 shows the same data as Fig. 14.5, but with additional information to make a complete phase diagram for the sulphur system. It is possible to heat rhombic sulphur above 96°C without it changing, because the rate of transition is slow. When it is outside its stable range of conditions, but is slow to change, it is said to be **metastable.** The dotted line *AB* shows the vapour pressure curve for metastable rhombic sulphur. Similarly the dotted line *AF* shows the vapour pressure of metastable monoclinic sulphur. *Note that the metastable allotrope has a higher vapour*

pressure than the stable form. The densities of the phases are in the order

rhombic > monoclinic > molten.

This explains why the lines *BD* and *FD* slope as they do (the slope is exaggerated in Fig. 14.6 for the sake of clarity). For instance, if molten sulphur at 140°C is subjected to about 1000 atm pressure, monoclinic sulphur will form because the volume is then less.

14.5 Enantiotropy

The property of elements whose allotropes possess a transition temperature is known as enantiotropy. Other examples of enantiotropic behaviour are:

(i) Grey tin ⇌ metallic tin, transition temperature 13·2°C. Grey tin is non-metallic in properties and objects made of tin disintegrate to powder if kept at prolonged low temperatures. Transition is more rapid once a nucleus of the new phase has become established to assist crystallization, giving the change the appearance of a "disease" afflicting the tin. Organ pipes, which in affluent times were often made of fairly pure tin, are said to have been subject to "tin plague" in cold churches. In more recent times, the leakage of fuel cans for Scott's antarctic expedition has been attributed to the failure of the soldered seams (Pb/Sn alloy) due to alloroptic change.

(ii) Iron loses its ferromagnetism at about 730°C. Above this temperature iron cannot be magnetized, nor is it attracted to a magnet.

14.6 Monotropy

Allotropes of many elements are metastable with respect to the stable form under all conditions. Such allotropes are said to be *monotropes,*

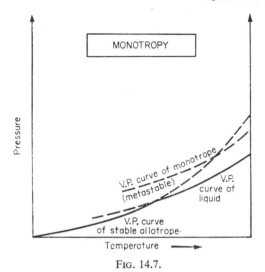

FIG. 14.7.

or to show monotropy. Thus white phosphorus has a vapour pressure greatly exceeding that of red, because it is composed of molecules (section 5.7). There can be no triple point at which red and white phosphorus are mutually in equilibrium with vapour (Fig. 14.7).

14.7 ΔG and ΔH for allotropic changes

When a metastable allotrope changes into its stable form, heat is usually evolved. It is often not practicable to measure this directly because of the slow rate of change, but Hess's law may be applied to the separate heats of combustion or reaction of the two forms. For instance:

$$C(\text{diamond}) + O_2(g) \longrightarrow CO_2(g);$$
$$\Delta H_d = -395 \text{ kJ}$$

$$C(\text{graphite}) + O_2(g) \longrightarrow CO_2(g);$$
$$\Delta H_g = -393 \text{ kJ}$$

Hence, $C(\text{diamond}) \longrightarrow C(\text{graphite});$
$$\Delta H = \Delta H_d - \Delta H_g$$
$$= -2 \text{ kJ (Fig. 14.8)}$$

The change from a metastable to a stable allotrope must be accompanied by a decrease in free energy (ΔG negative). The fact that ΔH is often observed to be negative as well is an

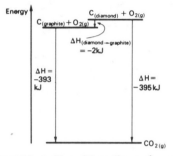

FIG. 14.8. Heat of transition: diamond → graphite (not to scale).

indication that the entropy changes are generally fairly small. The transition from one ordered crystal lattice to another of different order, does not involve very great change in the total disorder of the system.

14.8 Polymorphism of compounds

The arguments given above for elements apply equally well to compounds. Indeed the term allotropy can equally well be applied to compounds, though the term is not so common here. The term **polymorphism** (of many shapes) is a general term which can be taken to describe the property of any substance, element or compound, existing in more than one crystalline form.[*] It should not be thought that elements have some special property which enables them to exhibit phase transitions.

Chalk and limestone are crystalline forms of $CaCO_3$ which occur naturally as *sedimentary* rocks. Crystalline modifications have occurred

---

[*] Some books use the word allotropy even when no crystalline structures are involved, for instance to describe the various molecular states (S_2, S_4, S_8, etc.) which occur in molten sulphur.

due to the intense pressures and temperatures which prevailed, giving rise to *metamorphic* (changed shape) forms of calcium carbonate such as marble.

14.9 Vapour pressure of salt hydrates

The solubility curves of many salts in water show discontinuities analogous in some respects to the discontinuities in vapour pressure curves, such as those in Fig. 14.5. Closer investigation reveals that each part of the curve represents a different phase, generally due to the incorporation of varying numbers of moles of water in the crystalline phase. Many salts possess this **water of crystallization** and are known as **salt hydrates.** Where only one salt hydrate is formed, there will be no discontinuity in the solubility curve.

Figure 14.9 shows the solubility curve of sodium sulphate in water. There is a marked discontinuity at 32·4°C (point Q). Above this temperature sodium sulphate separates out as a solid crystalline phase of composition Na_2SO_4. Below 32·4°C the composition of the solid phase in equilibrium is $Na_2SO_4 . 10 H_2O$.

Figure 14.9 can be converted into a phase diagram, with the addition of a little more information. This is done in Fig. 14.10, in which the original solubility curve is shown in bolder lines.

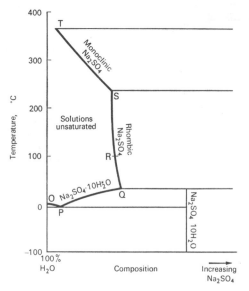

FIG. 14.10. Complete phase diagram for sodium sulphate.

Note how this phase diagram differs from the single-component diagrams shown in Fig. 14.3, 14.5 and 14.7: here there is an additional component, but pressure is no longer a significant variable so it is disregarded. Figure 14.10 plots temperature against composition instead of temperature against pressure. By convention composition is generally given the x-axis on phase diagrams, and so Fig. 14.9 has been turned through 90 degrees in making Fig. 14.10.

In a phase diagram, areas represent different compositions and these are labelled. The line $PQRST$ represents the composition of aqueous phase (solution) which is in equilibrium with the solid phase at various temperatures. The line OP is added because, if the solution is cooled sufficiently ice will separate out as the solid phase.

At the point Q, it is possible for the two different solid phases to be in equilibrium with the same solution. Assuming constant pressure, there is only one temperature at which this can occur. This is again called the *transition temperature.*

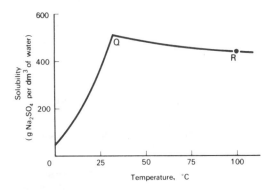

FIG. 14.9. Solubility of sodium sulphate in water.

14.10 The phase rule

A simple rule enables us to understand the effects of temperature, pressure, and composition upon a phase equilibrium. It is called the **phase rule**. The phase rule enables us to determine the number of **degrees of freedom** which a system in equilibrium possesses.

The number of degrees of freedom of a system is the number of variables (namely temperature, pressure and composition) which may be altered *independently* without altering the number of phases present.

The phase rule states that $P+F = C+2$

where C = number of components present,

F = number of degrees of freedom,

P = number of phases.

Referring back to Fig. 14.3, the number of degrees of freedom is zero at the triple point ($P = 3, C = 1, \therefore F = 0$) meaning that temperature and pressure, and of course composition, are uniquely determined.

Consider again the point Q in Fig. 14.10. Here $P=3$, and $C=2$, therefore $F=1$. This means that there is only one degree of freedom left: if we alter the pressure, the temperature will have to be altered or else one of the phases will disappear. Pressure and temperature cannot both be altered *independently*.

14.11 Efflorescence and deliquescence

Consider two different salt hydrates mixed together, and in equilibrium with water vapour (Fig. 14.11). Here $P = 3$ and $C = 2$, so again $F = 1$. If we now fix the temperature we have used up this degree of freedom and the vapour pressure becomes fixed as well. *If two salt hydrates are in equilibrium with each other and*

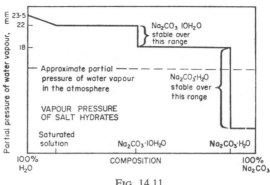

FIG. 14.11.

with water vapour, the vapour pressure must be fixed in value at a given temperature. If we attempt to alter the partial pressure of water vapour, equilibrium can only re-establish itself by removing one of the phases. For instance, if we increase the partial pressure of water vapour the higher hydrate will form at the expense of the lower, and vice versa.

If there is only one hydrate it can remain in equilibrium with water vapour over a range of pressures ($P = 2, C = 2, \therefore F = 2$, i.e. temperature and pressure can both be altered independently without causing a phase to disappear).

Figure 14.11 illustrates another type of phase diagram for a two-component system. This time temperature is assumed constant and omitted from the graph, and the y-axis represents pressure. The graph illustrates how the partial pressure of water vapour in equilibrium with sodium carbonate varies with the degree of hydration of the salt.

In normal atmospheric conditions washing soda, $Na_2CO_3 . 10H_2O$, produced by crystallizing sodium carbonate from aqueous solution, loses water. The decahydrate forms the monohydrate as follows:

$$Na_2CO_3 . 10\,H_2O(c) \underset{\text{above 18 mm}}{\overset{\text{below 18 mm}}{\rightleftharpoons}} Na_2CO_3 . H_2O(c) + 9\,H_2O(g).$$

The crystals take on a powdery appearance and the phenomenon is known as **efflorescence**.

If sodium carbonate decahydrate is subjected to a very high concentration of water vapour, it will undergo a phase change in the opposite direction, taking up water to form a solution. Figure 14.11 shows that this would only happen if the partial pressure of water vapour exceeded 22 mmHg; such conditions are not normally met in the atmosphere though they could be reproduced artificially. Many other salts, particularly highly soluble ones, take up water vapour to form a solution at a much lower partial pressure, and will do this if left to stand in the laboratory. The phenomenon is known as **deliquescence**. Sodium carbonate would never deliquesce in the laboratory, but many salts, for example $CaCl_2(c)$ have to be kept in a desiccator or well-stoppered bottle. To summarize:

Efflorescence is the loss of water of crystallization to form a lower hydrate or anhydrous salt.

Deliquescence is the absorption of water vapour by a solid to form a solution.

Both properties depend upon prevailing conditions of humidity.

The word *hygroscopic* is a general term applied to substances which absorb water vapour. For instance:

(i) Anhydrous copper sulphate (white) absorbs water to become $CuSO_4 \cdot 5H_2O$ (blue) on standing, but remains solid.

(ii) Many liquids, for example sulphuric acid and ethanol, absorb water while remaining liquid.

(iii) A common drying agent is *silica gel*; it can absorb considerable quantities of water without becoming wet.

14.12 Vapour pressure of miscible liquids and Raoult's law

When two liquids are shaken together the resultant system may consist of a single phase (liquids miscible) or two phases (liquids immiscible). Generally it is observed that "like dissolves like". If two liquids whose chemical properties are very similar are mixed it is found that each exerts a vapour pressure proportional to its **mole fraction.** Systems which obey this rule are said to obey **Raoult's law.** The mole fraction of a component is defined as

$$\frac{\text{number of moles of that component}}{\text{total number of moles of all components present}}$$

In a phase diagram of a liquid system containing two components it is not possible to show more than two of the degrees of freedom. Figure 14.12 shows a plot of vapour pressure against composition for a mixture which obeys Raoult's law, temperature being assumed constant.

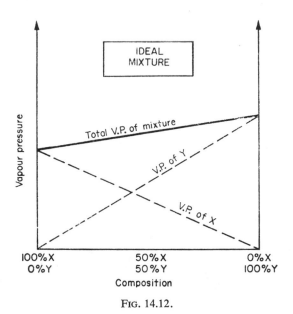

FIG. 14.12.

Typical examples of such a mixture, known as an **ideal mixture,** include the following:

(i) A mixture of $O_2(1)$ and $N_2(1)$ (liquid air) is nearly ideal.

(ii) A mixture of adjacent organic homologues, such as hexane and heptane, is almost ideal in behaviour.

14.13 Distillation of an ideal mixture of two liquids

Figure 14.13 shows the phase diagram corresponding to Fig. 14.12, but this time with temperature as a variable, pressure being assumed constant. The two components X and Y will be in equilibrium with vapour at a given temperature. In the diagram, Y is more volatile, and therefore the vapour composition will be richer in Y than the liquid from which it is derived. There are therefore two boundary curves on the phase diagram. Points between the two curves represent two phases of composition shown by the ends of the "tie-line". For instance, if liquid of composition represented by the point A at temperature T_1 is heated to T_2 it will form two phases, of composition B (liquid phase)

and C (vapour phase). On further heating to T_3 (point J) only one phase will be present, the whole of the mixture having been vaporized.

Figure 14.14 shows what will happen if the same mixture is distilled from a fractionating column. The labels on the diagram correspond to the stages listed below:

(1) The mixture of X and Y is heated until it just boils, at $T_4°C$ (point D).

(2) The mixture comes into equilibrium with its vapour at $T_4°C$, on the bottom plate of the column as shown. The vapour will be richer in component Y (point E).

(3) The vapour cools as it reaches the next plate of the column, reaching a temperature of $T_5°C$ (point F).

(4) Hotter vapour from below causes this condensed liquid to re-evaporate: the vapour is now richer still in Y (point G).

(5) This vapour again condenses on the next plate in the column at a temperature T_6.

And so the process continues. Equilibrium is eventually established with a temperature gradient existing up the column. If the column is a good one, practically pure Y will emerge from the top of it, and the mixture which drips back into the flask will become progressively richer in X.

The efficiency of a column is often expressed in terms of its **theoretical plate number.** This is the number of successive theoretical equilibrations which would produce a product of the observed purity. In Fig. 14.14 it is seen that only three theoretical plates produce a product which is better than 90% pure. A good laboratory fractionating column may have perhaps twenty theoretical plates, while a fractionation tower for the refinement of crude oil might have many more.

The process of fractionation has been studied in detail because of its great practical importance. *All* laboratory distillations should, if

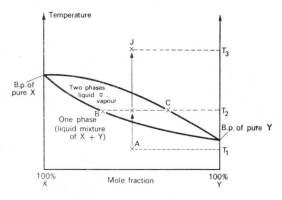

FIG. 14.13. Phase diagram for an ideal mixture of liquids (see text).

properly carried out, involve some degree of fractionation. Even in a simple distillation flask, the neck allows some refluxing to occur, and special care should be taken when using small-scale apparatus to ensure that heat is not too strongly applied, or an impure product will distil over.

Provided the mixture does not depart too far from Raoult's law, it will distil in the manner just indicated, enabling a fairly complete separation to be obtained.

If the two liquids are too close together in boiling point, distillation will not separate them because the number of theoretical plates which a fractionating column can achieve are insufficient. Liquids with boiling points less than 20 degrees apart require very efficient fractionation. For such cases a much more powerful method of separation is *gas chromatography* (section 14.21).

such a mixture might be expected to take place more readily than if the mixture was ideal. This phenomenon is observed in practice, and the curve of vapour pressure and composition no longer obeys Raoult's law. In many cases the deviations are so extreme as to lead to a maximum in the curve (Fig. 14.15(a)).

The point of maximum vapour pressure on this curve will correspond to a composition of *minimum boiling point*. If liquid of this composition is boiled its composition will not alter, and it will continue to boil at constant temperature. It is known as a **constant boiling mixture** or **azeotrope.** Figure 14.15(b) shows the temperature–composition phase diagram. It may be treated as two simple phase diagrams joined together—for instance the left-hand half of Fig. 14.15(b) behaves exactly like the system in Fig. 14.13.

14.14 Positive deviations from Raoult's law

Suppose two liquids are miscible in all proportions, but do not have great chemical affinity for one another. Evaporation of molecules from

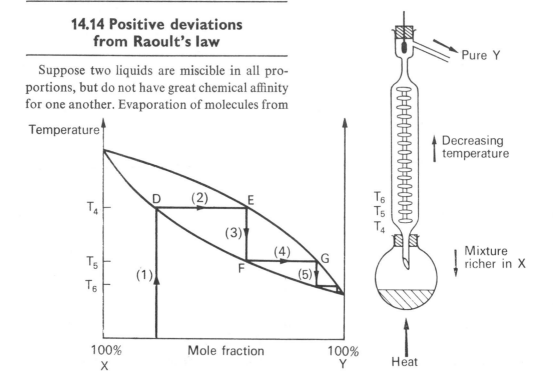

FIG. 14.14. Fractionation of an ideal mixture of liquids (see text).

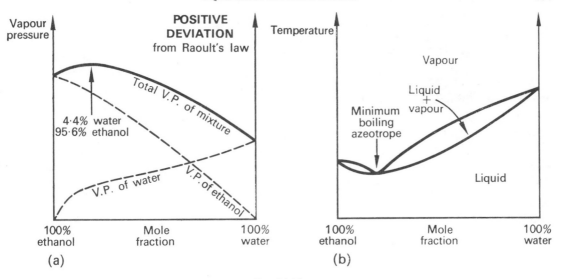

FIG. 14.15.

Ethanol and water form a well-known azeotropic system. No matter how efficient the distillation, the product distilling over will be the azeotrope, with composition 95·6% ethanol, 4·4% water at normal atmospheric pressure. If pure ethanol is required from a dilute aqueous solution, the final traces of water must either be removed chemically (for instance, with anhydrous calcium oxide), or by adding another component to the system (there is a method of preparing anhydrous alcohol by adding benzene to the system).

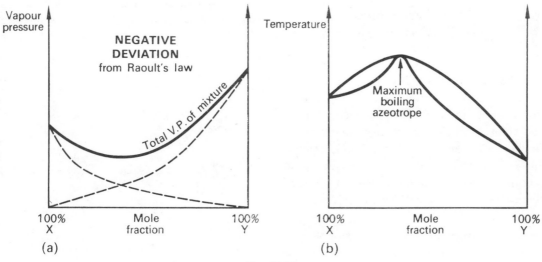

FIG. 14.16.

14.15 Negative deviations from Raoult's law

Where two miscible liquids have a strong chemical attraction for one another, perhaps showing a tendency towards loose compound formation, evaporation from the mixture will be *less* easy than from an ideal mixture. In extreme cases this commonly leads to another form of azeotrope, this time with a minimum vapour pressure, and maximum boiling point. Figures 14.16(a) and (b) illustrate this behaviour.

Systems such as nitric acid and water, or the hydrogen halides and water, exemplify this type of behaviour. Again it is easy to reason out what will happen when a given mixture is distilled, by treating Fig. 14.16(b) as two separate forms of Fig. 14.13 joined together.

As with the minimum boiling point system above, *if the azeotrope is distilled its composition will remain unchanged*, since the two curves for composition of liquid and vapour meet at this point.

14.16 Azeotropes and bond formation

Systems which show a negative deviation from Raoult's law do so because extra chemical bonds form which prevent such ready vaporization of the molecules. For instance, acetone and chloroform show a negative deviation due to the formation of hydrogen bonds (section 18.7). This interpretation is borne out by the fact that, when two liquids which show a negative deviation are mixed, heat is evolved. The amount of heat evolved per mole is not very great (of the order 20 kJ mol⁻¹) showing that the bonds formed are much weaker than covalent or ionic forces.

Systems which show a positive deviation from Raoult's law, such as alcohol and benzene, have a positive (endothermic) enthalpy of mixing showing that bonds are being broken.

14.17 Immiscibility

The formation of two separate phases instead of one, when two liquids are shaken together, is an extreme case of a positive deviation from Raoult's law. The phenomenon can be illustrated by reference to the aliphatic alcohols, Table 14.1.

14.18 Steam distillation

If a mixture of two immiscible liquids is boiled, each phase will exert its own vapour pressure separately, since neither phase is diluting the other. The boiling point of a system of two immiscible liquids is therefore always below that of both the pure liquids.

In organic preparations (for example, the preparation of aniline, $C_6H_5NH_2$) volatile components immiscible with water are readily removed from the tarry mixture in the preparation flask by blowing steam through the mixture (Fig. 14.17). Distillation will proceed because the temperature of the steam is higher than the boiling point of the mixture. The weight of each component distilled will be proportional to its

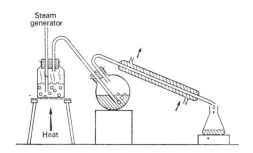

Fig. 14.17. Steam distillation.

TABLE 14.1

Name	Formula	Miscibility with water, etc.
Methanol	CH_3OH	Positive deviation from R.L. but no maximum. Miscible with water in all proportions
Ethanol	C_2H_5OH	Positive deviation from R.L., maximum V.P. observed. Miscible with water in all proportions
Propanol-1	C_3H_7OH	Pronounced maximum in V.P. curve. Miscible with water in all proportions
Butanol-1	C_4H_9OH	Two phases formed on mixing, each component being fairly soluble in the other
Pentanol-1	$C_5H_{11}OH$	Two phases formed, solubility being less than for butanol-1, etc.

molar concentration in the vapour. Provided the components can be assumed *completely* immiscible, we can say

$$\frac{\text{number of moles of X in distillate}}{\text{number of moles of water in distillate}}$$

$$= \frac{\text{s.v.p. of X}}{\substack{\text{s.v.p. of water at the distillation} \\ \text{temperature}}}$$

The distillate will consist of an oily emulsion, which is generally difficult to separate by mechanical means such as with a separating funnel. The usual procedure is to add diethyl ether (or some other solvent immiscible with water but miscible with the non-aqueous phase) and separate the ether layer in a separating funnel. The ether layer may then be stood over a desiccant such as anhydrous calcium chloride to remove traces of dissolved water, and then redistilled.

In theory the method of steam distillation can be used to determine the molecular weight of X (see Study Question 7), but in practice the method is rather inaccurate. This is partly due to the fact that many so-called "immiscible" liquids are in fact partially miscible, and partly because true equilibrium between the phases may not be established.

14.19 Eutectics and alloys

The melting point of a pure solid is lowered by adding an impurity. For instance the melting point of pure camphor could be lowered by adding another solute such as naphthalene. The converse is also true: the melting point of pure naphthalene will be lowered by adding small quantities of camphor. Similar behaviour is observed with mixtures of molten metals, and these provide convenient examples for study. One of the simplest is a mixture of lead and tin. It is found that each lowers the melting point of the other, and if a graph of freezing point against composition is plotted, the result is like Fig. 14.18. Careful plotting reveals a

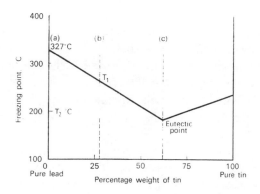

FIG. 14.18. Eutectic formation.

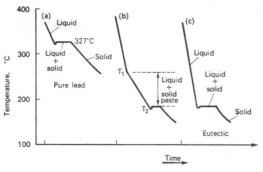

FIG. 14.19.

sharp minimum in the curve. The alloy at this composition is called a **eutectic** and the point on the curve is called a **eutectic point**.

A close examination of the alloys as they solidify on cooling shows that solidification is gradual, and this can be further verified by plotting a cooling curve for the various compositions. Figure 14.19(a) shows the sharp melting point associated with pure lead (or pure tin). Figure 14.19(b) shows what happens with an alloy other than the eutectic. Crystallization commences at a temperature T_1, and cooling then slows down. When T_2, the freezing point of the eutectic, is reached the temperature remains constant until the whole has become solid. Thereafter normal cooling of the solid continues.

The explanation of this behaviour is as follows: at T_1, pure lead crystallizes out. The melt becomes richer in tin and its freezing point is correspondingly lowered. This continues until the eutectic composition is reached. The eutectic has the lowest possible freezing point, and thereafter both metals separate out together as two distinct solid phases.

If the eutectic itself is cooled (line (c) in Fig. 14.18) then the freezing point is as sharp as for a pure substance (Figure 14.19 (c)). Sharpness of melting point or freezing point is generally taken to be one of the tests for deciding

whether a substance is pure or a mixture, but here the test breaks down. A eutectic is not a pure substance, i.e. not a compound of lead and tin in this case, as the following evidence shows:

(a) It is non-stoichiometric; (this alone would not be a sufficient piece of evidence).

(b) Separate phases of the components can be seen by microscopic examination. The arguments against eutectics being compounds are similar to those for azeotropes. Both eutectics and azeotropes are phenomena which result from the peculiar behaviour of phases in equilibrium.

14.20 Phase diagrams of eutectics

The information plotted in Figs. 14.18 and 14.19 enables a complete phase diagram for the lead–tin system to be constructed. Figure 14.20 shows how this is done, and the areas are labelled to show their significance.

The system sodium sulphate–water (Fig. 14.10) shows a eutectic point, where the "ice" curve meets the $Na_2SO_4 . 10 H_2O$ curve. Most salts form eutectics with water in this way.

The phase diagrams so far considered are the result of pure **solid phases** separating out.

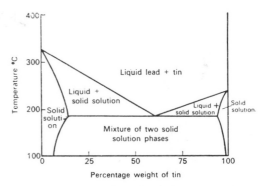

FIG. 14.20. Complete phase diagram for the lead–tin system.

Occasionally, particularly with metals, a **solid solution** may be formed. A solid solution is a *homogeneous* solid phase containing a mixture of two or more substances. Such phase diagrams will not be considered further here.

14.21 Applications of eutectics

(1) Low-melting alloys are used as safety devices, for instance as plugs in automatic fire sprinklers and as "fail-safe" devices in boilers. By a suitable choice of metals, very low melting points can be achieved. Wood's metal (50% Bi, 25% Pb, $12\frac{1}{2}$% Sn, $12\frac{1}{2}$% Cd) even melts below the boiling point of water (m.p. 65°C).

(2) Freezing mixtures are made by mixing ice and salt, or better ice and ammonium chloride. If the ratio of the two components is near to the eutectic composition, the solids will come to equilibrium with melted liquid phase at a temperature near to the freezing point of the eutectic. The temperature will drop until equilibrium is reached, the heat being taken up as latent heat of fusion.

14.22 Partition of a solute between two phases

Section 10.11 showed that, if a solute is added to two immiscible solvents, it will obey the equilibrium law at low concentrations. This fact is put to practical use in a number of ways, for example **solvent extraction** and **partition chromatography.**

Ether is immiscible with water, and organic substances can be extracted from an aqueous emulsion by shaking the emulsion up with ether. For example, in the steam distillation of aniline (section 14.18) the distillate contains a small quantity of aniline together with a relatively large amount of water in the form of an emulsion which only partly separates out. On adding ether the aniline partitions itself between the ether layer (less dense upper layer) and the aqueous layer (lower layer), being very much more soluble in the ether. On passing through a separating funnel, containing most of the aniline the lower layer can be obtained separately. A further quantity of ether can then be added to the remaining aqueous layer, enabling most of the remaining aniline to be extracted from it.

Worked Example 1. The partition coefficient of a solute X between ether and water is 3, the substance being more soluble in ether. 100 cm³ of an aqueous solution containing 10 g of X is shaken with 100 cm³ of ether. Calculate the weight of X left in the aqueous solution.

Since the volumes of the layers are the same,

$$\frac{\text{weight of X in ether}}{\text{weight of X in water}}$$

$$= \frac{\text{concentration of X in ether}}{\text{concentration of X in water}} = 3$$

If w_1 = weight of X remaining in the aqueous layer, then $(10 - w_1)$ = weight of X in the ether layer

$$\therefore \ \frac{10 - w_1}{w_1} = 3$$

Whence $w_1 = 2 \cdot 5$ g.

A further 50 cm³ of ether are added to the aqueous layer, after the first separation. Calculate the weight of X which now remains in the aqueous layer.

Here the two volumes are unequal, and

$$\frac{\text{weight of X in ether}}{\text{weight of X in water}}$$

$$= \frac{\text{vol. of ether} \times \text{conc. of X in ether}}{\text{vol. of water} \times \text{conc. of X in water}}$$

$$= \frac{50}{100} \times \frac{3}{1} = \frac{3}{2}$$

Let w_2 = weight of X now remaining in the aqueous layer, then $(2\cdot5 - w_2)$ = weight of X in the ether layer

$$\therefore \frac{2\cdot5 - w_2}{w_2} = \frac{3}{2}$$

$$\therefore w_2 = 1\cdot0 \text{ g.}$$

A succession of extractions will enable a solute to be extracted from one solvent into another, provided the partition coefficient is reasonably favourable. In the worked example above, with a partition coefficient of only 3, several extractions would be needed to effect a good separation, but in the ether extraction of aniline two or three extractions are quite sufficient.

14.23 Countercurrent distribution

Countercurrent distribution is an automatic procedure, used industrially and in research laboratories, for separating two closely related solutes by making use of their slightly different partition coefficients in two immiscible liquid solvents. The technique is used when other methods such as distillation are inadequate.

The apparatus consists of a number of identical tubes (about $100-200$) and the lower heavier solvent is placed in each of them. The sample is dissolved in a portion of the upper, lighter, solvent and placed in tube 1, which is shaken, and allowed to settle. The upper solvent is then automatically transferred to tube 2, while a new portion of upper solvent is added to tube 1. The process is repeated with tubes 2 and 3, and so on.

The effect of this is to cause both the solutes in the sample to be gradually transferred along the sequence of tubes, but they will undergo a gradual separation. Suppose that component A favoured the upper solvent more than component B did. Component A would then be transferred from tube to tube in slightly greater amounts than B, and after a number of operations would leave component B behind.

Countercurrent distribution can handle involatile mixtures readily, and can cope with up to 20 g of solute. It can be operated as a closed cycle to save on the number of tubes used: for instance, the sample from tube 50 of a fifty-tube unit can be returned to tube 1, and the whole operation continued.

14.24 Partition chromatography

Paper chromatography operates by a mechanism analogous to countercurrent distribution. We will consider first **partition chromatography** in which we use a strip of chromatography paper (specially prepared absorbent paper rather like filter paper) and place the solute mixture, generally dissolved in water, at one end of the strip. A second solvent is now allowed to travel along the strip, and it will extract the solute in a manner analogous to countercurrent distribution. Figure 14.21 shows two typical arrangements for separation by paper chromatography.

The analogy with countercurrent distribution is as follows: the paper itself holds a quantity of moisture, which can act as solvent for the sample. Moreover, the paper fibres themselves will attract the solute molecules by dipolar forces. The paper is referred to as the **stationary phase.** It is analogous to the lower layer of heavy solvent in countercurrent distribution. The solvent which travels across the paper is known as the **moving phase,** or mobile phase: It is analogous to the upper liquid in countercurrent distribution. Strictly the solvent in the two phases should not mix, though in practice some diffusion of solvent between the phases is bound to occur. Indeed the moving phase used in paper chromatography is very often water-

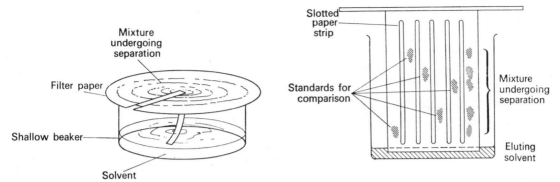

FIG. 14.21. Paper chromatography.

miscible and the "partition" which occurs is really between the liquid moving phase and the polar surface of the paper fibres. In fact elementary experiments, such as the separation of inks, can be carried out on blotting paper or filter paper without a non-aqueous liquid being used: the solutes in ink distribute themselves between the paper fibres and the moving water phase.

Partition chromatography was invented by Martin and Synge in 1941, so it is a relatively recent technique. **Adsorption chromatography** has been known for much longer, however, the first systematic studies being done by Tswett, a Polish botanist whose main work was done at the beginning of this century, in connection with the components of the green pigmentation in plants (chlorophylls and xanthophylls). Adsorption chromatography uses a solid stationary phase, such as alumina, and a moving solvent which may or may not be aqueous. Martin and Synge suggested partition chromatography using the following:

stationary phase: water, adsorbed on silica gel;

moving phase: a liquid immiscible with water.

Paper chromatography proceeds by a mechanism which is partly partition and partly adsorption.

Adsorption chromatography, and partition chromatography of the type used by Martin and Synge, can be carried out using a column, (Fig. 14.22). The advantage of a column over paper is that it can handle greater weights of solute, but the disadvantages are cost and slowness of operation.

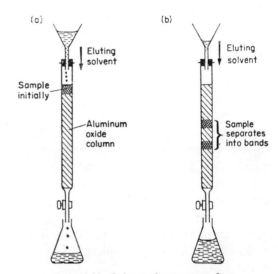

FIG. 14.22. Column chromatography.

For rapid separations where identification of the components of a mixture is all that is required, paper chromatography is usually the most convenient. For instance the sugars obtained in the breakdown of starch by hydrolysis can be identified[*] or the method can be used for resolving the mixture of amino-acids obtained from proteins.

14.25 Gas–liquid partition chromatography (GLPC)

Martin and Synge, in their 1941 paper on partition chromatography, mentioned the possibility of using a gas as the moving phase instead of a liquid. The idea was not developed by anyone until James and Martin published the first paper on gas–liquid chromatography in 1952. The stationary phase is a liquid, such as molten vacuum grease, or silicone oil, impregnated on a powder support and packed in a column. The stationary liquid must be involatile at the temperature used. The moving phase is a carrier gas, such as nitrogen, argon or helium. It is important that the moving phase does not react with the sample, so a carrier gas such as oxygen would generally be unsatisfactory. The whole column is placed in a thermostatically controlled oven or vapour jacket (Fig. 14.23).

The sample may be gaseous, liquid or solid, provided that it is appreciably volatile at the column temperature. For convenience most samples are introduced into the column as vapour or liquid, with a syringe. The emerging sample is passed through a detector whose electrical response depends on the concentration of sample in the gas stream. Various types of detector are in use, two typical arrangements being:

(a) **A katharometer,** shown in Fig. 14.23. The temperature of a heated platinum resistance wire varies with the thermal conductivity of the gas which surrounds it, and will thus alter if the carrier gas contains sample. The wire is

connected to a Wheatstone's bridge, and any change in its temperature will alter the resistance, throwing the bridge out of balance. The out-of-balance voltage is fed to an automatic chart recorder.

(b) **An ionization detector.** Some detectors use a hydrogen flame, and a carrier gas of nitrogen. The ionization energy of nitrogen is high and it is not ionized as it passes into the flame, but the ionization energy of organic molecules is generally low. The flame in the detector becomes a weak electrical conductor when organic molecules are passed through it. The current is suitably amplified and fed to a chart recorder as before.

The trace produced by the pen recorder is known as a chromatogram. A typical gas chromatogram is shown in Fig. 14.24. Gas chromatography is used largely as an analytical tool, where it can handle samples as small as 10^{-9} g and effect remarkable separations. Substances which fail to separate even in the most elaborate fractionating column will generally yield to gas chromatographic separation. For instance, a typical, "theoretical plate number" (section 14.13) for fractional distillation is 20. For gas–liquid chromatography it is more like 1000, and by using columns constructed from fine capillary tubing, plate numbers of more than 100 000 have been achieved.

[*]See, for instance, *Nuffield O-level Chemistry,* Experiment 21.1 (Longmans, 1966).

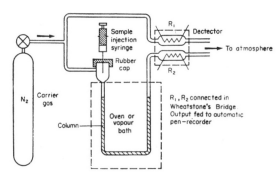

FIG. 14.23. Gas–liquid chromatography.

Gas chromatography cannot handle such large masses of sample as fractional distillation or countercurrent distribution, though samples of the order 2–3 g can be separated on special, wide-bore **preparative scale columns.** The method is restricted to samples which can be volatilized.

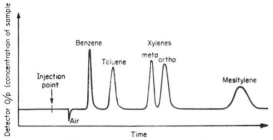

FIG. 14.24. A typical gas chromatogram.

Factors affecting separation. If the column is a non-polar one, the components of the solute mixture will emerge roughly in the order of their boiling points. The more volatile ones will remain in the vapour phase preferentially and are therefore swept through faster.

If the column is polar it will tend to retard polar samples preferentially, by dipolar attraction. Components in a mixture can be roughly characterized by chromatographing them on two contrasting columns, one polar and one non-polar.

Applications of gas chromatography. The technique has grown enormously since its invention in 1952, and already there are special journals dealing exclusively with published work on the subject. Some industrial applications are:

(1) Petroleum mixtures, which are far too complex for complete chemical analysis, are readily separated by gas–liquid chromatography. The various components are identified by comparing their *retention times* on the column with known standards.

(2) The complex chemical mixtures present in foodstuffs can, when volatile, be identified. Much research has been done into such problems as the odour of coffee and the "bouquet" of wine, presumably to develop ways of improving on the natural product.

(3) Reaction products of industrial processes can be rapidly monitored, and automatic control can be applied to chemical plant.

(4) Alcohol in blood and urine is measured by gas chromatography.

Study Questions

1. Are the following systems homogeneous or heterogeneous?

    (a) Sand and water.   (b) Salt solution.   (c) Mist.
    (d) Petrol.   (e) Salad **cream.**   (f) Smoke.
    (g) Granite.   (h) Jelly.

2. (a) How many phases and components are present in the **following** systems?

        (i) Ether (1) $\rightleftharpoons$ Ether (g)
        (ii) Solid tin/lead eutectic
        (iii) $NH_4Cl(c) \rightleftharpoons NH_3(g) + HCl(g)$
        (iv) $CaCO_3(c) \rightleftharpoons CaO(c) + CO_2(g)$

    (b) How many degrees of freedom does each system possess?

3. In Fig. 14.3:

    (a) What do the lines TC and TY represent?
    (b) What causes the slope of the line TY?
    (c) Why does the line TC stop at C?
    (d) Is it possible to obtain ice above 0·1°C at pressures of less than 220 atm?
    (e) Water, obtained below 0°C by supercooling is metastable. What would you expect to happen if a small crystal of ice was added to the water?
    (f) Can you suggest how water below 0°C could be obtained, other than by supercooling?

4. The vapour pressure of some salt hydrates and their saturated solutions at 20°C are as follows:

Substance	v.p. of hydrate	v.p. of sat.soln.
$CaCl_2.H_2O$	2·5 mm	7·5 mm
$CuSO_4.5 H_2O$	5·0 mm	16·0 mm
$Na_2SO_4.10 H_2O$	16·2 mm	16·6 mm

What will happen to these three salts if they are exposed to an atmospheric water vapour pressure of (a) 4 mm, (b) 10 mm, (c) 17 mm?

5. (a) Ethanol boils at 78·4°C; ethyl acetate at 77·2°C; an azeotropic mixture containing 31% of the alcohol boils at 71·8°C. How can you account for this behaviour? What will happen if a 50% solution is distilled?

  (b) Water boils at 100°C; HCl at −85°C; a maximum boiling mixture of these substances containing 20·2% HCl boils at 108·6°C. How can you account for this behaviour? What will happen if a 50% solution is distilled?

†**6.** Explain the following observations as far as you can:

  (a) Above 66°C, phenol and water are miscible in all proportions. Below this temperature, they are miscible only in certain proportions.

  (b) Nicotine and water are miscible in all proportions only below 61°C and above 208°C.

7. At 98·5°C, the vapour pressures of aniline and water are 42 and 718 mm Hg respectively. In an actual experiment at this temperature, the steam distillate was found to contain 11·02 g of aniline and 38·81 g of water. ($H_2O = 18$.)

  (a) Calculate the molecular weight of aniline.

  (b) Comment on the accuracy of this method as a way of determining molecular weights.

8. The following results refer to the systems (a) α-naphthol/naphthalene, (b) β-naphthol/naphthalene, (c) phenol/aniline. The freezing-points (°C) are given for mole fractions of the first mentioned component.

Mole fraction	0·0	0·1	0·2	0·3	0·4	0·5	0·6	0·7	0·8	0·9	1·0
(a)	80°		71°		62°		75°		87		96°
(b)	80°		89°		99°		107°		114°		121°
(c)	−6°	−11°	−3°	16°	28°	31°	28°	20°	15°	30°	41°

(i) Sketch a phase diagram for each system.
(ii) Comment on the features shown by each diagram.

9. (a) Draw the temperature–composition phase diagram for two solids A and B that form a compound AB, and label each area.

  (b) Suggest a pair of substances that might show this behaviour.

  (c) How many eutectic points would you expect for two substances that form n compounds?

10. The distribution coefficient of iodine between water and CCl_4 is 86, the iodine being more soluble in the latter. A solution of 1 g of iodine in 1 dm³ of water is to be extracted with 100 cm³ of CCl_4.

  (a) How much will be extracted if all the CCl_4 is used at once?

  (b) How much will be extracted if the CCl_4 is used as two 50 cm³ portions?

  (c) How would you use the CCl_4 in order to extract the maximum amount of iodine from the water?

11. Suggest methods for separating the following:

  (a) A mixture of amino-acids (the breakdown products of proteins).

  (b) The hydrocarbons present in a commercial petrol.

  (c) $SOCl_2$ (b.p. 78°C) and $POCl_3$ (b.p. 107°C).

  (d) Iodobenzene from the sludge after an organic preparation.

CHAPTER 15

Colligative properties

15.1 Lowering of vapour pressure by an involatile solute

If an *involatile* solute is added to a solvent, the vapour pressure of the solvent will be lowered. Figure 15.2 shows the effect on the vapour pressure curve of water when adding a solute to it. The effect is a consequence of Raoult's law: adding solute lowers the mole fraction of water present. Pure water has a mole fraction of 1, and if sufficient sugar is added to lower the mole fraction of water to 0·99, then the vapour pressure of water will be lowered to 99% of what it was before.

In the above case, if the mole fraction of water is 0·99, it follows that the mole fraction of solute sugar, is 0·01, since the total mole fraction must add up to unity. The relative lowering of the vapour pressure is 1% of the original vapour pressure.

When applied to an involatile solute dissolved in a solvent, we may therefore state Raoult's law in one of the following special forms:

(1) *The relative lowering of vapour pressure is equal to the mole fraction of the solute;* or,

(2) *The relative lowering of vapour pressure is directly proportional to the molar*

concentration of solute, provided the solution is dilute.

In practice it is found that, provided the solution is dilute, Raoult's law is obeyed, and (2) provides a valuable means of determining the molecular weight of substances in solution. In principle, we could make up a solution of known concentration in gdm^{-3}, and determine the relative lowering of vapour pressure. We count the number of moles by comparing this solution with one for which the number of moles per dm^3

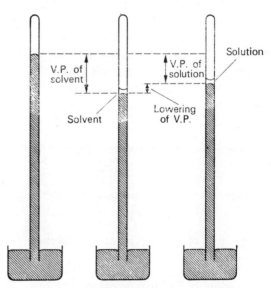

FIG. 15.1. Lowering of vapour pressure by an involatile solute.

is already known. We then use the relationship

$$\text{molecular weight} = \frac{\text{number of grams per dm}^3}{\text{number of moles per dm}^3}.$$

Figure 15.1 shows how this might be done by introducing the solution into the vacuum above a mercury barometer, but it would not be very accurate. There are accurate methods of measuring vapour pressure to determine molecular weights, but a much more accurate method is to use a property which depends upon lowering of vapour pressure such as:

(a) *elevation of boiling point of the solution;*

(b) *depression of freezing point of the solution;*

(c) *osmotic pressure.*

These properties of solutions are known as **colligative properties**—properties determined by the *number* of particles present (in this case particles, or moles, of solute).

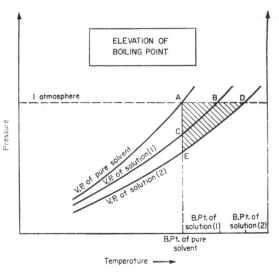

FIG. 15.2.

15.2 Elevation of boiling point

The boiling point of a liquid is the temperature at which the vapour pressure reaches atmospheric. If an involatile solute is lowering the vapour pressure, the boiling point will become higher. Figure 15.2 shows that, provided the solutions are dilute, the elevation of boiling point is proportional to the lowering of vapour pressure. In Fig. 15.2,

lowering of vapour pressure for solution (1)
$$= AC$$

lowering of vapour pressure for solution (2)
$$= AE$$

elevation of boiling point for solution (1)
$$= AB$$

elevation of boiling point for solution (2)
$$= AD$$

If the solutions are dilute, triangles ABC and ADE will be similar, as we may assume the lines BC and DE to be small and parallel. Therefore,

The elevation of boiling point of a solution is proportional to the molar concentration of the solute, provided the solution is dilute.

Measurement of the boiling point of a solution therefore affords a convenient way of counting the number of moles of solute. Elevation is determined solely by the number of moles present and not upon their nature. In order to measure a small temperature change, a thermometer is required which can register precisely very small *differences* in temperature. Such a thermometer need not necessarily give actual values in °C, but it must be able to measure the difference in boiling point between a pure solvent and a dilute solution. A good thermometer for this purpose is the Beckmann thermometer, which has a large bulb but a very fine column, giving perhaps 1 degree per inch, or better. The scale is divided into hundredths of a degree, and at the top of the thermometer is a small mercury reservoir which enables the total quantity of mercury in the operating part of the thermometer to be varied. In this way, a Beckmann thermometer can be set for, say, boiling point elevations using water in the range 99°–104°C, or freezing point measurements in the region 0°C and below.

When taking the boiling point of a solution, the thermometer must be placed in the solution, *not* in the vapour above. If the bulb is in the vapour, it will register the boiling point of pure solvent which condenses on it. A device is needed which bathes the bulb in solution, and Cottrell's apparatus, shown in Fig. 15.3, is one way of doing this. The force of boiling solution sends some of it up the tubes by the side of the thermometer bulb.

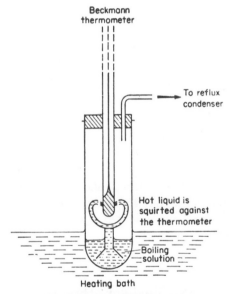

FIG. 15.3. Cottrell's apparatus.

The elevation of boiling point is measured:

(i) with a solution of known molarity, solution (A), and

(ii) with a solution whose concentration in g dm⁻³ is known, but whose molar concentration is unknown, solution (B).

Hence,

$$\frac{\text{molar concentration of (A)}}{\text{molar concentration of (B)}}$$
$$= \frac{\text{elevation of b.p. of (A)}}{\text{elevation of b.p. of (B)}}$$

In Cottrell's apparatus the weight of solvent is best found by observing the *volume* of solution actually present when the measurement is taken, knowing its density. A graduated tube is useful for this purpose. The weight of solute is subtracted from the weight of solution thus determined.

Solvents other than water are also used for measurements of boiling point elevation. It is not necessary to take a second measurement on a liquid of known molar concentration, provided the ebullioscopic constant of the solvent is known (see below).

15.3 Depression of freezing point

When a solution is cooled to freezing point, crystals of *pure* solvent nearly always separate first, provided the solution is dilute. The freezing point of the system is that temperature at which the vapour pressure curve of the liquid phase (the solution) intersects that of the solid phase (pure solvent)—see section 14.2. Fig. 15.4 illustrates what happens in the case of aqueous solutions. From this diagram, by a similar argument to that used for elevation of boiling point, the triangles *ABC* and *ADE* are similar. Hence

Depression of freezing point ∝ lowering of vapour pressure,

∴ **Depression of freezing point of a solution is proportional to the molar concentration of the solute, provided the solution is dilute.**

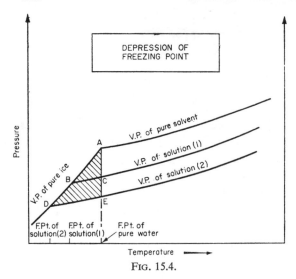

FIG. 15.4.

Various solvents are useful for cryoscopic (freezing point) measurements, and generally they lead to a more accurate determination of molecular weight than ebullioscopic (boiling point) measurements. This is partly because the temperature changes are larger, and partly because the experimental techniques are often simpler.

A typical apparatus for cryoscopic measurements is shown in Fig. 15.5. A Beckmann ther-

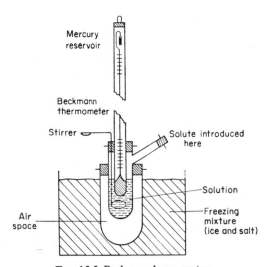

FIG. 15.5. Beckmann's apparatus.

mometer is shown, but for quick measurements a simple thermometer accurate to one-tenth of a degree will do. The freezing point of a known weight of pure solvent is first taken, with vigorous stirring to prevent supercooling. For maximum accuracy a cooling curve is plotted (Fig. 15.6(a)). After re-warming, a weighed

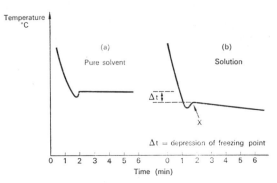

FIG. 15.6. Cooling curves obtained in freezing point determination.

amount of solute is dissolved and a new freezing point taken (Fig. 15.6(b)). The correct freezing point is at X in the figure: once much solvent has crystallized out, the concentration will increase, and the freezing point will gradually drop as shown. For maximum accuracy, a given solution should be warmed and refrozen two or three times, taking the average. Further solute can then be added and the whole process repeated.

15.4 Rast's method for determining molecular weights

Rast's method of determining molecular weights is suitable for organic substances which can be dissolved in molten *camphor*, (m.p. 177°C). Camphor shows a remarkably large freezing point depression for a given molar concentration of solute, making the Beckmann thermometer superfluous. An ordinary thermometer reading to one-tenth of a degree is quite adequate, and the method is very quick.

The melting point of a small quantity of *pure* camphor (labelled "for molecular weight determinations") is first determined, either in a melting point bath, or alternatively on a heating block. A weighed quantity of solute and a weighed amount of pure camphor are then mixed and melted together, and stirred until homogeneous. The mixture is cooled until solid, ground to a fine powder, introduced into the melting point apparatus, and heated until completely molten. The temperature at which solidification begins on cooling is carefully noted. The whole process is repeated using a standard of known molecular weight, such as naphthalene, $C_{10}H_8$. The molecular weight of solute is calculated as before. Note that the molecular weight of camphor is *not* required, and is not relevant to the problem.

15.5 Ebullioscopic and cryoscopic constants

The depression of freezing point or elevation of boiling point of a solvent is independent of the nature of solute molecules, and is determined solely by their concentration in moles. For convenience in calculating, tables have been prepared listing:

(1) The elevation of boiling point of typical solvents, in K mol⁻¹, for 1 kg of solvent—this is the **ebullioscopic constant** of the solvent.

(2) The depression of freezing point, in K mol⁻¹—this is the **cryoscopic constant** of the solvent.

Table 15.1 lists some typical constants for well-known solvents. It does *not* follow that solutions of one mole per kg of solvent would have the exact boiling points and freezing points predicted, because solutions of this concentration are too concentrated for ideal behaviour to apply. The constants are purely constants of proportionality, devised to facilitate calculations.

TABLE 15.1
EBULLIOSCOPIC AND CRYOSCOPIC CONSTANTS
(K MOL⁻¹ FOR 1 kg SOLVENT)

Solvent	Cryoscopic constant	Ebullioscopic constant
Water	1·86	0·52
Ethanol	—	1·15
Chloroform	—	3·66
Benzene	5·12	2·67
Acetic acid	3·9	2·53
Phenol	7·5	—
Aniline	—	3·22
Camphor	40·0	—

Worked Example 1. A solution of 1·35 g of urea in 72·3 cm³ of water boiled at 100.162°C, (pressure = 760 mm Hg). The ebullioscopic constant of water is 0·52 K mol⁻¹ for 1 kg of water. Calculate the molecular weight of urea. Urea may be assumed involatile at 100°C.

1 mole of urea in 1000 g water would theoretically elevate b.p. 0·52 K.

An elevation of b.p. of 0·162 K will be produced by $\dfrac{0·162}{0·52}$ mole of urea in 1 kg water, since elevation $\propto$ concentration.

$\therefore$ Number of moles in 72·3 cm³ of water

$$= \frac{0.162}{0·52} \times \frac{72·3}{1000}.$$

Let molecular weight $= M$

$$M = \frac{\text{number of grams present}}{\text{number of moles present}}$$

$$= \frac{0·52}{0·162} \times \frac{1000}{72·3} \times 1·35 = 60.$$

(Analysis shows the empirical formula to be N_2H_4CO which has a formula weight of 60, therefore this is also the molecular formula. The structural formula is $H_2N.C.NH_2$.)

$$\overset{\|}{O}$$

Warning: ebullioscopic and cryoscopic constants are sometimes quoted for 100 g of solvent. In this case the value will be ten times greater than the above, e.g. the ebullioscopic constant for water is 5·2 K mol^{-1} for 100 g of solvent. It is unwise to memorize a standard formula for calculating molecular weights from cryoscopic and ebullioscopic constants, as this can lead to mistakes.

15.6 Boiling point and freezing point data for "counting moles"

Elevation of b.p. and depression of f.p. are colligative properties, and essentially a way of *counting* solute particles, or moles of solute. It frequently happens that a solute *dissociates* completely or partially, and this fact is revealed by freezing point or boiling point data.

Worked Example 2. One-tenth of a mole of the following solutes were each dissolved in 1 kg of solvent with freezing point depressions as quoted in Table 15·2. What conclusions can you draw about the molecular state of the solutes in each case?

　(a)　sugar in water: f.p. −0·186°C;
　(b)　sodium chloride in water: f.p. −0·372°C;
　(c)　calcium chloride in water: f.p. −0·558°C,
　(d)　acetic acid in benzene: depression of f.p. 0·256 K.

(a) *Sugar in water.* Number of moles of sugar added = 0·1 per kg of water.
Number of moles per kg of water, calculated from cryoscopic constant:

$$= \frac{0·186}{1·86} = 0·1$$

∴ sugar does not dissociate.

(b) *Sodium chloride in water.* Number of moles NaCl(c) added = 0·1 per kg of water.

Number of moles solute per kg of water, calculated from cryoscopic constant:

$$= \frac{0·372}{1·86} = 0·2$$

∴ sodium chloride has dissociated, one mole of NaCl(c) giving two moles of ions.

$$aq + NaCl(c) \longrightarrow Na^+(aq) + Cl^-(aq).$$
　　　　　　　　　　　one mole　　one mole

(c) *Calcium chloride in water.* Number of moles of solid added = 0·1 per kg of water. Number of moles of solute per kg of water, calculated from cryoscopic constant

$$= \frac{0·558}{1·86} = 0·3$$

∴ calcium chloride has dissociated, one mole of $CaCl_2$(c) giving three moles of ions.

$$aq + CaCl_2(c) \longrightarrow Ca^{2+}(aq) + 2\,Cl^-(aq)$$
　　　　　　　　　　　one mole　　two moles
　　　　　　　　　　　　(Total: three moles)

(d) *Acetic acid in benzene.* Again there is 0·1 mole added per kg of solvent. Number of moles per kg of benzene, calculated from cryoscopic constant

$$= \frac{0·256}{5·12} = 0·05 \text{ mole.}$$

This result is interpreted by postulating *association*. Double molecules, called *dimers*, are formed:

$$\underbrace{CH_3COOH + CH_3COOH}_{\text{two moles}} \longrightarrow$$

one mole of dimer

15.7 Osmosis

Many membranes in living systems, particularly the very thin membranes which constitute the walls of living cells, allow water to pass fairly freely through them, but act as a barrier to dissolved substances in the water. These are called **semi-permeable membranes,** and the passage of solvent through a semi-permeable membrane is called **osmosis.**

Osmosis is extremely important to the working of living systems—the uptake of water by plants largely depends upon it for instance—though it can be observed for non-organic systems as well, and synthetic semi-permeable membranes can be made.

Considerable pressure has to be applied to a system undergoing osmosis, in order to restore equilibrium and prevent osmosis from occurring. This pressure is termed **osmotic pressure.** Osmotic pressure is a *colligative* property of a solution, and provided an accurate method of measuring it can be devised, it forms a method for determining molecular weights of dissolved solutes.

15.8 The mechanism of osmosis

The exact mechanism by which semi-permeability works is often not known, but a process analogous to osmosis can be made to take place using a "membrane" of phenol separating a solution of calcium nitrate from pure water (Fig. 15.7). Saturated calcium nitrate is a convenient choice because it is denser than phenol. A half-inch layer of phenol is placed over a few ml of saturated calcium nitrate in a test-tube, and water is carefully added. The level of the phenol layer is marked and the tube left to stand for a day or so. After a time pure water will have diffused through the phenol, and the

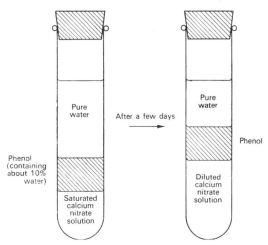

FIG. 15.7. Experiment to illustrate "osmosis".

phenol layer will have risen up the tube. This will continue until finally the phenol layer is resting on top of calcium nitrate solution. The mechanism here is understandable. Water dissolves in phenol to the extent of about 10% by weight, and the phenol becomes saturated with water from A. The water in C has a lower mole fraction than pure water, and hence it is not in equilibrium with the phenol layer B. Water moves down from B to restore equilibrium. Water then moves down from the A to re-establish equilibrium between A and B.

15.9 Simple experiments in osmosis

If a cell from a living system is placed in a solution which is more concentrated than its own internal fluid, it will shrink and shrivel up through loss of water. This effect is called **plasmolysis.** Conversely a cell placed in a too dilute solution will swell until it finally bursts. It is possible to find a solution whose concentration exactly balances the concentration of solutes inside the cell, such a solution is said to be **isotonic** and a cell placed in an isotonic solution will remain stable. The "normal saline" solution used for injections is a dilute salt solution, designed to be isotonic with the fluid in blood corpuscles. The effects of plasmolysis are readily observed with blood under a microscope.

A large-scale demonstration of the effect of solutions on cells, and of osmosis generally, can be shown using eggs. It is first necessary to remove the shell without disturbing the inner skin membrane, and dilute hydrochloric acid will do this. Take two eggs whose shells have been dissolved, and place one in pure water and the other in concentrated brine. After a few days the contrast in sizes of the two eggs will be quite evident.

An idea of the size of the forces operative in osmosis can be obtained by setting up an inverted thistle funnel with a long extension tube, and a semi-permeable membrane stretched over it. Living systems are well adapted to producing semi-permeable membranes, and pig's bladder is often recommended for the thistle funnel, although "cellophane" will do equally well provided it is fixed without any leakage. Water will diffuse through the membrane into solution placed into the funnel and will continue to rise almost indefinitely, the limit often being set in practice by mechanical failure of the membrane.

15.10 Accurate determination of osmotic pressure—osmometry

The forces required to prevent osmosis are indeed often very large, and simple apparatus as described above is inadequate for accurate work, due to leakage of solvent, and mechanical failure of the membrane. A mechanically reliable membrane was devised by Berkeley and Hartley (1909), who used an inorganic precipitate as membrane. They set up an electrolytic cell, and passed current to cause $Cu^{2+}(aq)$ and $Fe(CN)_6^{4-}(aq)$ to diffuse towards each other within the walls of porous pot. The following precipitation reaction occurs:

$$2\,Cu^{2+}(aq) + Fe(CN)_6^{4-}(aq) \longrightarrow$$
$$\longrightarrow Cu_2[Fe(CN)_6](c)$$

Such a membrane is supported by the porous pot, and can withstand applied hydrostatic pressures of several hundred atmospheres.

Measurement of osmotic pressure — **osmometry** — is a reliable and cheap method of determining high molecular weights, and is particularly important in the measurement of the molecular weights of polymers.

A modern osmometer uses different membranes from those used by Berkeley and Hartley: cellulose, polyvinyl alcohol, and non-aqueous solvents are frequently employed. The method of calculating molecular weight is given in section 15.13 below.

15.11 The laws of osmosis

Measurements on the osmotic pressure of solutions have led to three general laws, which only apply when the solutions are dilute. The failure of the laws to deal with concentrated solutions is due to the fact that osmotic pressures are related to the *activities* of the solutes, not to their concentrations.

(1) The osmotic pressure of a solution is directly proportional to its molar concentration, provided temperature is constant. (This law is closely analogous to Boyle's law for gases, section 8.3.)

(2) The osmotic pressure of a solution of given concentration is directly proportional to its absolute temperature. (This is comparable to Charles' law.)

(3) At a given temperature, solutions of the same molar concentration have the same osmotic pressure. (This is the law which enables us to make molecular weight determinations.)

The analogy with the gas laws is evident. In fact it is even closer than the above laws would suggest. It is found experimentally that **the osmotic pressure of a solution is the same as that pressure which the solute alone would exert if it were a gas occupying the same volume at the same temperature.**

$$\Pi V = nRT$$

where V = volume of the solution,

n = number of moles of solute dissolved in this volume,

T = temperature in K,

R = a constant, found to be equal to the gas constant,

Π = the osmotic pressure.

15.12 Osmotic pressure and lowering of vapour pressure

It may be wondered why osmosis occurs, why it ceases at a certain opposing pressure—the osmotic pressure—and why it is a colligative property like lowering of the vapour pressure. We cannot always say exactly *how* osmosis occurs—clearly the mechanism of diffusion described for phenol cannot apply exactly to a solid membrane, though a combination of dissolving and adsorption may be operating. We can apply an equilibrium argument to osmosis however, which is quite independent of mechanism, to show that osmotic pressure is related to lowering of vapour pressure of a solution.

The argument is as follows: in Fig. 15.8, the system in (a) cannot possibly represent an equilibrium, because we have two liquids exerting different vapour pressures in contact with the same vapour. Since pure water exerts the higher vapour pressure (p), it would "distil" over and

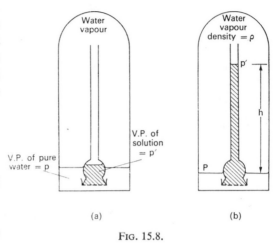

Fig. 15.8.

condense in the solution, by evaporation (vapour pressure of solution = p'). This process would continue until the vapour pressures became equal.

Now consider (b). Here we may imagine that the liquid in the inner tube has risen through osmosis. If we neglect dilution, its vapour pressure will be still p'. However, the pressure of water vapour is now slightly less, because of the increased height.

Suppose that the vapour pressure outside the tube is exactly equal to the pressure inside. We have now reached equilibrium. At this point the hydrostatic pressure due to the liquid column is equal to the osmotic pressure of the solution. If osmosis were occurring, it would stop at this level when equilibrium was reached between the vapours.

The mathematical relationship is simple provided the solution is very dilute, and we assume that the density of solvent vapour does not change too much over the height considered, and that we can take an average, ρ.

Let ρ = average density of the solvent vapour,

  D = density of the solution,

  p = vapour pressure of pure solvent,

  p' = vapour pressure of the solution,

  h = height of the column of solution when equilibrium is established,

  Π = osmotic pressure,

  g = acceleration due to gravity.

Π is equal to the hydrostatic pressure at equilibrium.

$$\therefore \; \Pi = gDh$$

The pressure exerted by the column of water vapour, height h, is equal to the pressure difference at the two heights.

$$\therefore \; p - p' = g\rho h.$$

Combining these equations,

$$\frac{\Pi}{p-p'} = \frac{D}{\rho} = \text{constant, if solution is dilute.}$$

Therefore, **osmotic pressure is directly proportional to lowering of vapour pressure.**

15.13 Osmotic pressure as a method for molecular weight determination

Osmotic pressure, like other colligative properties, depends upon *molar* concentrations of a solute, not its concentration in grams. Essentially then, osmotic pressure enables us to *count the number of moles of solute* present in a solution, using the relationship $\Pi V = nRT$ where n = number of moles present. One or two worked examples should make this clear:

Worked Example 3. 4·07 g of a substance dissolved in 82·4 cm³ of water gave an osmotic pressure of 46·74 mm Hg at 22°C. Calculate the molecular weight.

We have, $\Pi V = nRT$, and we are first required to calculate n. Since we are not given a value for R in the units we require it, we can derive it by using the relationship

$$R = \frac{p_0 V_0}{T_0} \text{ for 1 mole of ideal gas at s.t.p.}$$

$$= \frac{760 \times 22\,400}{273} \text{ cm}^3 \text{ mm Hg K}^{-1}.$$

Now, $T = 22 + 273 = 295$ K.

Substituting, number of moles

$$n = \frac{\Pi V}{RT} = \frac{46 \cdot 74 \times 82 \cdot 4 \times 273}{760 \times 22\,400 \times 295}$$

Now molecular weight

$$= \frac{\text{number of grams}}{\text{number of moles}}$$

$$= \frac{4 \cdot 07 \times 760 \times 22\,400 \times 295}{46 \cdot 74 \times 82 \cdot 4 \times 273} = 19\,450.$$

Note the procedure in the above calculation:

(i) obtain R in the required units;

(ii) calculate the number of moles using $\Pi V = nRT$;

(iii) calculate molecular weight using $M = $ number of grams/number of moles.

Worked Example 4. A solution of M NaCl(aq) was found to have approximately twice the osmotic pressure of a solution of M glucose at the same temperature. What can you deduce from this?

The fact that the sodium chloride shows twice the expected osmotic pressure indicates that there are twice as many particles or moles present, despite the fact that the solutions are both nominally molar. Two conclusions are theoretically possible, either (i) the sodium chloride has dissociated into ions or (ii) the glucose has associated into double molecules. In the light of our knowledge of the properties of sodium chloride, (i) would seem to be the correct explanation for the fact observed.

Note that osmotic pressure has been used to *count moles*. If the number of moles turns out to be greater than predicted by a formula, this suggests *dissociation*. If it turns out to be less, this suggests *association*.

The term **degree of dissociation** (degree of ionization) is applied to an electrolyte which appears to reach an intermediate stage. Suppose 1 mole of a binary electrolyte gave $(1+\alpha)$ moles of particles, as measured by osmotic pressure. α is called the degree of dissociation.

Osmotic pressure is the only practicable colligative property of a solution which can be used to measure very large molecular weights, such as those of polymers in the region 100 000–1 000 000. For instance a 1% solution of a polymer of molecular weight 100 000 in benzene would show the following colligative properties:

elevation of b.p.—0·00026 K ⎫ not measurable
depression of f.p.—0·00051 K ⎭ urable

osmotic pressure—2·8 cm of benzene; measurable.

Study Questions

1. (a) Calculate the mole fractions of solvent and solute in the following:

  (i) 180 g of glucose ($C_6H_{12}O_6$) in 882 g water.
  (ii) 1·8 g glucose in 900 g water.
  (iii) 12·8 g naphthalene ($C_{10}H_8$) in 70·2 g of benzene (C_6H_6)
  (iv) 0·512 g sulphur in 760 g of CS_2.

  (b) Why can the molarities of these solutions *not* be calculated from these figures?

2. (a) The cryoscopic constant for water is 1·86 K mol^{-1} for 1 kg of the solvent. Calculate the freezing temperature you would expect if the solutes had remained undissociated in each of the following cases:

  (i) A solution of 8 g of NH_4NO_3 in 10 dm^3 of water froze at −0·0372°C.
  (ii) A solution of 0·49 g of sulphuric acid (H_2SO_4) in 1 kg of water froze at −0·0279 °C.

  (iii) A solution of 3·32 g of KI in 1 kg of water froze at −0·0744°C.
  †(iv) 4·54 g of HgI_2 were dissolved in a litre of solution (c). The freezing-point was then −0·0558°C.

  (b) Into how many moles has each solute dissociated?

  (c) Suggest an equation for the dissociation in each case.

  (d) At what temperature would a solution of 0·45 g of glucose in 500 g of water freeze?

3. (a) At 18°C, the osmotic pressure of a solution containing 8·20 g dm^{-3} of a substance was 307 mm Hg. What is the molecular weight of the substance?

  (b) Why do you think that this method has been found to be unsuitable for substances of low molecular weight?

CHAPTER 16

Surfaces

16.1 Adsorption

When a substance becomes concentrated at the surface dividing two phases, it is said to undergo adsorption. There is a distinction between *ad*sorption and *ab*sorption (Fig. 16.1).

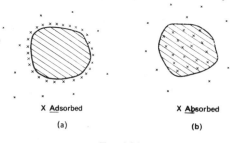

X Adsorbed   X Absorbed

(a)   (b)

FIG. 16.1.

Adsorption implies that the substance is concentrated at a surface, whereas absorption implies penetration of a solute into the bulk of a phase. A great many examples of adsorption could be quoted, for instance:

(a) *Adsorption of gases by solids.* One of the most powerful adsorbents known is charcoal, which is thought to be a microcrystalline form of graphite with a very high surface area. The large surface area is responsible for the adsorbent properties to a great extent. Gases are

readily adsorbed—a fact that was made use of in manufacturing gas-masks for protection against poison gas—and the readiness of a gas to be adsorbed seems to be very approximately in proportion to its readiness to condense to a liquid phase (Table 16.1).

TABLE 16.1

ADSORPTION OF GASES BY CHARCOAL

Gas	Volume adsorbed by 1 g charcoal at 15°	Boiling point (°C)
Sulphur dioxide	380 cm³	−10
Chlorine	285 cm³	−35
Ammonia	181 cm³	−33
Hydrogen sulphide	99 cm³	−62
Hydrogen chloride	72 cm³	−85
Nitrous oxide	54 cm³	−88
Carbon dioxide	48 cm³	−78
Methane	16 cm³	−160
Carbon monoxide	9·3 cm³	−190
Oxygen	8·2 cm³	−183
Nitrogen	8·0 cm³	−196
Hydrogen	4·7 cm³	−253

Heats of adsorption of gases on charcoal are low, of the order 20 kJ mol⁻¹. This is comparable to latent heats of vaporization rather than to the energy needed for forming chemical bonds. It is thought that adsorption generally

178

involves weak attachment of molecules to a surface by van der Waals forces.

Some naturally occurring inorganic solids, for instance the *zeolites* which are complex silicates, have very porous structures favourable to adsorption. These and other substances prepared artificially can be used as **molecular sieves,** since the channels in their structures are of molecular dimensions (0·3—0·5 nm). Different grades of molecular sieve are available commercially as adsorbents for removing impurities from gas streams. For instance, it is possible to obtain a grade which will allow straight chain hydrocarbons through while impeding and adsorbing the more bulky branched chains.

Adsorption of gases in charcoal is generally reversible. The gas can be desorbed by heating. With some adsorbents, a more powerful form of adsorption takes place, where chemical bonds are formed between the adsorbed substances and the surface. This is **chemisorption,** which is generally less easy to reverse. Chemisorption is characterized by a higher heat of adsorption.

The adsorption of gases on solids has recently been applied to chromatography, in **gas–solid chromatography.** Components of the gas mixture are separated by virtue of their differing affinities for the mobile gas phase and the stationary adsorbent surface.

(b) *Adsorption of solutes from solution.* Charcoal is also effective for adsorbing solutes from solutions. Activated charcoal (charcoal which has been rid of adsorbed gases by heating) will extract coloured impurities from a discoloured solution of an organic compound. This is often necessary in the final stages of purification of an organic substance.

(c) *Ion exchange.* A special form of adsorption occurs with some solids which have ionic groups in their structure. The phenomenon was first noticed with the zeolites, and is called **ion exchange.** Ion exchange has important technical uses, such as water softening, and nowadays synthetic resins have replaced the zeolites as ion exchangers. A typical modern ion exchange resin is an organic macromolecule, such as polystyrene, which has been treated chemically to embody sulphonic acid groups, $-SO_3^-$, in the lattice. The resultant structure, known as sulphonated polystyrene, is a sort of "macro-anion" which must have a corresponding number of cations present to neutralize the charge. Most cation exchange resins are sold as their sodium salts (Fig. 16.2).

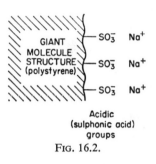

Acidic
(sulphonic acid)
groups
FIG. 16.2.

The resin may be converted quantitatively and reversibly into another salt by allowing a solution of some other cation to flow over it. For instance, if a solution of calcium ions is passed down a column containing beads of the sodium salt of a resin, ion exchange occurs as follows:

$$Ca^{2+}(aq) + 2NaR \text{ (resin)}$$
$$\text{mobile} \qquad \text{stationary}$$
$$\rightleftharpoons CaR_2 \text{ (resin)} + 2Na^+(aq)$$
$$\text{stationary} \qquad \text{mobile}$$

The adsorption is reversible, as the sign $\rightleftharpoons$ indicates. If a solution of $Na^+(aq)$ is now passed over the converted resin calcium salt, the sodium salt is regenerated and calcium ions are washed out, or *eluted.*

A resin for exchanging anions can be made, containing basic groups such as quaternary ammonium groups, NR_3^+. Such a resin will exchange quantitatively anions such as OH^-, NO_3^- and SO_4^{2-} (Fig. 16.3).

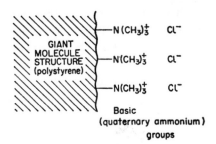

FIG. 16.3.

Simple applications of ion exchange adsorption are:

(i) *Deionization of water.* A mixture of resins containing the hydrogen "salt" of a cation exchanger plus the hydroxide "salt" of an anion exchanger, will remove all the ions present in tap water, substituting instead $H^+(aq)$ and $OH^-(aq)$ ions. Water of such extremely high purity has very low electrical conductivity, and is known as **conductivity water.**

(ii) *Estimation of the hardness of water.* If tap water is passed through a cation exchanger in its hydrogen form (acid form), it will quantitatively displace hydrogen ions equivalent to the cations it contains (mainly Ca^{2+} and Mg^{2+}). The solution which emerges can be titrated with standard alkali.

Number of moles
of Ca^{2+} = 2 × number of moles of
                  H^+ displaced
        = 2 × number of moles of
                  OH^- required for titration.

16.2 Ion exchange chromatography

A column of ion exchange resin can be used for the chromatographic separation of mixtures of anions or cations, making use of the selective adsorption properties of the resin for the dif-

ferent ions. Any ion-pair will come to equilibrium, concentrations being related by an equilibrium constant. Provided the column has sufficient theoretical plates (section 14.9), ions may be separated even when the equilibrium constants are very close; the process is analogous to a sort of countercurrent adsorption. Two typical applications of ion exchange chromatography are:

(i) A mixture of $Cl^-(aq)$, $Br^-(aq)$ and $I^-(aq)$ can be separated on a column containing the nitrate salt of an anion exchanger, eluting with a concentrated solution of $NO_3^-(aq)$. The separation depends upon the different equilibrium constants of the halide equilibria:

$$NO^-(aq) + RX \text{ (resin)}$$
$$\rightleftharpoons X^-(aq) + RNO_3 \text{ (resin)}.$$

The effluent is collected in 5 or 10 cm³ portions, and these are titrated with silver nitrate solution from a burette:

$$Ag^+(aq) + X^-(aq) \longrightarrow AgX(c).$$

The results are plotted graphically (Fig. 16.4).

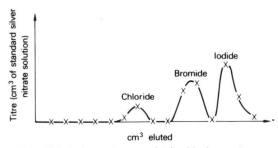

FIG. 16.4. A chromatogram obtained by ion exchange.

(ii) The first straightforward process for separating the lanthanide metals was by ion exchange. Before this, laborious procedures of fractional crystallization had to be resorted to, and even then it was difficult to obtain a pure product. A mixture of lanthanide ions, $M^{3+}(aq)$, is placed on an *anion* exchanger, and eluted with citrate ions. Anionic complex ions are formed, the stabilities of which vary along the series of metals. The separation depends upon these different stabilities: the more stable citrate complexes are formed by the smaller ions at the end of the series, and these are therefore retarded least by the column. Pure samples of the lanthanide elements are made industrially in this way. It is interesting to note that the element promethium, Pm, does not occur in nature and was first made artificially. It was identified by the fact that it fitted into the vacant space in the series on an ion exchange column (Fig. 16.5). Elements 99 (einsteinium) and 100 (fermium) were similarly detected in the dust from a thermonuclear explosion in the Pacific in 1953—on the 100-atom scale!

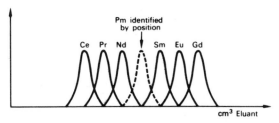

FIG. 16.5. Chromatography of the lanthanide elements.

16.3 Colloids

A **colloid** is a heterogeneous system intermediate between a suspension and a true solution. A true solution contains solute particles less than about 10^{-9} m in diameter (single molecules or ions) dispersed in a solvent. In a colloid the particle size is greater (up to 10^{-6} m), though not so great as to allow the particles to settle out under gravity. A system in which the particles settle slowly out is known as a **suspension.**

The colloidal particles constitute what is called the **disperse phase,** and the "solvent" is known as the **dispersion medium.** Generally the dispersion medium is a liquid, though it would be quite correct to describe a gaseous system such as smoke as a colloid, in which the dispersion medium is a gas. The colloidal particles are usually clumps of molecules or ions, though in the case of substances of very high molecular weight, such as starch and proteins, each colloidal particle may be a single molecule.

CLASSIFICATION OF COLLOIDS

Disregarding colloids in which the dispersion medium is not a liquid, there are various ways in which colloids may be classified.

(a) *Sols and gels.* One way of classifying colloids is by their physical properties, notably viscosity. Take as an example a colloidal solution of gelatin in water: in very dilute solutions the viscosity approaches that of water itself, and the particles of disperse phase are effectively single gelatin molecules. Such a system is called a **sol,** the term **hydrosol** being sometimes used to indicate that the dispersion medium is water. If the concentration of disperse phase is increased, cross-linking between particles begins to occur, and the structure begins to become more rigid—in fact jelly-like. Such a system is called a **gel.** If a gel is heated it melts over a short temperature range, in a manner analogous to an ordinary solid.

(b) *Lyophobic and lyophilic colloids.* Colloids may also be classified in terms of their stability. In lyophobic (literally, solvent hating) colloids there is a tendency for the particles to coagulate. The system is therefore not stable, but tends to revert to a suspension. A number of factors can accelerate this. For instance metals like gold and platinum form lyophobic hydrosols which tend to precipitate in the presence of aqueous metal ions.

Lyophilic (solvent-liking) colloids are stable and do not tend to cluster into particles which settle out. Gelatine forms a lyophilic hydrosol—

the particles may cross link, with the occlusion of water molecules, but a jelly will not settle out. Lyophobic colloids on the other hand do not form gels, and their viscosity is always very close to that of the pure dispersion medium. Colloids in which the particles are macromolecules, such as hydrated silica and albumen, are generally lyophilic.

16.4 Properties of colloids

(a) *Optical properties.* If the particles are comparable in size to the wavelength of light, a beam of light passed through the colloid will show up rather like the beam of light in a dusty atmosphere. This is known as the **Tyndall effect.** Many lyophobic colloids have extremely small particles which fail to show the Tyndall effect, although most colloids do show **Brownian motion** (section 8.1). It is Brownian motion which prevents sedimentation of the particles.

(b) *Colligative properties.* Colligative properties depend upon the number of moles of particles present, and if the particles have a very high average "molecular weight", of the order 1 000 000 as is generally the case, such effects as elevation of boiling point and depression of freezing point will be very slight indeed. Some colloids do have a measurable osmotic pressure however, and this has been used to determine the molecular weight of colloidal particles (section 15.13).

(c) *Electrical properties.* Colloidal particles have a large area in proportion to their volume, so that the properties of their surfaces are often more important than their bulk properties. One important surface phenomenon is the *adsorption* of ions on the particles of a disperse phase. Such adsorption is generally accompanied by the secondary adsorption of further particles of the

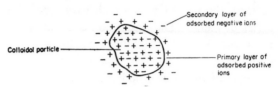

FIG. 16.6. Electrical double layer on a colloid particle.

opposite charge, forming what is known as an **electrical double layer** (Fig. 16.6).

The particles in a colloid carry an overall charge, either positive or negative due to adsorbed ions, and repulsion between like charges stabilizes the colloid and prevents coagulation. If an electric field is applied, a process analogous to electrolysis occurs, called **electrophoresis.** The colloidal particles are deposited on the electrode and lose their charge. This technique has recently been applied to painting: the article forms an electrode in a colloidal paint bath.

The addition of a quantity of highly charged ions to a hydrosol may cause coagulation to occur by destroying the electrical double layer and hence the charge on the particles. For instance, blood is coagulated by Al^{3+}(aq) (this is made use of in styptic pencils) and colloidal mud at the mouth of a river is coagulated by the salts in sea water (the formation of river deltas relies on this effect). If a negatively charged sol is added to a positively charged one, both will precipitate out.

16.5 Dialysis

The process of separating colloidal particles from the smaller particles which form the solute in a true solution is called **dialysis.** Dialysis is necessary when preparing many lyophobic colloids—it is necessary to remove electrolyte which would otherwise coagulate the particles. (A small amount of electrolyte is necessary to stabilize the colloid.) Dialysis is really a form of filtration, where the filter pores are extremely small. A cellophane membrane will serve, and the arrangement is shown schematically in Fig. 16.7.

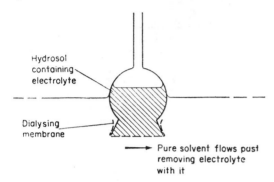

FIG. 16.7. Dialysis.

Dialysis has recently assumed great importance as a means of treating kidney failure. Blood is a colloidal system, in which the disperse phase consists of cells—red and white corpuscles—suspended in an aqueous medium called plasma. Plasma contains dissolved salts, and also traces of waste products such as urea which are normally removed by the kidneys and excreted as urine. In cases of kidney failure, death can follow rapidly due to the build-up of impurities in the blood. An "artificial kidney machine" enables the patient's blood to be dialysed. The high cost of these machines is largely caused by the complex instrumentation needed to monitor all the components in the bloodstream, and keep them in correct balance.

16.6 Preparation of colloids

The methods may be classified as **dispersion methods** (in which larger particles are broken down to a colloidal size) and **aggregation methods** (in which the colloid particles are built up from molecules and ions).

DISPERSION METHODS

(i) **Colloid mill.** The disperse phase is disintegrated mechanically by feeding it between two metal plates suspended in the dispersion medium very close together and rotating in opposite directions.

(ii) **Peptization.** Some precipitates undergo the reverse of coagulation when shaken with water, for instance a suspension of silver chloride in water containing electrolyte becomes colloidal when shaken up. The particles of the suspension adsorb ions and become charged, thereby breaking up.

(iii) **Heating.** Lyophilic sols can be made by heating together two suitable substances, such as gelatin and water. Lyophobic sols cannot be made in this way.

AGGREGATION METHODS

(i) **Precipitation of ions.** A very high rate of precipitation, produced by mixing ions in very high concentrations, will produce particles of colloidal size. For instance colloidal barium sulphate can be obtained by mixing 4 M solutions of Ba^{2+}(aq) and SO_4^{2-}(aq).

(ii) **Reduction.** Many metals can be made as hydrosols by chemical reduction of their salts. Such sols are generally coloured, the colour depending upon the particle size.

(iii) **Oxidation.** Colloidal sulphur is made, and frequently appears when it is not wanted, by oxidation of H_2S or S^{2-}(aq), or by hydrolysis of $S_2O_3^{2-}$(aq).

(iv) **Bredig's arc method.** A platinum hydrosol can be made by striking an electric arc between platinum electrodes held under water. Hydrosols of other precious metals can be prepared similarly.

Study Questions

1. (a) Why does a beam of light passing through smoke show the Tyndall effect?

    (b) Colloidal As_2S_3, a negative sol, is readily precipitated by a small amount of $AlCl_3$(aq, M): it is also precipitated by about seven times the amount of M $BaCl_2$ or M $ZnCl_2$, but only by several hundred times as much M HCl or M NaCl. Can you see any significance in these results?

    (c) Is colloidal As_2S_3 a lyophobic or a lyophilic colloid?

    (d) Fe_2O_3 forms a positive sol. Use part (b) to predict whether K_2SO_4 or $MgCl_2$ would be the more effective in precipitating Fe_2O_3 from a colloidal solution.

    (e) What would happen if colloidal solutions of Fe_2O_3 and As_2S_3 were mixed?

2. (a) Suggest a method for separating compounds of Am, Cm, Bk and Cf.

    (b) If you only had very small quantities, how would you show that a separation of these elements had been achieved?

CHAPTER 17

Reaction rates

17.1 Introduction

This chapter is concerned with the various factors which affect the rate of chemical reactions. So far in this book we have not considered this problem very much, and previous chapters have mainly dealt with equilibria. It was seen that measurement of ΔG could tell us *how far* a reaction was capable of going, though it was incapable of saying *how fast*.

The kinetic theory of matter (Chapter 7) is consistent with experiment insofar as it predicts that, in general, raising the temperature will increase the number of collisions and hence raise the reaction rate. It can be shown that, for molecules in a gas, the number of collisions between molecules rises in proportion to $\sqrt{T}$, provided volume is kept constant. This means that for a rise in temperature of 10 K, the number of collisions only increases a few per cent.

However, it is found by experiment that the rate of most chemical reactions increases *by a factor of two or three* for a 10 K rise in temperature. Evidently the rate of a reaction does not depend simply upon the number of collisions which occur.

We shall return to this important theoretical point in a later section, but first we shall present some experimental facts.

17.2 Definition of reaction rate

Reaction rate, or reaction velocity may be defined as the rate of change of concentration of a stated reactant or product. It is generally convenient to express concentration in mol dm^{-3}, and the time unit might be seconds, minutes or even hours or days. Alternatively, rate may be expressed in terms of *number* of moles, or even number of grams, rather than in terms of concentration of a reactant: such a procedure is more realistic if the total volume of the reactants is changing.

17.3 Factors which can influence reaction rate

Simple experiments show that the following factors must be considered:

(a) **Accessibility of reactants.** By this is meant the ease with which two or more species of reacting molecule can come together in order to react. In the case of a heterogeneous system, in which the reactants occupy different phases the surface area of the interface is an important factor. In a homogeneous reaction system there is no such limitation on reaction rate.

(b) **Temperature.** Reactions increase in rate as the temperature is increased.

(c) **Concentration.** Increasing the concentration of the reactants can have a variety of effects on reaction velocity. Generally an increase in concentration of a given reactant produces an increase in overall reaction velocity, though this is not always the case. In some cases the rate is found to be directly proportional to the concentration of a given reactant, and in other cases rate may be proportional to the square of concentration, or to concentration raised to some other power. The effect of reactant concentration on rate cannot be predicted theoretically (except sometimes by analogy with other known reactions of a similar nature), but must be determined *experimentally*.

(d) **Presence of catalysts.** Many reactions proceed at an increased rate in the presence of another substance called a catalyst. A catalyst is not consumed in a reaction, though it may take part in the formation of intermediate compounds, and certainly forms temporary chemical bonds with reactant molecules at some stage in the process.

17.4 Experimental measurement of reaction velocity

The rate of a chemical reaction is determined by analysis of the reaction mixture at suitable time intervals. The choice of analytical method depends on the reaction being considered.

(a) **Reactions in solution.** These are very often studied most conveniently by titration. Consider for instance the rate of hydrolysis of an ester like methyl acetate by $OH^-(aq)$ ions. The equation is:

$$CH_3COOCH_3(aq) + OH^-(aq)$$
methyl acetate

$$\longrightarrow CH_3OH(aq) + CH_3COO^-(aq);$$
methanol　　　　acetate ions

This reaction may be followed by extracting small portions of the reaction mixture by pipette at intervals, and running them into an excess of dilute acid. This effectively stops the reaction proceeding further, and as a further precaution the reaction mixture may be cooled in ice. The excess acid is then titrated with standard alkali to give the amount of alkali that has been consumed.

(b) **Reactions in the gas phase.** These are often more conveniently studied by observing some physical property which does not involve taking samples from the reaction mixture. For instance, the gas phase decomposition of t-butyl iodide could be followed by sampling the mixture at intervals and titrating the liberated hydrogen iodide with $OH^-(aq)$, but it is much more easily followed by plotting a graph of pressure against time, at constant volume:

$$(CH_3)_3CI(g) \longrightarrow \underbrace{(CH_3)_2C{=}CH_2(g) + HI(g)}$$
1 mole　　　　　　　total 2 moles

(c) **Heterogeneous reactions.** The experimental difficulties involved in making theoretical studies of heterogeneous reactions are often considerable, and the results complex, particularly when the reaction takes place at a catalytic surface. Thus a reaction like

$$C_2H_4(g) + H_2(g) \longrightarrow C_2H_6(g)$$

which takes place readily at the surface of finely divided nickel is heterogeneous, but very many variables—state of surface, area of surface and method of preparation as well as temperature, concentration and pressure—affect the overall rate.

Even in a non-catalytic heterogeneous process such as

$$CaCO_3(c) \longrightarrow CaO(c) + CO_2(g)$$

the structure and nature of the solid surface can have a great effect on the rate.

Measurements do not give the rate of reaction directly: they can only give the concentration of a reactant at a particular time. A graph of concentration against time is then plotted. The rate of reaction is the gradient of the tangent at a given point (Fig. 17.1).

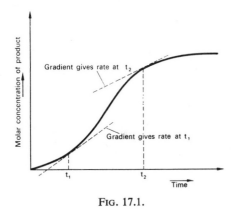

FIG. 17.1.

From these data further graphs may be plotted, for instance rate against time, or rate against concentration. Experiments show that the rates of reactions depend upon the concentration of the reactants, though they are not necessarily directly proportional (Fig. 17.2).

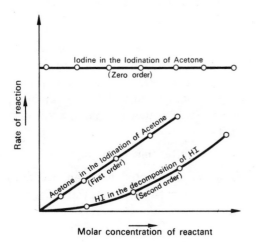

FIG. 17.2.

17.5 Order of reaction

The dependence of rate of a reaction upon concentration of a given reactant can be expressed by the experimentally determined relationship

$$\text{rate} \propto [\text{reactant}]^n$$

where n is defined as the order of reaction with respect to that particular reactant, and we say the reaction follows *nth order kinetics*. The order of a reaction can be determined by plotting log[reactant] against log(rate). The gradient will give n, the order. Where two reactant concentrations affect the rate, we may write

$$\text{rate} \propto [A]^m[B]^n$$

and $m = $ order with respect to reactant A, and $n = $ order with respect to reactant B. The overall order of reaction is $(m+n)$.

Order of reaction may be defined as the number of concentration terms in the experimentally determined rate equation.

First order reactions. The thermal decomposition of t-butyl iodide, section 17.4(b), is first order. Another first order reaction is the isomerization of cyclopropane to propene:

$$\begin{array}{c} CH_2 \\ \diagup \quad \diagdown \\ CH_2\!\!-\!\!CH_2 \end{array} \longrightarrow CH_3\!\!-\!\!CH\!\!=\!\!CH_2$$

Rate $= k[\text{cyclopropane}]$
where [ ] denotes concentration of reactant, and
$k = $ the **rate constant** of the reaction.

The decay of a radioactive element, though not strictly a chemical process, is first order since the rate of decay, measured by counting the intensity of emitted α- or β-particles, depends upon the amount of the radioactive element present.

Second order reactions. These are much more common. One of the first to be thoroughly studied was the gaseous decomposition of hydrogen iodide, and the reverse reaction (combination of hydrogen and iodine vapour) by Bodenstein in 1896. The reaction proceeds to equilibrium and the results for the equilibrium constant were quoted in Chapter 10. Bodenstein's procedure was to seal the reactants in glass bulbs and place them in a vapour bath at constant temperature. After a given time interval the bulbs were removed, cooled rapidly —quenched—to prevent further reaction, and their contents analysed.

We may express the results of the reaction between hydrogen and iodine by writing

$$- R(H_2) = - R(I_2) = k_1[H_2][I_2]$$

where the notation $R(H_2)$ denotes "rate of reaction with respect to hydrogen concentration" and the minus sign denotes rate of *disappearance* of hydrogen. Using this notation the rate could equally well have been expressed in terms of the rate of formation of the product $HI(g)$:

$$R(HI) = k_2[H_2][I_2]$$

Note that $k_2 = 2k_1$, because two moles of hydrogen iodide are formed for every mole of hydrogen or iodine consumed.

However the rate is expressed, the reaction is second order overall, and first order with respect to each separate reactant.

The backward reaction is also second order:

$$- R(HI) = k_3[HI]^2,$$

where $- R(HI)$ denotes rate of disappearance of hydrogen iodide.

Of all the hydrogen halides, Bodenstein was fortunate in choosing hydrogen iodide. The rate equations for the formation and decomposition of HBr are considerably more complex, while HCl and HF do not decompose according to any clear mathematical law.

It is not possible to *deduce* the kinetics of a reaction from its stoichiometric equation, or from its equilibrium constant: the hydrogen halides all follow the same stoichiometry,

$$H_2 + X_2 = 2HX$$

but their rate equations are entirely different.

If one of the reactants in a chemical reaction is present in great excess, its concentration may be almost unaffected by the progress of the reaction. In this case its concentration term may be omitted from the rate equation, and the experimentally determined order will be lower. For instance, in the above case, if the reaction were carried out with a small concentration of hydrogen and a very large excess of iodine vapour it would have an overall order of one— first order with respect to hydrogen, and *zero order* with respect to iodine.

Order of reaction then is not an absolutely fundamental property of a reaction system, it is simply a convenient way of describing the rate equation which best fits the facts.

Some reactions are found to be of *fractional order*. For instance, an order of 3/2:

$$CH_3CHO(g) \longrightarrow CO(g) + CH_4(g);$$
$$-R(CH_3CHO) = [CH_3CHO]^{3/2}$$

17.6 First order rate equation

Most chemical changes gradually slow down as they proceed, since the molar concentration of the reactants falls off as they are consumed. Consider a first order reaction A → products, and let the initial value of $[A] = a$ mol dm^{-3}. Suppose that after time t, x mol dm^{-3} of A have reacted.

$$\therefore \text{ after time } t, [A] = a - x$$

Using the differential notation, we express rate of change of [A] with time (rate of reaction) as follows:

$$- \frac{d[A]}{dt} = - \frac{d(a-x)}{dt}$$

The minus sign indicates that, although rate is a positive quantity, [A] decreases as t increases.

For a first order reaction,

$$-\frac{d(a-x)}{dt} = +\frac{dx}{dt} = k(a-x)$$

$$\therefore \frac{dx}{a-x} = k\, dt$$

$$\therefore \int \frac{dx}{a-x} = k \int dt \quad (+ \text{ integration constant})$$

At the beginning of the reaction, $x = 0$ and $t = 0$, and we may therefore write in limits to the integrals, and eliminate the constant of integration:

$$\int_0^t \frac{dx}{a-x} = k \int_0^t dt$$

$$\therefore \log_e\left(\frac{a}{a-x}\right) = kt$$

$$\therefore k = \frac{1}{t}\log_e\left(\frac{a}{a-x}\right)$$

or,

$$k = \frac{2\cdot303}{t}\log_{10}\left(\frac{a}{a-x}\right).$$

This is the equation for the graph shown in Fig. 17.3, and enables the rate constant k to be evaluated.

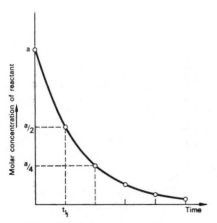

FIG. 17.3. First order reaction.

17.7 Half-life for a first order change

Let $t_{0\cdot5}$ = the time taken for exactly half of the original reactant to decompose. That is, when $t = t_{0\cdot5}$, $x = a/2$.

Substituting,

$$k = \frac{2\cdot303}{t_{0\cdot5}}\log_{10} 2$$

Note that this equation does not include the concentration of the reactant. Therefore, for a first order reaction, the time taken for exactly half the reactant to decompose is independent of the initial concentration. In other words, a first order process has a definite **half-life** period. This fact has already been commented on for radioactive decay. The form of curve shown in Fig. 17.3 is known as an **exponential decay** curve.

The time for any other definite fraction of reactant to decompose is also constant. For instance, a first order reaction has a constant "quarter-life" period.

17.8 Rate equations for other orders

For a second order reaction in which rate $\propto$ [A]², we may write

$$\frac{dx}{dt} = k(a-x)^2$$

$$\therefore \int_0^x \frac{dx}{(a-x)^2} = k \int_0^t dt$$

$$\therefore \frac{x}{a(a-x)} = kt$$

$$\therefore k = \frac{1}{t}\cdot\frac{x}{a(a-x)}$$

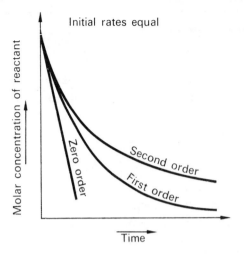

FIG. 17.4.

The half-time for a second order change does depend upon the initial concentration, as the following shows:

$$t = \frac{1}{k} \cdot \frac{x}{a(a-x)}$$

$$t_{0\cdot5} = \frac{1}{k} \cdot \frac{a/2}{(a-a/2)a}$$

$$= \frac{1}{k} \cdot \frac{1}{a} = \frac{1}{ka}.$$

A second order process of the type rate $\propto k[A]^2$ therefore has a half-life period inversely proportional to the initial concentration of A. Examination of the half-life period of a reaction, after plotting a graph, is often a suitable way of determining the order of reaction. In general, for an nth order reaction,

$$\text{half-life} \propto \frac{1}{a^{n-1}}.$$

If the rate of reaction is constant, and independent of concentration [A], this can be expressed mathematically as rate $\propto [A]^0$, and the reaction is of zero order. A zero order reaction will produce a straight line plot for concentration against time (Fig 17.4).

†17.9 Effect of temperature on rate of reaction

Increasing the temperature has the effect of increasing the rate of most reactions. A typical organic reaction shows a doubling or trebling of the rate constant for a rise of 10 degrees. It is found by experiment that the logarithm of the rate constant is inversely proportional to the absolute temperature T, the relationship being

$$\log_e k = \text{constant} - \frac{E}{RT}$$

or, $\quad E = -2\cdot303 \, RT \log_{10} k + \text{constant}$

where R = the gas constant, in J mol^{-1} K^{-1}. E has the dimensions of energy, and is termed the **energy of activation** of the reaction. This equation is strikingly similar to that previously given for the relation between free energy ΔG and the equilibrium constant K:

$$(\Delta G = -2\cdot303 \log_{10} K).$$

If $\log_{10} k$ is plotted against $1/T$, a straight line is obtained, Fig. 17.5. The intercept at $1/T = 0$ is the constant above, which may be written

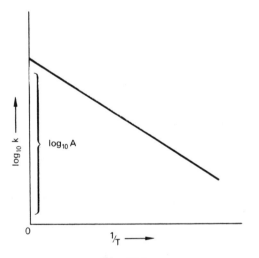

FIG. 17.5.

$\log_{10} A$. The above equations are collectively referred to as the Arrhenius equation, and the constant A is termed the Arrhenius constant or the "A factor".

$$k = A e^{-E/RT}.$$

17.10 Energy of activation

The energy term E in the Arrhenius equation is the minimum energy which molecules need to acquire before they can react by collision. For instance, in the reaction between $H_2(g)$ and

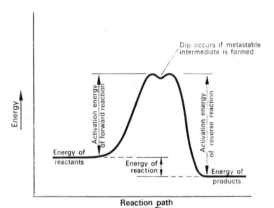

FIG. 17.6. Activation energy.

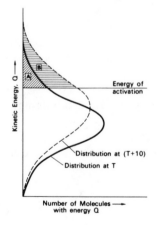

FIG. 17.7.

$I_2(g)$ only a very tiny fraction of the collisions actually lead to the production of $2 HI(g)$. In most cases the collision energy is insufficient and the molecules simply rebound elastically. However, if they collide with sufficient energy they may coalesce to form a **transition state,** otherwise known as an activated complex, in which H—H and I—I bonds are breaking and H—I bonds are forming.

$$\begin{array}{ccccc} \text{H—H} & & \text{H----H} & & \text{H H} \\ & \rightleftharpoons & \vdots \quad \vdots & \rightleftharpoons & | \quad | \\ \text{I—I} & & \text{I----I} & & \text{I I} \end{array}$$

The energy required to form this transition state is the activation energy. This is depicted graphically in Fig. 17.6.

At a given temperature the number of molecules possessing a given energy is shown by a distribution curve; at higher temperatures the "spread" of energies is larger and far more molecules possess very high energies (Fig. 17.7).

The figure also shows how the total number of molecules whose energy exceeds E varies with temperature. If area A represents the number of molecules with energy greater than E at temperature T, then area B represents the number of additional molecules with energy greater than E at $(T+10)$. When E has a value of about 40–60 kJ mol⁻¹, the number of molecules with energies greater than E is approximately doubled for a temperature rise of 10 K.

Complex reactions which proceed in a series of steps often do not obey the Arrhenius equation. Others may do, by a combination of mathematical factors, though the actual value of E does not have any very real physical significance. Nevertheless the concept of energy of activation is a great help to the chemist in learning how to control reactions and reaction rates.

Energy of activation is comparable in magnitude with the values of bond energy terms

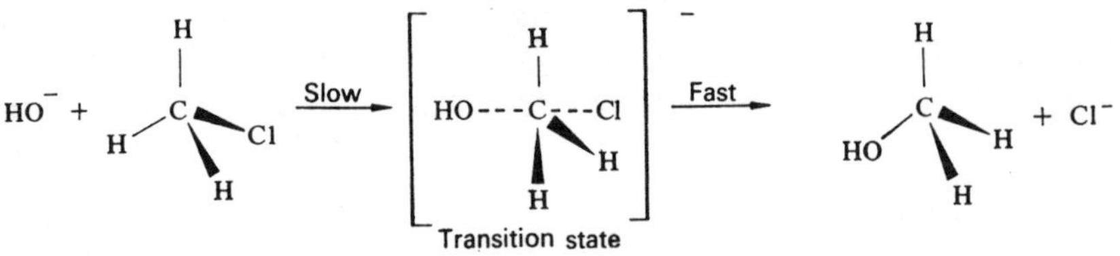

Transition state

FIG. 17.8.

(Chapter 9). This is to be expected, since formation of the transition state involves breaking bonds.

17.11 Molecularity of a reaction step

Some confusion has existed in school text books concerning the meaning of the terms *unimolecular, bimolecular*, etc., with regard to chemical reactions. We shall define **molecularity** as *the number of molecules (or atoms or ions) which form the transition state in a reaction.* For instance the molecularity of the reaction $H_2(g) + I_2(g) \rightarrow 2HI(g)$ is two (bimolecular) because the transition state is formed from two molecules.

The confusion probably arises because very many bimolecular processes follow second order kinetics, and unimolecular reactions often follow first order kinetics. This is by no means always true however, and the concepts of order and molecularity should be carefully distinguished:

(a) *Order of reaction* is an experimentally determined quantity, and is nothing more than a mathematical convenience—a number in a rate equation which can have all sorts of values from zero upwards, including fractional values;

(b) *Molecularity* is a theoretical concept—the number of molecules which are *postulated* as participating in formation of the activated

complex of a particular *reaction step*. We can refer to the order of a complex reaction, as the observed kinetics of its many steps. We can only refer to the molecularity of each individual stage in a reaction proceeding as a series of steps. Molecularity can only have integral values, usually 1 or 2. A molecularity of 3 would be very rare indeed since the chances of a three-body collision of a suitable type occurring are very remote.

An example of a bimolecular reaction is the attack of an alkyl halide such as CH_3Cl by $OH^-(aq)$, (hydrolysis) (Fig. 17.8). This reaction follows second order kinetics, and is described as an SN2 reaction.

Alkyl halides such as t-butyl iodide on the other hand are found experimentally to follow first order kinetics, and the above mechanism does not apply. The C—I bond is far weaker than the C—Cl bond, and the **rate-determining step** in the reaction is thought to be independent of [OH $^-$], being merely the rate at which C—I bonds, activated by random collision, break up. This postulated mechanism accounts for the observed experimental fact that rate $\propto$ [t-butyl iodide]:

(i) First stage is rate determining:

$$(CH_3)_3C—I \longrightarrow$$

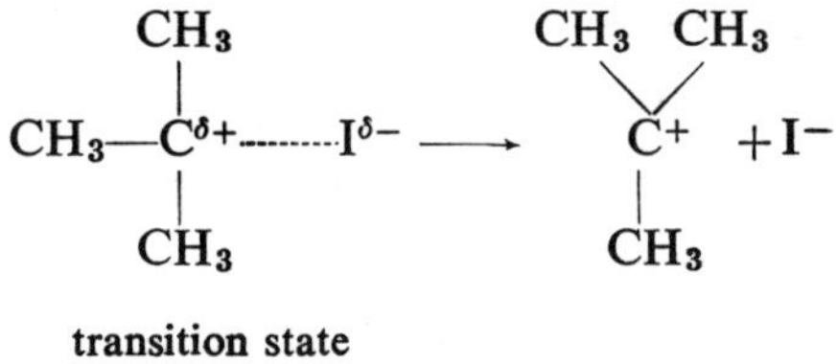

transition state

(ii) Second stage is rapid, because negligible energy of activation is required for the combination of oppositely charged ions as they attract one another:

$$(CH_3)_3C^+ + OH^- \longrightarrow (CH_3)_3C.OH.$$

Here we have an example of two reactions which are identical from a "stoichiometric" point of view, yet follow different kinetics and have different mechanisms on account of the different structures of the reactants.

The fact that the slowest step determines the overall rate is readily appreciated by considering the "washing up" analogy. Imagine three people washing up plates, one washing, one drying, and one stacking them into piles in the cupboard. The rate determining step in the process will undoubtedly be the drying stage. The third stage cannot proceed any faster than the supply of plates allows it to!

17.12 The effect of radiation on reaction velocity

If radiation is to affect the velocity of a reaction, its wavelength must be such that it is absorbed by one of the components of the reaction mixture. Ultraviolet radiation generally has the most effect, and indeed the results are often very dramatic. For instance, at room temperature in the dark, little observable reaction takes place between $H_2(g)$ and $Cl_2(g)$. In subdued sunlight smooth reaction occurs, forming hydrogen chloride, while in strong ultraviolet light, the mixture will explode.

The presence of radiation is essential to the process of photosynthesis, since here many stages in the process involve an increase in ΔG, and will not take place at all without light.

One of the most important consequences of the absorption of radiation in the visible and ultraviolet regions of the electromagnetic spectrum is the breaking of bonds. The reason why the hydrogen–chlorine reaction becomes explosive in the presence of ultraviolet light is that free atoms of chlorine are formed in the process

$$Cl_2(g) \longrightarrow 2\,Cl(g);$$
$$\Delta H = +240 \text{ kJ mol}^{-1} \qquad (1)$$

From the relationship, $E = h\nu$

$$E = \frac{Nhc}{\lambda}$$
$$= \frac{1 \cdot 19 \times 10^{-4}}{\lambda}$$
$$\text{kJ mol}^{-1}$$

where h = Planck's constant = $6 \cdot 63 \times 10^{-34}$ N s

c = velocity of light = $2 \cdot 998 \times 10^8$ m s^{-1}

N = Avogadro number = $6 \cdot 023 \times 10^{23}$,

the energy of a quantum of visible light at 400 nm will just break a bond of 300 kJ mol^{-1} energy—a relatively weak bond.

The process by which bonds are artificially ruptured by radiant energy is called **photolysis.** A schematic apparatus is shown in Fig. 17.9. The reaction vessel is generally quartz, which is transparent to ultraviolet radiation, and a mercury lamp is the energy source.

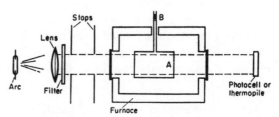

FIG. 17.9. Photolysis apparatus. A, quartz reaction vessel; B, connection to gas handling apparatus.

17.13 Chain reactions

The remarkable thing about the hydrogen–chlorine explosion reaction is its remarkably high **quantum yield**. We may define quantum yield as

$$\frac{\text{number of molecules transformed by reaction}}{\text{number of quanta of radiation absorbed}}.$$

In the above case, quantum yields of up to 10^7 have been observed. This suggests that absorption of a single quantum of energy "triggers off" a whole process, rather in the manner that a small spark can trigger off a combustion reaction.

The process is thought to occur by a whole sequence of reactions, starting with (1) above. Further reactions can occur, such as

$$Cl + H_2 \longrightarrow HCl + H \qquad (2)$$
$$H + Cl_2 \longrightarrow HCl + Cl \qquad (3)$$

Reactions (2) and (3) can proceed indefinitely, as long as the free atoms produced are able to react further.

The presence of free atoms and free radicals is often important in chemical reactions. Polymerizations to form "Perspex", polyvinyl chloride and polystyrene are examples.

17.14 Catalysis

A catalyst may be defined as a substance which can increase the rate of a reaction, without itself undergoing permanent chemical change. Catalysts have the following general properties:

(a) A small number of moles of catalyst can catalyse a very large number of moles of reactant. For instance, hydrogen peroxide can be decomposed into water and oxygen by as little as one part per million of colloidal platinum.

(b) A catalyst may undergo intermediate physical changes, and it may form temporary chemical bonds with the reactants, but it will be unchanged in amount and composition at the end of the reaction.

(c) A catalyst cannot affect the ΔG of a reaction, and hence it cannot change the equilibrium position. It may accelerate the attainment of an equilibrium, and its function can be likened roughly to "lubricating the mechanism". Since it is chemically unchanged overall, it cannot add energy to, or remove energy from, the system.

Catalysts may be homogeneous (same phase as the reactants) or heterogeneous (different phase, usually solid), but heterogeneous catalysts have wider applications in practice. Transition metals and their compounds are very often effective. This is generally due to their ability to form intermediate bonds and assist the formation of transition states (activated complexes), though the exact mechanism of a particular catalysis is often obscure. The following are some examples of transition metals catalysts used industrially:

(i) Finely divided nickel catalyses the hydrogenation of the olefinic bond $C = C$. This is used in the conversion of vegetable oils to margarine.

(ii) A finely divided mixture of iron and aluminium oxide is the catalyst for the Haber process for ammonia manufacture.

(iii) Platinized asbestos, and vanadium(V) oxide (vanadium pentoxide) are both used for the reaction $SO_2(g) + \frac{1}{2} O_2(g) \rightarrow SO_3(g)$.

Catalysts work by lowering the energy of activation of the reaction, by providing an alternative path with a lower "hump" to be overcome. (Fig. 17.10). Consequently more molecules are present with sufficient energy to react.

Very often a different choice of catalyst will lead to different products from a given reaction

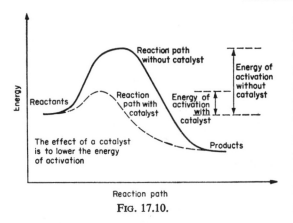

Reaction path
without catalyst

Energy of
activation
without
catalyst

Reactants

Reaction
path with
catalyst

Energy of
activation
with
catalyst

Products

The effect of a catalyst
is to lower the energy
of activation

Reaction path

FIG. 17.10.

mixture. Thus the Fischer–Tropsch synthesis of hydrocarbons uses $CO(g)$ and $H_2(g)$ over a mixture of oxides of cobalt and thorium. The same gases passed over zinc oxide yield methanol. Catalysts are highly specific, and the principles underlying their operation are even today far from being understood. The choice of a catalyst for an industrial organic synthesis may often be a combination of science and "cookery", or inspired guesswork, and the recipe for a good catalyst is often a closely guarded industrial secret.

Study Questions

1. Suggest how the rates of the following reactions might be followed experimentally:
  (a) The reaction of iodine solution with acetone in the presence of acid.
  (b) The thermal decomposition of acetaldehyde vapour to give methane and carbon monoxide.
  (c) The action of hydrochloric acid on sodium thiosulphate solution. (One of the products is sulphur.)
  (d) The catalytic decomposition of hydrogen peroxide solution into water and oxygen.

2. The decomposition of $N_2O_5(g)$ into $NO_2(g)$ and $O_2(g)$ was followed by observing the change in the pressure. The partial pressure of N_2O_5 in the mixture was as follows:

Time	Pressure (mm)	Time	Pressure (mm)
0 min	348	40 min	105
10 min	247	50 min	78
20 min	185	60 min	58
30 min	140		

Calculate (a) the order of the reaction, (b) the rate constant for this reaction.

3. The thermal decomposition of acetaldehyde was followed at 770 K, the overall pressure varying as follows:

Time (seconds)	Pressure (mm)
0	363
42	397
105	437
242	497
480	557
840	607

Calculate (a) the order of the reaction, and (b) its rate constant.

4. Comment on the following observations:
  (a) Hydrogen and oxygen do not normally react at room temperature, but if a trace of platinum black is present, an explosion occurs.
  (b) The substance urease causes urea, $CO(NH_2)_2$ to be hydrolysed, yet it has no effect on methyl urea, $CONH_2(NHCH_3)$.

5. Ammonia is made by passing nitrogen and hydrogen over an iron catalyst. Will the addition of iron speed up, or retard, the decomposition of ammonia into its elements?

6. Classify the industrial catalysts mentioned in section 17.14 as homogeneous or heterogeneous.

†**7.** $d-C_2H_5(CH_3)CHBr$ can be hydrolysed using NaOH solution.
What shape and configuration will the product possess if the reaction follows (a) first-order, (b) second-order, kinetics?

8. (a) The hydrolysis of methyl acetate (section 17.4) with alkali using equimolar quantities of the reactants follows second order kinetics, but in the presence of a large excess of alkali first order kinetics are observed. Why is this?
  (b) Will the molecularity of the process also alter?

†**9.** The reaction of $NO(g)$ with $H_2(g)$ is thought to proceed by the following mechanism:

$2NO \longrightarrow (NO)_2;$		(Fast)
$(NO)_2 + H_2 \longrightarrow$ Initial products;		(Slow)
Initial products $\longrightarrow$ Final products;		(Fast)

What will be
  (a) The reaction orders with respect to NO and H_2?
  (b) The overall reaction order?
  (c) The molecularity of the rate-determining step?

CHAPTER 18

Hydrogen

18.1 Occurrence

Of all the atoms in nature, the hydrogen atom is the simplest. Its lighter isotope consists only of a single proton surrounded by a single electron. Isolated hydrogen atoms, H(g), do not exist under normal conditions at the Earth's surface, though they are present in the Sun and in interstellar space, and can form at high temperatures, for example, in flames. The free element does not occur in the atmosphere except in trace quantities, though large quantities are manufactured for industrial use. The element hydrogen occurs commonly in the following forms:

(a) **Positive ions.** The simplest positive ion is the proton itself, which can exist in a discharge tube but not as a free chemical species in solution. The hydrogen ion in aqueous solution is a proton combined with one or more molecules of water, written for convenience $H^+(aq)$. Crystalline solids containing the ion H_3O^+, the hydroxonium ion, have been prepared, but in an aqueous solution it is difficult to establish the exact composition of $H^+(aq)$, and species such as H_3O^+, $H_5O_2^+$, etc., may all be present in varying concentrations.

The hydrogen ion may exist in other solvents as the solvated proton. For instance, in liquid ammonia we might write $H^+(am)$ for the various species present. The ammonium ion, NH_4^+, is one possible species.

(b) **Negative ions.** Hydrogen occasionally forms negative ions, H^-, and this ion is isoelectronic with He and Li^+, containing a filled $1s$-orbital. In fact the alkali metals combine with hydrogen gas to form salt-like structures. Sodium hydride, $Na^+H^-(c)$, is in many ways analogous to sodium chloride, $Na^+Cl^-(c)$, and has a simple cubic (6 : 6) structure. Evidence for the H^- ion is provided by the fact that hydrogen is evolved at the *anode* when a molten alkali metal hydride is electrolysed.

(c) **Covalently bonded atoms.** Hydrogen can also solve its energy level requirements by forming a single covalent bond with other atoms.

TABLE 18.1

ENERGIES OF SOME BONDS WITH HYDROGEN ATOMS (kJ mol^{-1})

			H—H 435
C—H 415	N—H 390	O—H 465	F—H 565
Si—H 293	P—H 318	S—H 338	Cl—H 430
	As—H 247	Se—H 276	Br—H 338
		Te—H 238	I—H 297

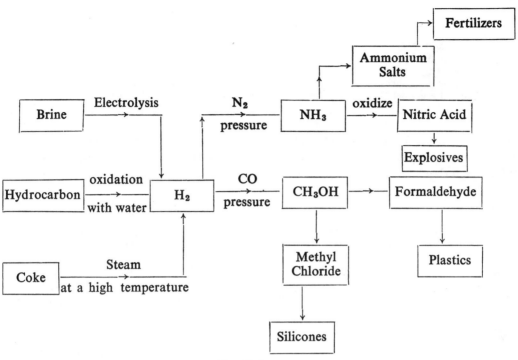

Most non-metals (but not the noble gases) and some metals will form such bonds.

The simplest bond is that formed in the hydrogen molecule itself, $H_2(g)$, (Chapter 3). Table 18.1 shows a selection of some bonds between elements and hydrogen, together with their bond energies. Metal–hydrogen bonds have low bond energies, and hence hydrides like SnH_4 and SbH_3 are thermally unstable.

18.2 Properties and uses of hydrogen gas

Though a small amount of hydrogen is used for filling balloons, and some pure hydrogen is consumed in the use of the oxy-hydrogen flame for cutting and welding metals, the major bulk of commercially produced hydrogen is used in chemical reactions. Figure 18.1 is a flow-sheet illustrating some of the uses. Impure hydrogen, as a constituent of town gas and water gas, is used extensively as a fuel.

The naturally occurring isotope ^2H, deuterium (often given the symbol D) is used when labelled hydrogen atoms are required. The isotope ^3H, tritium, was first made in 1934 by bombarding deuterium compounds with high energy deuterium nuclei:

$$^2_1D + {}^2_1D \longrightarrow {}^3_1T + {}^1_1H;$$

$$\Delta H = -3{\cdot}8 \times 10^8 \text{ kJ.}$$

As tritium emits β-particles, it is also a useful form of labelled hydrogen. It has been used in studying reaction mechanisms. This nuclear reaction liberates a very large amount of energy, and is an example of **nuclear fusion**. It is a possible source of power in future nuclear reactors using controlled fusion processes though the

TABLE 18.2

Ionization energy	H $\longrightarrow$ H$^+$+e$^-$;	$\Delta H = +1310$ kJ g-atom^{-1}
Electron affinity	H$^- \longrightarrow$ H+e$^-$;	$\Delta H = +75$ kJ g-atom^{-1}
Bond energy	H$_2 \longrightarrow$ 2 H;	$\Delta H = +435$ kJ g-equation^{-1}
Internuclear distance		0·074 nm
Half-cell potential	H$^+$(aq)+e$^- \rightleftharpoons \frac{1}{2}$ H$_2$(g);	$E° = 0·000$ V

technical obstacles to obtaining continuous energy in this way are proving enormous. Table 18.2 summarizes some data for hydrogen.

18.3 The production of hydrogen gas

From the flow-sheet (Fig. 18.1) it is clear that considerable quantities of hydrogen are required industrially. This hydrogen comes from three main sources:

(i) **Electrolysis.** It is rarely necessary to set up special plant for making hydrogen electrolytically, since it is a by-product of the process for making sodium hydroxide and chlorine from brine, NaCl(aq). Chlorine is liberated at the anode, while at the cathode

$$H^+(aq)+e^- \longrightarrow \tfrac{1}{2} H_2(g).$$

Deuterium is manufactured in countries where electrical power is cheap, such as Norway. Its separation from ordinary hydrogen depends on the fact that the rate of liberation of D$_2$(g) is lower than that of H$_2$(g). A solution of sodium hydroxide is electrolysed until its relative bulk has been considerably reduced and it will then be enriched in the heavier isotope.

(ii) **Water gas.** This somewhat misleading name refers to the mixture of carbon monoxide and hydrogen produced by blowing steam over white-hot coke:

$$C(c)+H_2O(g) \longrightarrow CO(g)+H_2(g);$$
$$\Delta H = +130 \text{ kJ.}$$

In a process called the *Bosch process*, water gas is mixed with more steam and passed over a catalyst consisting mainly of iron(III) oxide at about 500°C:

$$CO(g)+H_2O(g) \longrightarrow CO_2(g)+H_2(g);$$
$$\Delta H = -42 \text{ kJ}$$

The carbon dioxide is absorbed by dissolving it in water under pressure. The resultant hydrogen is not all that pure but is pure enough for making ammonia.

(iii) **Petroleum.** When high molecular weight hydrocarbons are *cracked* (broken up catalytically into smaller molecules) some hydrogen is formed. In countries where natural gas is abundant, methane is converted into hydrogen:

$$\text{Ni 900°C}$$
$$CH_4(g)+H_2O(g) \longrightarrow CO(g)+3 H_2(g);$$
$$\Delta H = +205 \text{ kJ}$$

Some hydrogen is present in the mixture called *coal gas* evolved when coal is heated in the absence of air. The domestic gas popularly referred to as coal gas, is often derived from petroleum as well as from coal, and contains rather less hydrogen. Its more correct name is *town gas*.

Small laboratories which are not equipped with hydrogen cylinders generally use the following reaction to generate hydrogen, hydrochloric acid being dropped on to impure granulated zinc:

$$Zn(c)+2 H^+(aq) \longrightarrow H_2(g)+Zn^{2+}(aq).$$

Hydrogen is produced in a great many reactions involving the reduction of hydrogen ion or water, generally by a metal, for instance:

oxidized (electron removed)

$$Na(c) + H_2O(l) \longrightarrow Na^+(aq) + OH^-(aq) + \tfrac{1}{2} H_2(g).$$

reduced (electron added)

Miscellaneous reactions which produce hydrogen are:

(i) the action of alkali on some elements:

$$Al(c) + 3\,OH^-(aq) + 3\,H_2O$$
$$\longrightarrow \tfrac{3}{2} H_2(g) + Al(OH)_6^{3-}(aq)$$

(ii) the action of water on saline (salt-like) hydrides:

$$CaH_2(c) + 2\,H_2O \longrightarrow Ca(OH)_2(aq) + H_2(g)$$

18.4 Chemical properties of hydrogen gas

Despite the fact that hydrogen burns, and mixtures with oxygen explode, $H_2(g)$ is not as reactive a gas as one might suppose. Its bond energy is fairly high (435 kJ) which makes it difficult to break. Its readiness to give electrons and form hydrogen ions such as $H^+(aq)$ make it a fairly powerful *reducing agent*. Its reactions may be classified under the following headings:

(a) **Reductions in aqueous solution.** Half-cell potential data predict that hydrogen gas will reduce any couple with a positive ε° value. Frequently, however, reductions do not occur, in the absence of a catalyst, due to the high energy required to break the H–H bond. For instance, $H_2(g)$ fails to reduce $Fe^{3+}(aq)$ to $Fe^{2+}(aq)$ ($\varepsilon^\circ = +0.771$ V), and fails to reduce $Ag^+(aq)$ to $Ag(c)$ ($\varepsilon^\circ = +0.799$ V).

(b) **Reduction of metal oxides.** The conditions under which it becomes possible for a metal oxide to be reduced by gaseous hydrogen can be determined from studying a free energy diagram like Fig. 13.5. However, a simple aid to memory is that metals below iron (approximately) in the electrochemical series are readily produced by passing hydrogen gas over the heated oxide.

$$CuO(c) + H_2(g) \rightleftharpoons Cu(c) + H_2O(g);$$
$$\Delta G \text{ negative at } 500°C$$

In the above case of copper(II) oxide, the equilibrium lies heavily to the right. Iron and steam tend to form an equilibrium with hydrogen and iron(II, II, III) oxide:

$$Fe_3O_4(c) + 4\,H_2(g) \rightleftharpoons 3\,Fe(c) + 4\,H_2O(g);$$
$$\Delta G \simeq \text{zero at } 500°C$$

The reaction may be made to proceed in either direction. This behaviour is quite common with equilibria: an equilibrium is readily "driven to completion" by sweeping the products away from the reaction site so that balance is never achieved.

Hydrogen is rarely used for extracting metals on any scale, as it is too expensive compared with, say, coke. Tungsten and germanium are extracted using hydrogen, where quantities are not large and the cost of reductant not an overriding factor.

(c) **Direct combination with elements.** Hydrogen has never, to date, been persuaded to combine with a noble gas, nor does this seem likely from energy considerations. Direct combination has been observed between hydrogen and all other non-metals, however, though in some cases the conditions are rather extreme. For instance at one end of the scale, acetylene is formed in small amounts by blowing hydrogen gas through an arc struck between carbon electrodes. This reaction probably involves the

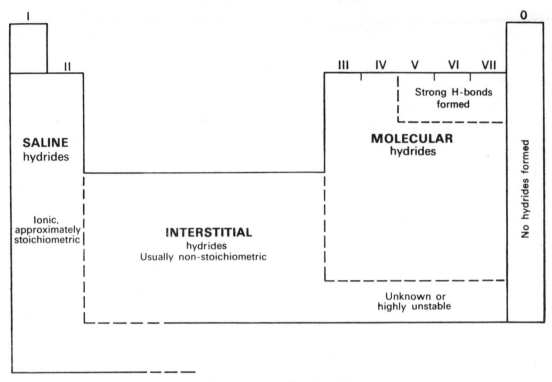

FIG. 18.2. The properties of hydrides.

dissociation of hydrogen into atoms which are able to attack the heated carbon.

$$\text{2 C(c)} + \text{H}_2\text{(g)} \xrightarrow{\text{high temperature}} \text{C}_2\text{H}_2\text{(g)};$$
$$\Delta H = +343 \text{ kJ at } 25°\text{C}.$$

At the other end of the reactivity scale, hydrogen and fluorine explode spontaneously even in the dark at very low temperatures:

$$\text{H}_2\text{(g)} + \text{F}_2\text{(g)} \longrightarrow \text{2 HF(g)};$$
$$\Delta H = -535 \text{ kJ at } 25°\text{C}.$$

The equilibrium between hydrogen and iodine was considered in Chapter 10.

Metals of Groups IA and IIA combine directly with hydrogen. With the very electropositive metals in Group IA, the formation of the H⁻ ion is important.

Most transition metals form non-stoichiometric hydrides described as **interstitial** since the hydrogen atoms can fit into the interstices (gaps) in the original metal lattice without much alteration to the metallic structure. Such hydrides are often more akin to solid solutions than to true chemical compounds, and they are electrical conductors, though their conductivity is usually impaired. A remarkable example is the palladium–hydrogen system, of variable composition PdH_x, where $x = 0$ to 0.75. Palladium powder is used for selectively "mopping up" traces of hydrogen in a gas mixture, and can absorb several hundred times its own volume at room temperature.

The formulae of simple hydrides provide a good illustration of the periodic law (Fig. 1.11), though the heavier hydrides have empirical formulae which do not fit the pattern. Figure 18.2 shows how the properties of hydrides depend upon position in the periodic table.

18.5 Chemical properties of the aqueous hydrogen ion

When we say that an aqueous solution is *acidic*, we are in fact making rather loose use of the term. In the next chapter, where the theory of acids and bases will be developed in more detail, the word *acid* will be given a more specialized theoretical meaning. For practical purposes, however, as long as we consider aqueous solutions, a solution is said to be acidic when the concentration of $H^+(aq)$ is greater than in pure water.

Pure water dissociates slightly into ions by a reaction known as auto-ionization. At room temperature, the product $[H^+]\times[OH^-]$ is about 10^{-14}, and hence the concentration of $H^+(aq)$ is equal to 10^{-7} mol dm^{-3}.

$$H_2O \rightleftharpoons H^+(aq)+OH^-(aq);$$

Concentrations: 10^{-7} 10^{-7}

(mol dm^{-3})

Activities: 10^{-7} 10^{-7}

Note that, at a concentration of 10^{-7} mol dm^{-3}, concentration and activity can be taken as numerically equal.

The reactions of dilute aqueous acids, such as hydrochloric acid, $HCl(aq)$, nitric acid, $HNO_3(aq)$ and sulphuric acid, $H_2SO_4(aq)$, are all essentially reactions of the hydrogen ion. Similar reactions are shown by other substances which produce a relatively high concentration of $H^+(aq)$ when added to water. Thus solutions of metal ions low in the electrochemical series show the reaction of acids:

$$Al^{3+}(aq)+H_2O \rightleftharpoons H^+(aq)+Al(OH)^{2+}(aq)$$
hydroxo-complex

The reactions of $H^+(aq)$ are closely bound up with equilibria, and a proper treatment of them will have to be deferred to the next chapter, but a few of them are summarized here for convenience.

(a) **Indicators.** Indicators are coloured substances which undergo changes of colour depending upon the concentration of $H^+(aq)$ present. Typical reactions of indicators with $H^+(aq)$ are shown in Table 18.3, though it should be borne in mind that the exact concentration of $H^+(aq)$ required to effect the colour change depends upon the indicator chosen (Chapter 19).

TABLE 18.3

Indicator	Colour in acid	Colour in alkali
litmus	red	blue
methyl orange	red	yellow
phenol phthalein	colourless	red

The concentration of $H^+(aq)$ in a solution can be roughly measured by adding **universal indicator,** which is a mixture of several acid-sensitive dyes.

(b) **Metals.** Many metals can reduce hydrogen ions to $H_2(g)$, being oxidized to an aqueous ion. A typical example of this reaction type is

$$Mg(c)+2 H^+(aq) \longrightarrow Mg^{2+}(aq)+H_2(g).$$

Metals whose couples with aqueous ions have positive values—those below hydrogen in the electrochemical series—cannot liberate hydrogen from acids.

(c) **Alkalis and bases.** Alkalis are essentially aqueous solutions of OH^- ion, and the neutralization of an alkali like calcium hydroxide, $Ca(OH)_2$, with an acid is essentially

$$H^+(aq)+OH^-(aq) \longrightarrow H_2O;$$
$$\Delta H = -58 \text{ kJ}$$

Many metal oxides and hydroxides are insoluble in water but nevertheless dissolve in acids with the evolution of heat, by reactions analogous to the above:

$$2\,H^+(aq) + Cu(OH)_2(c)$$
$$\longrightarrow Cu^{2+}(aq) + 2\,H_2O(l).$$
$$6\,H^+(aq) + Fe_2O_3(c)$$
$$\longrightarrow 2\,Fe^{3+}(aq) + 3\,H_2O(l).$$

Some metal oxides, for instance chromium(III) oxide, Cr_2O_3, are such strongly-bound structures that they are not attacked by acids appreciably, except with prolonged boiling.

Substances which react with hydrogen ions are known as **bases,** though the term base, like acid, has acquired a fairly special meaning (Chapter 19).

(d) **Catalysis.** Reactions are frequently catalysed by $H^+(aq)$, for instance the hydrolysis of esters.

$$CH_3COOEt + H_2O \overset{H^+}{\rightleftharpoons}$$

ethyl acetate

$$CH_3COOH + EtOH$$

acetic acid   ethanol

$$(Et = C_2H_5, \quad ethyl)$$

18.6 Preparation of non-metal hydrides

(a) DIRECT COMBINATION OF ELEMENTS

Although most non-metal hydrides *can* be made by direct combination of the elements this is not always the most convenient way. Indeed some hydrides are of little more than theoretical interest and are rarely prepared. Two very important direct combinations ought to be mentioned:

(i) Ammonia, formed from nitrogen and hydrogen in the Haber process. Details are given in Chapter 23.

(ii) Hydrogen chloride, $HCl(g)$ is manufactured largely from electrolytically produced hydrogen and chlorine which react together by burning. Hydrogen chloride gas is converted into hydrochloric acid by simply dissolving in water.

$$HCl(g) + aq \longrightarrow H^+(aq) + Cl^-(aq).$$

(b) HYDROLYSIS OF A SUITABLE "X-IDE"

The hydride of an element X can often be made by hydrolysing suitable "X-ide". For instance, hydrogen sulphide can be made by the hydrolysis, either by water or dilute acid, of a suitably chosen metal sulphide. Similarly phosphine, $PH_3(g)$, can be made by the hydrolysis of a metal phosphide. The method is fairly general, in the sense that suitable reactants can usually be found, e.g.

many metal sulphides + dilute acid → hydrogen sulphide, H_2S

calcium nitride + water or dilute acid → ammonia, NH_3.

calcium carbide + water or dilute acid → acetylene, C_2H_2, and other hydrocarbons.

magnesium silicide + dilute acid → silane mixture, SiH_4, Si_2H_6, Si_3H_8, etc.

(c) REDUCTION OF HALIDES WITH LITHIUM ALUMINOHYDRIDE

Lithium aluminohydride, $LiAlH_4(c)$, is a very vigorous reducing agent used particularly for specific organic reductions. It can be used to prepare fairly pure hydrides, though it would not be used where cheaper, large-scale, methods are available. An example of its use is:

$$SiCl_4(l) + LiAlH_4(c) \longrightarrow$$
$$SiH_4(g) + LiCl + AlCl_3.$$

(d) MISCELLANEOUS REACTIONS

Where small laboratory supplies of certain gaseous hydrogen compounds are required, the common methods do not fall into the above categories. The following is worth mentioning:

$$Ca(OH)_2(c) + 2NH_4Cl(c) \xrightarrow{heat}$$
$$CaCl_2(c) + 2H_2O(g) + 2NH_3(g)$$

Any solid ammonium salt will do here. The gas is alkaline and can be dried over any basic drying agent, e.g. CaO(c), quicklime.

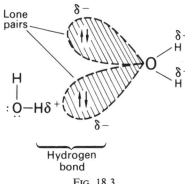

FIG. 18.3.

18.7 The hydrogen bond

The hydrogen bond is the name given to the force which attracts hydrogen atoms to electronegative atoms in certain circumstances. Although it is termed a bond, it is very much weaker than the chemical forces which hold atoms together in molecules and lattices, yet noticeably stronger in most cases than the dipole attractions which come under the general heading of van der Waals forces. The situation is summarized in Table 18.4.

Although the hydrogen bond is essentially an electrostatic attraction, it is directional. It may be illustrated by examining the structure of ice. The water molecule has bonds which are strongly dipolar, and two lone pairs. The lone pairs occupy regions of negative charge which can attract hydrogen atoms from other molecules which are themselves positively charged due to attachment to an electronegative atom. Figure 18.3 illustrates this.

It is found that hydrogen bonds are strongest where the atoms are attached to the most electronegative first row elements, N, O, and F. Here the dipolar force will be largest, giving the H atoms a strong charge $\delta+$. Note that a hydrogen atom attached to, say, carbon or silicon is not sufficiently "$\delta+$" to be attracted to molecules such as H_2O, NH_3 or HF.

It thus appears that the commonest hydrogen bonds are *bridges* between any two of the atoms N, O, and F.

Hydrogen bond formation accounts for the

TABLE 18.4

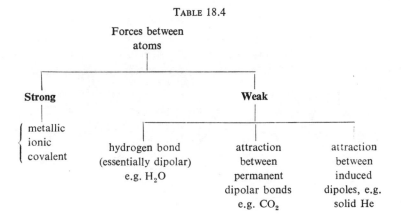

Forces between atoms		
Strong	**Weak**	
metallic ionic covalent	hydrogen bond (essentially dipolar) e.g. H_2O	attraction between permanent dipolar bonds e.g. CO_2	attraction between induced dipoles, e.g. solid He

fact that the three first row hydrides, water, ammonia and hydrogen fluoride, have melting and boiling points which are somewhat higher than would be expected (Fig. 18.4).

Hydrogen bond formation also accounts for many solubility effects. For instance methane is almost insoluble in water, while ammonia is highly soluble (one volume of water can dissolve about 1000 volumes of gas), and liquid hydrogen fluoride is miscible in all proportions.

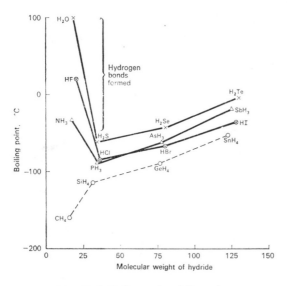

FIG. 18.4. Hydrogen bond formation.

Water and hydrogen fluoride, and to a lesser extent ammonia, show strong evidence of **association** in the liquid state. Association is the joining together of two or more molecules, and is really a similar effect to the behaviour

of H_2O molecules in ice. Liquids are best regarded as *broken-down solids*, and liquid water for instance consists of small associated units $(H_2O)_x$, where $x =$ about 6, which have structures akin to ice. Hydrogen fluoride also associates in the liquid state, and $(HF)_x$ chains exist in the vapour just above the boiling point.

Hydrogen bonding between hydrogen and fluorine leads to the formation of an ion HF_2^- in solid salts, and trace amounts in liquid hydrogen fluoride. This ion is found to be linear, with equal H—F bond lengths:

$$[F\ldots\ldots H\ldots\ldots F]^-$$

This can be regarded as a structure in which a single electron pair is shared between three atoms. The bond order of these bonds is therefore only one-half. Salts such as $KHF_2(c)$ contain this ion.

18.8 Importance of the hydrogen bond to living systems

Although the hydrogen bond is relatively weak (about 20 kJ mol^{-1} in most cases), it has a considerable effect upon the structure of substances, especially the macromolecules of living systems. Moreover when a large number of hydrogen bonds exist between molecules, the overall force can become very large indeed, making it virtually impossible to separate the molecules without breaking them up.

Many examples could be quoted concerning the structure of biological molecules, but we will consider just one, namely the structure of

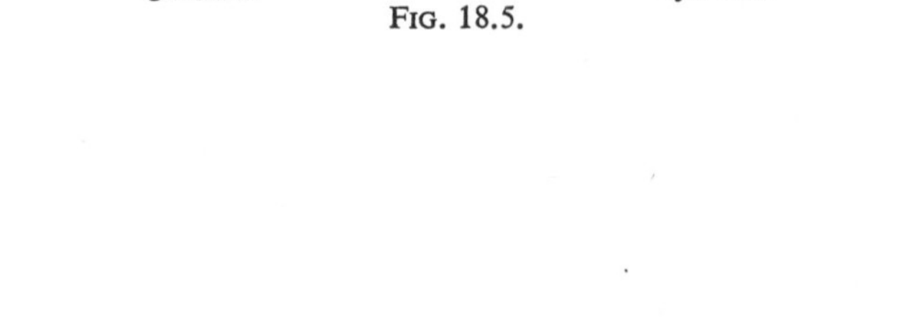

guanine   cytosine

FIG. 18.5.

the molecule of DNA (deoxyribonucleic acid), which contains all the "genetic information" required for the development of living cells. The quest for unravelling its structure still continues, but the essential features are now known, and hydrogen bonds play a very important part.

The overall shape of DNA is a double helix, that is, rather like a piece of two-core electrical "flex" (the cloth-covered variety). The links between the two molecular strands are provided by hydrogen bonds between pairs of organic bases (Fig. 18.5).

Although a single hydrogen bond is weak, it is very difficult to pull apart the two strands of DNA chain from the double helix. If one or other side is momentarily broken away, the structure soon repairs itself as suitable molecules come along which can "fit" the structure. Provided both strands of the helix do not break simultaneously at the same base pair, the molecule will hold together. Thus the mechanism of life itself is seen to depend on such simple things as the hydrogen bond.

Study Questions

1. Starting with heavy water, D_2O, suggest how the following compounds might be prepared

(a) NaD, (b) ND_3, (c) C_2D_2, (d) DCl, (e) $LiAlD_4$.

2. What would you expect the formulae and properties of the simplest hydrides of the following elements to be?

(a) K, (b) Ni, (c) Ge, (d) Se, (e) Ne.

3. Using Chapter 13 and this chapter, suggest whether molecular hydrogen might be an effective reducing agent for the oxides of
(a) silver, (b) sodium, (c) zinc, (d) nickel.

4. Why is the boiling point of water (100°C) so much greater than the boiling points of HF (19°C) and NH_3 (−33°C)?

5. Account for the following observations:
(a) When fused calcium hydride is electrolysed, hydrogen is obtained at the anode.
(b) Calcium hydride reacts with water, giving off a gas.
(c) HCl(1) is a poor conductor of electricity, yet HCl(aq) is an excellent electrical conductor.

6. Like the alkali metals (Chapter 20), hydrogen only has one outer electron. In what ways are the chemistries of hydrogen and the alkali metals (a) similar, (b) different?

7. The element boron forms a number of hydrides: one of these contains 21·8% H and 78·2% B and has a molecular weight of 27·65.

(a) What is the empirical formula of the hydride?
(b) What is the molecular formula of the hydride?

(c) Can you suggest a structure for this compound that is consistent with your theories of bonding?

8. When monosilane, SiH_4, is treated with strong alkali, a gas that burns with a slight explosion is formed. 0·8 g of silane gave 2240 cm^3 of the gas (at s.t.p.).

(a) How many moles of silane gave how many moles of the gas?
(b) What is the gas?
(c) Write an equation for the reaction between SiH_4 and NaOH(aq).

9. Write down equations for the formation of:

(a) H_2S from CaS.
(b) NH_3 from magnesium nitride.
(c) Acetylene from calcium carbide.
(d) Si_3H_8 from magnesium silicide.

10. When hydrogen was passed over a heated red powder, the powder turned into a liquid looking like mercury, which however solidified as it was cooled. 6·85 g of the powder gave 6·21 g of the metal. As the reaction took place, condensation was observed on a cold surface of the reaction tube.

(a) Suggest what the metal was. How many g-atoms of it were formed?
(b) Suggest what the condensate was.
(c) What is the empirical formula of the red powder?
(d) What is the trivial name of the red powder?
(e) What is the systematic name of the red powder?

11. In what ways does life depend on the phenomenon of hydrogen bonding?

CHAPTER 19

Acids

19.1 Acids as solutions of $H^+(aq)$

In the previous chapter an aqueous solution was said to be acidic if its hydrogen ion concentration, $[H^+(aq)]$, was greater than that in pure water, ($>$ about 10^{-7} mol dm^{-3} at room temperature). Such as definition of *acidic* is commonly accepted, but a more general definition is required in order to deal with non-aqueous and heterogeneous systems. The purpose of this chapter is first, to examine some of the properties of acids, and second to consider some of the theories of acid–base behaviour.

The familiar dilute acids in the laboratory have one factor in common: they are all the result of adding a substance to water, giving rise to an equilibrium involving hydrogen ions. They vary in equilibrium position however, the so-called strong acids having an equilibrium composition consisting almost entirely ($>$ about 99·9%) of ions. The weak acids, on the other hand, may be less than about 1% ionized yet still give rise to a concentration of $H^+(aq)$ noticeably higher than that of pure water. Table 19.1 gives a selection of familiar acids in aqueous solution.

A substance which dissolves in water to give a high percentage of ions is termed a **strong electrolyte.** A substance which only ionizes to a small extent is termed a **weak electrolyte.** The equilibrium constant here is termed the **dissociation constant.** For acids, it is generally given the symbol K_a.

TABLE 19.1

Name	Formula of acid	Equation for reaction with water	Approximate value of K_a at room temperature	Approximate % present as ions in M solution	Classified as
Nitric acid	$HNO_3(l)$	$HNO_3+aq \rightleftharpoons H^+(aq)+NO_3^-(aq)$	large	100	strong
Hydrogen chloride	$HCl(g)$	$HCl(g)+aq \rightleftharpoons H^+(aq)+Cl^-(aq)$	large	100	strong
Sulphuric acid	$H_2SO_4(l)$	$H_2SO_4+aq \rightleftharpoons H^+(aq)+HSO_4^-(aq)$	large	100	fairly strong
Hydrogen fluoride	$HF(g)$ or (l)	$HF(g)+aq \rightleftharpoons H^+(aq)+F^-(aq)$	$6·7\times10^{-4}$	2·7	weak
Acetic acid	CH_3COOH (l) or (c)	$CH_3COOH(l)+aq \rightleftharpoons H^+(aq)+ CH_3COO^-(aq)$	$1·8\times10^{-3}$	0·4	weak
Hydrogen sulphide	$H_2S(g)$	$H_2S(g)+aq \rightleftharpoons H^+(aq)+HS^-(aq)$	$1·0\times10^{-7}$	0·03	very weak

19.2 Ionic product of water

Water auto-ionizes (Chapter 18) to an extent which varies with temperature. The product $[H^+]\times[OH^-]$ is a constant termed the ionic product of water, and is given the symbol K_w. Figure 19.1 is a plot which shows how K_w

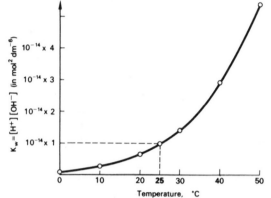

FIG. 19.1. Variation of K_w with temperature.

varies with temperature. The variation is consistent with le Chatelier's principle, the ionization being endothermic.

$$H_2O(l) \rightleftharpoons H^+(aq)+OH^-(aq);$$
$$\Delta H = +58 \text{ kJ.}$$

$[H^+]$ denotes the *activity* of $H^+(aq)$ ions in this expression, but this is numerically equal to the concentration provided the latter is low. At higher concentrations, it is only permissible to treat the two quantities as numerically equal to within an order of magnitude. A few worked examples will make this point clearer.

Worked Example 1. Calculate $[H^+]$ in a solution of M HF(aq), at room temperature.

From the data in Table 19.1,

$$K_a = \frac{[H^+][F^-]}{[HF]} = 6.7\times10^{-4}$$

Inspection of this expression shows the numerator to be far smaller than the denominator,

indicating that the hydrogen fluoride is present mainly as molecules rather than ions. We may therefore write $[HF] \simeq 1$ mol dm^{-3}.

$$\therefore [H^+][F^-] = 6.7\times10^{-4}.$$

For every positive ion formed, there will be one negative ion, and therefore

$$[H^+] = [F^-].$$
$$\therefore [H^+]^2 = 6.7\times10^{-4}$$
$$[H^+] = \sqrt{6.7\times10^{-4}}$$
$$= 2.6\times10^{-2} \text{ mol dm}^{-3}.$$

Notes:

(a) Although hydrogen fluoride is a weak acid, $[H^+]$ has a considerably greater value than in pure water.

(b) When substituting concentration values in the expression for K_a, we have made the assumption that activity = concentration. The calculation is therefore only approximate.

(c) The fact that the calculation is in any case approximate, is some justification for the initial assumption that $[HF] = 1$. Working back, we now see that a value of $[HF] = 1 -(2.6\times10^{-2})$ $= 0.974$ would have been more valid.

Worked Example 2. Calculate $[H^+]$ for a solution of M $H_2SO_4(aq)$, assuming that both H_2SO_4 and HSO_4^- are strong acids.

The above assumptions state that the following equilibria lie mainly to the right:

$$H_2SO_4+aq \rightleftharpoons H^+(aq)+HSO_4^-(aq)$$
$$HSO_4^-(aq) \rightleftharpoons H^+(aq)+SO_4^{2-}(aq)$$

Hence one mole of H_2SO_4 produces *two* moles of $H^+(aq)$ in the equation. Such an acid is called a **dibasic acid.**

Since the equilibria are almost 100% to the right,

$$[H^+] = 2 \text{ mol dm}^{-3}$$

Note: the **basicity** of an acid is defined as the number of moles of hydrogen ion given by one mole of the acid in the equation.

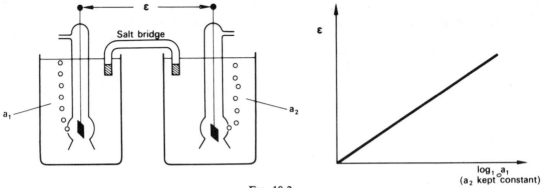

FIG. 19.2.

19.3 Hydrogen ion concentration and activity

It is not easy to obtain a direct measurement of the hydrogen ion *concentration in* a solution, but it is easy to measure its *activity* by means of a cell. If two hydrogen electrodes are coupled together as in Fig. 19.2, the e.m.f. ε is related to the ratio of activities of hydrogen ion, a_1 and a_2 (section 11.7).

$$\varepsilon = \frac{RT}{nF} \log_e \frac{a_1}{a_2} = \frac{2 \cdot 303\,RT}{nF} \log_{10} \frac{a_1}{a_2}$$

In conditions where the ionic concentration is very low, we may write $a_1 = c_1$ and $a_2 = c_2$

(where c_1, c_2 are the concentrations of hydrogen ion). Provided the activity or concentration of one dilute solution is known, a value for the other solution may be calculated from e.m.f. measurement.

If the concentration of solution in one cell is kept constant, and the other varied, a graph of e.m.f. against logarithm of concentration can be plotted. This is a straight line, on account of the expression above, provided concentrations are kept low.

19.4 The normal hydrogen electrode and the glass electrode

The normal hydrogen electrode was referred to in Chapter 11. It is the reference electrode to which all quoted values for ε° are related. Measurements of e.m.f. show that for a solution of dilute hydrochloric acid to have a hydrogen ion activity of 1, the concentration of $[H^+] = 1 \cdot 18$ mol dm^{-3}. This is a good illustration of how widely numerical values of concentrations and activities can differ. The divergence here is 18 per cent.

Hydrogen electrodes are inconvenient to operate for routine measurements of $[H^+]$, and a much simpler alternative is the **glass electrode** (Fig. 19.3).

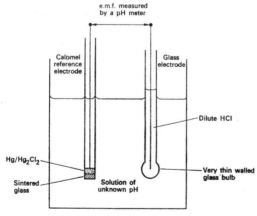

FIG. 19.3. pH measurement with a glass electrode.

Glass has the property of being slightly permeable to hydrogen ions, and the potential difference between the inner and outer surfaces depends upon the relative concentrations of hydrogen ion in contact with these surfaces. It is found experimentally that if a glass electrode is coupled up with a reference electrode unaffected by hydrogen ions, a cell is formed whose e.m.f. is directly proportional to $\log_{10}$ [H⁺], just as with an ordinary hydrogen electrode. This is very convenient, for it means that a glass electrode can be used in place of a hydrogen electrode for routine measurements of [H⁺]. It does however need calibrating before use, and this is described in a later section.

19.5 The pH scale

Since e.m.f. measurements are related to $\log_{10}$ [H⁺], and not directly to [H⁺] itself, hydrogen ion concentrations are customarily expressed as pH values, where

$$\text{pH} = -\log_{10} [\text{H}^+].$$

The term pH was coined by Sörensen in 1909, and the notation originally stood for "potential of hydrogen". The "p" notation can be applied to other quantities and always means "minus $\log_{10}$ of". For instance,

$$\text{pOH} = -\log_{10} [\text{OH}^-]$$
$$\text{p}K_w = -\log_{10} K_w, \quad \text{and so on.}$$

Worked Example 3. Express the following values of [H⁺] as pH.

(a) [H⁺] = 1 mol dm⁻³
    pH = $-\log_{10} 1 = 0$
(b) [H⁺] = 10^{-14} mol dm⁻³
    pH = $-\log_{10} 10^{-14} = -(-14) = +14$
(c) [H⁺] = 2 mol dm⁻³
    pH = $-\log_{10} 2 = -0.3010$
(d) [H⁺] = 2×10^{-5} mol dm⁻³
    pH = $-(5.3010)$
       = $+5 - 0.3010$
       = $+4.699.$

Worked Example 4. Express the following in 'p' notation.

(a) [OH⁻] = 10^{-3} mol dm⁻³
    pOH = $-\log_{10} [\text{OH}^-] = +3.$
(b) $K_w = 10^{-14}$ at room temperature.
    p$K_w = -\log_{10} (10^{-14}) = +14.$
(c) K_a for acetic acid = 2×10^{-5} mol dm⁻³ at room temperature.
    p$K_a = 4.699$ (cf. Example 3(d) above).

Worked Example 5. Calculate the pH of the following solutions at room temperature.

(a) M H₂SO₄(aq).
    From Worked Example 2,
    [H⁺] = 2 mol dm⁻³.
    From Worked Example 3(c), pH = -0.301

(b) Pure water.
    $K_w = [\text{H}^+] [\text{OH}^-] = 10^{-14}$, and
    [H⁺] = [OH⁻].
    $\therefore$ [H⁺] = $\sqrt{K_w} = 10^{-7}$
    $\therefore$ pH = $+7.$

(c) M NaOH(aq).
    Sodium hydroxide is a strong electrolyte, $\therefore$ [OH⁻] = 1 mol dm⁻³
    Using the "p" notation,
    pK_w = pH + pOH
    Now, pOH = 0 in this example, and pK_w = 14.
    $\therefore$ pH = 14. Note that even this solution, which is strongly alkaline, has still some H⁺(aq) ions present, although [H⁺] = 10^{-14} mol dm⁻³

(d) 0.01 M Ca(OH)₂(aq).
    Calcium hydroxide is a strong electrolyte, and from its formula, will give two moles of OH⁻(aq) per mole of Ca(OH)₂.
    [OH⁻] = 0.02 mol dm⁻³
    pOH = $-(\bar{2}.301) = +1.699$
    pH = pK_w − pOH = $14 - 1.699$
       = $12.301.$

(e) M $CH_3COOH(aq)$.

($K_a = 2 \times 10^{-5}$ mol dm^{-3} approximately, at room temperature).

$$K_a = \frac{[H^+][CH_3COO^-]}{CH_3COOH]} = 2 \times 10^{-5}$$

$[H^+]^2 = 2 \times 10^{-5}$ (cf. the same argument in Worked Example 1).

$\therefore$ pH $= \frac{1}{2}pK_a = \frac{1}{2}(4.699) = 2.35$.

19.6 Buffer solutions

A **buffer solution** is one whose pH is relatively insensitive to the addition of small quantities of acid or alkali. Solutions vary widely in their pH sensitivity. Pure water, and pure NaCl(aq), are very sensitive: a litre of pure water would show a marked change (± 2 pH units) on adding only 0.01 cm^3 of molar H^+(aq) or OH^-(aq). In some chemical reactions it is necessary to stabilize the pH, and this can be done by adding substances which can absorb variations in $[H^+]$.

A typical acidic buffer solution is a mixture of a weak acid with a solution of its anion, for example acetic acid plus sodium acetate. The concentrations of $[CH_3COOH]$ and $[CH_3COO^-]$ are relatively high compared to $[H^+]$, the three concentrations being related by the expression for the dissociation constant of the acid:

$$K_a = \frac{\overset{\text{relatively large}}{[CH_3COO^-]}\,\overset{\text{small}}{[H^+]}}{\underset{\substack{\text{relatively} \\ \text{large}}}{[CH_3COOH]}} = 2 \times 10^{-5} \text{ mol dm}^{-3}$$

approximately, at room temperature.

Small additions of H^+(aq) are immediately taken up by a shift of the equilibrium

$$CH_3COOH(aq) \rightleftharpoons CH_3COO^-(aq) + H^+(aq)$$

to the left, and small additions of OH^-(aq) by a shift to the right. As long as $[CH_3COOH]$ and $[CH_3COO^-]$ are relatively large, their magnitude will not be much affected by the added acid

or alkali, and hence when equilibrium is re-established, pH will be almost the same as it was initially.

Worked Example 6. Calculate the pH of the following buffer solutions.

(a) A mixture which is molar with respect to both $CH_3COOH(aq)$ and $CH_3COONa(aq)$.

Sodium acetate is a strong electrolyte, and therefore $[CH_3COO^-] = 1$. Acetic acid is a weak electrolyte, and therefore $[CH_3COOH] = 1$. Substituting,

$$\frac{1 \times [H^+]}{1} = K_a$$

$\therefore$ pH $= pK_a$

$= 4.699$ (cf. Worked Example 4).

(b) A mixture which is 0.1 M with respect to $CH_3COOH(aq)$ and molar with respect to CH_3COONa.

$$\frac{1 \times [H^+]}{0.1} = K_a$$

$\therefore$ pH $= pK_a + 1 = 5.699$.

A buffer which has an alkaline pH can be made up by mixing a weak alkali such as ammonia with one of its salts, such as ammonium chloride (section 19.13).

19.7 Methods of measuring pH

(a) BY pH METER

The most precise way of measuring pH is by means of a glass electrode. This is set up with a reference electrode whose half-cell potential is not affected by pH. One of the features of a glass electrode is its very high electrical resistance ($\sim 10^8$ Ω), rendering measurements with a conventional voltmeter useless. A d.c. amplifier is needed, embodying also a potentiometer which enables the e.m.f. to be checked against that of a standard cell (usually a mercury cell).

The e.m.f. of the cell is amplified and fed to the voltmeter. Voltage is related linearly to pH. Simple meters, accurate to about ± 0.1 pH unit, have a direct readout of pH on the scale, while other systems use a null method which can give an accuracy of three places of decimals.

The procedure for determining pH with a pH meter is as follows:

(i) Place the pH electrodes in a standard buffer solution, and standardize its e.m.f. against the standard cell provided. The exact procedure varies from instrument to instrument, but the purpose of this operation is to allow for day to day variations in the characteristics of the glass electrode. The meter is set so that the pH it reads corresponds to that of the standard buffer.

(ii) The meter is now ready for use. The electrodes can now be placed in a solution of unknown pH.

(b) INDICATOR METHOD

This method is generally only accurate to about ± 0.5 pH unit. With special precautions, about ± 0.1 pH unit can be achieved. The simplest application is the use of **universal indicator** which is a mixture of indicators of varying pK_a values, and different colours. A universal indicator shows a different colour for each pH value over a wide range.

The action of indicators is best understood by taking a familiar example, methyl orange, which is yellow in alkaline solution but turns red when $H^+(aq)$ is added*, due to the equilibrium

$$HX(aq) \rightleftharpoons H^+(aq) + X^-(aq);$$
$$\text{red} \qquad\qquad\qquad\qquad \text{yellow}$$

$$K_a = \frac{[H^+][X^-]}{[HX]} = 2 \times 10^{-4} \text{ mol dm}^{-3} \text{ approx.}$$

---

* In the laboratory it is more usual to use screened methyl orange, which is a mixture of two dyes, methyl orange and xylene cyanol FF. This mixture is designed to give a neutral grey at the point where $pH = pK_a$.

Substituting in the expression for K, we see that, to obtain a neutral colour, where $[HX] \simeq [X^-]$, it is necessary for $pH = pK_a$ (cf. Worked Example 6(a)). Methyl orange shows its halfway tint at about $pH = 3.7$.

If the pH is altered by one unit in the acidic direction, we now have $[HX] = 10 \times [X^-]$ if the expression for K_a is to be satisfied (cf. Worked Example 6(b)). Greater change in pH than this will not result in much visible change as the eye cannot detect changes of one coloured species in the presence of a large excess of the other. A similar argument applies to pH change in the alkaline direction. At a pH of 4.7, $[X^-] = 10 \times [HX]$ and the solution will be almost completely yellow. The pH range of methyl orange is therefore approximately from 2.7 to 4.7.

Table 19.2 shows the data for some commonly used indicators.

TABLE 19.2

Indicator	Colour		pK_a	pH range approximately
	acid	alkali		
Methyl orange	red	yellow	3·7	2·7— 3·7
Methyl red	red	yellow	5·1	4·1— 6·1
Bromothymol blue	yellow	blue	7·0	6·0— 8·0
Phenol-phthalein	colourless	red	9·4	8·4—10·4

19.8 Alkalis and bases

In just the same way as a solution with $pH < 7$ is defined loosely as acidic, a solution with $pH > 7$ is said to be **alkaline.** Such a solution will have a concentration of $OH^-(aq)$ greater than the value for pure water at room temperature, namely 10^{-7} mol dm^{-3}. That is, $pOH < 7$. (Remember that $pH + pOH = pK_w = 14$.)

Hydroxides of the s-block metals, Group IA and IIA, are soluble in water forming metal ions and hydroxide ions (with the exception of beryllium hydroxide which is almost insoluble), and hence their solutions are alkaline.

Group IA: $MOH(c) \rightleftharpoons M^+(aq) + OH^-(aq)$

$(M = Li, Na, K, Rb, Cs \text{ or } Fr)$

Group IIA: $M(OH)_2(c) \rightleftharpoons$

$M^{2+}(aq) + 2OH^-(aq)$

$(M = Mg, Ca, Sr, Ba \text{ or } Ra)$

In fact magnesium hydroxide is only sparingly soluble, with a solubility product, K_s (section 10.12), of only 2×10^{-11} at 25°C. This means that a saturated solution of magnesium hydroxide has a relatively low pH at room temperature, and it may be classed as a weak alkali.

If an alkali is added to an aqueous acid, the reaction which takes place is termed **neutralization.** It is essentially

$$H^+(aq) + OH^-(aq) \longrightarrow H_2O(l)$$

This reaction is commonly used to produce salts, for when it occurs the *spectator ions* remain in solution and can be crystallized out. For example:

(i) $HCl(aq) + NaOH(aq)$

$\equiv \underbrace{H^+(aq) + OH^-(aq)}$

$\downarrow$

H_2O

$+ \underbrace{Na^+(aq) + Cl^-(aq)}_{\text{spectator ions}}$

$\downarrow$

$NaCl(c)$ on crystallization

(ii) $2 HNO_3(aq) + Ba(OH)_2(aq)$

$\equiv \underbrace{2H^+(aq) + 2OH^-(aq)}$

$\downarrow$

$2 H_2O$

$+ \underbrace{Ba^{2+}(aq) + 2NO_3^-(aq)}$

$\downarrow$

$Ba(NO_3)_2(c)$

Many metal hydroxides and oxides are insoluble in water, but can nevertheless neutralize acids in a similar way. Such substances are termed **bases** (basic hydroxides and basic oxides). Alkalis are therefore soluble bases. An example of a neutralization with an insoluble base would be

(iii) $H_2SO_4(aq) + Cu(OH)_2(c) \longrightarrow$

$2 H_2O(l) + \underbrace{Cu^{2+}(aq) + SO_4^{2-}(aq)}$

$\downarrow$

$CuSO_4 . 5H_2O(c)$

on crystallization

The only true spectator ion here is SO_4^{2-}—it is the only one which does not actually participate in the reaction—though it is conveniently written into the equation. Any aqueous acid will react with a base producing the ions of the metal. The above can be written

$2 H^+(aq) + Cu(OH)_2(c) \longrightarrow$

$2 H_2O(l) + Cu^{2+}(aq)$

Worked Example 7. Calculate the pH of the following aqueous solutions of bases:

(a) M NaOH(aq)

$[OH^-] = 1 \text{ mol dm}^{-3}$

$\therefore \ pOH = 0$

$\therefore \ pH = 14 - 0 = 14$

(b) 0·1 M $Sr(OH)_2(aq)$.

One mole of $Sr(OH)_2$ gives two moles of $OH^-(aq)$, therefore

$[OH^-] = 2 \times 10^{-1} \text{ mol dm}^{-3}$

$\therefore \ \log_{10}[OH^-] = \bar{1}·301$

$\therefore \ pOH = 1 - 0·301 = 0·699$

$\therefore \ pH = 14 - 0·699 = 13·301$

Worked Example 8. Calculate the pH of the saturated solution formed by adding MgO(c) to water at 25°C.

When the hydroxide is appreciably soluble in water, the oxide will react with water thus:

$$MgO(c) + H_2O(l) \longrightarrow Mg^{2+}(aq) + 2\ OH^-(aq)$$

$$\begin{array}{cc} x & 2x \\ \text{mole dm}^{-3} & \text{mole dm}^{-3} \end{array}$$

From the data given,

$$K_s = [Mg^{2+}][OH^-]^2 = x.(2x)^2 = 4x^3$$
$$= 2 \times 10^{-11}$$

Hence $x = 1 \cdot 7 \times 10^{-4}$ mol dm^{-3}

Hence pOH $= -\log_{10}[OH^-]$
$$= -\log_{10} 1 \cdot 7 \times 10^{-4} = 3 \cdot 77$$

Hence pH $= 14 - \text{pOH}$
$$= 10 \cdot 23.$$

Many other substances react with water to give solutions which are alkaline, for instance ammonia.

$$NH_3(aq) + H_2O(l) \rightleftharpoons NH_4^+(aq) + OH^-(aq)$$
$$\begin{array}{cc} \text{ammonia} & \text{ammonium} \\ & \text{ions} \end{array}$$

"Household ammonia" is in fact an aqueous solution of ammonia, in which the above species are present at equilibrium. *Ammonium hydroxide* is the name often given to this solution, but strictly this name should only apply to the species on the right-hand side of the equation.

Worked Example 9. Calculate the pH of a 0·1 M solution of NH$_3$(aq), given that the equilibrium constant, K_b, is 2×10^{-5} mol dm^{-3} at room temperature.

$$K_b = \frac{[NH_4^+][OH^-]}{[NH_3]} = 2 \times 10^{-5}, \quad \text{where square}$$

brackets here denote activities.

The expression may therefore be rewritten,

$$\frac{[NH_4^+][OH^-]}{[NH_3]} = \frac{[NH_4^+][OH^-]}{0 \cdot 1}$$
$$= \frac{[OH^-]^2}{0 \cdot 1} = 2 \times 10^{-5}$$

$$\therefore [OH^-]^2 = 2 \times 10^{-6}$$
$$\therefore [OH^-] = (2 \times 10^{-6})^{1/2}$$
$$\therefore \text{pOH} = -\log_{10} [OH^-]$$
$$= -\tfrac{1}{2} \log_{10} (2 \times 10^{-6})$$
$$= -\tfrac{1}{2} (\bar{6} \cdot 3010)$$
$$= +2 \cdot 85 \text{ approximately}$$
$$\therefore \text{pH} = 14 - 2 \cdot 85$$
$$= 11 \cdot 15$$

A 0·1 M solution of NH$_3$(aq) is therefore less strongly alkaline than a 0·1 M solution of NaOH(aq) (Worked Example 7).

The constant K_b is termed the **dissociation constant** of the weak base, ammonia.

19.9 Applications of pH measurement

(a) **Acid–base titrations.** The end-point of an acid–base titration occurs when there is an abrupt change of pH on adding solution from the burette. An indicator registers the colour change. It is essential to choose an indicator with the correct pK value, except when titrating a strong acid with a strong alkali, when the pH swing is large. The theory of indicators for titrations is dealt with further in section 19.14.

(b) **Soil testing.** Most plants will only grow in soils with pH values between about 6 and 8. Within these limits, however, pH can have a marked effect on the type of plant which is favoured. Soil testing is most conveniently carried out using universal indicator paper or solution, with a colour range specially chosen for the pH values encountered. A portable pH meter, incorporating a glass electrode, is sometimes used, and the soil is shaken with distilled water before testing.

(c) **Reactions in solution.** For some aqueous reactions the pH must be adjusted to within certain limits, such as in the preparation of some complex salts. Universal indicator paper is usually quite accurate enough for this.

(d) **Measurement of dissociation constants.** If the pH of a solution of a weak acid or base is accurately measured with a pH meter, the dissociation constant can be calculated.

Worked Example 10. A 0·1 M solution of formic acid, HCOOH(aq), is found to have a pH of 2·4 at 25°C. Calculate its dissociation constant at that temperature.

$$K_a = \frac{[H^+][HCOO^-]}{[HCOOH]} = \frac{[H^+]^2}{[HCOOH]}$$

$$\therefore \ pK_a = 2pH - p[HCOOH]$$

$$\therefore \ pK_a = 4·8 - 1 = 3·8$$

$$\therefore \ \ K_a = 1·6 \times 10^{-4} \ mol \ dm^{-3}$$

This method is not practicable for measuring the dissociation constant of a strong acid, beyond establishing that the value is large. For instance it makes very little difference to the pH of a solution of an acid HX(aq) whether $K_a = 10^{10}$ or 10^{20}. In both cases, for practical purposes the acid is fully (>99·99%) ionized in solution.

19.10 Oxo-acids and oxo-anions

Most of the non-metals form **oxo-acids.** With some elements, such as sulphur and phosphorus, a large number of different oxo-acids can form with various structures and oxidation numbers, but in other cases fewer species are known. The well-known mineral acids HNO_3, nitric acid and H_2SO_4, sulphuric acid, are oxo-acids of nitrogen and sulphur respectively.

Table 19.3 shows oxo-acids in which the oxidation number of the central non-metallic element is equal to the number of the group in the periodic table to which that element belongs.

When an element forms oxo-acids of more than one oxidation number, the ending *-ic* indicates a higher oxidation state, and *-ous* a lower oxidation state. In the case of elements like chlorine, which form a whole series of oxo-acids, prefixes *per-* and *hypo-* have to be added. Table 19.4 illustrates this for the oxo-acids of chlorine.

The strengths of oxo-acids appear to be governed by an empirical rule. If the formula is written out structurally, as $(HO)_n XO_m$, then

$m = 0$: weak acid, pK_a generally about 10;

$m = 1$: fairly weak acid, pK_a generally between 0 and 5;

$m = 2$: moderately strong acid, pK_a generally between -3 and 0;

$m = 3$: a strong acid, practically fully ionized in aqueous solution.

TABLE 19.3

	Group III	Group IV	Group V	Group VI	Group VII
First short period	boric acid $HO-B\begin{smallmatrix}\diagup OH\\ \diagdown OH\end{smallmatrix}$ planar	carbonic acid $O{=}C\begin{smallmatrix}\diagup OH\\ \diagdown OH\end{smallmatrix}$ planar	nitric acid $O{=}N\begin{smallmatrix}\diagup O\\ \diagdown OH\end{smallmatrix}$ planar	–	–
Second short period	–	silicic acid $SiO_2, \ xH_2O$ non-stoichiometric, complex structure	phosphoric acid $\begin{smallmatrix}O\diagdown \ \diagup OH\\ P\\ HO\diagup \diagdown OH\end{smallmatrix}$ tetrahedral	sulphuric acid $\begin{smallmatrix}O\diagdown \ \diagup OH\\ S\\ O\diagup \diagdown OH\end{smallmatrix}$ tetrahedral	perchloric acid $\begin{smallmatrix}O\diagdown \ \diagup O\\ Cl\\ O\diagup \diagdown OH\end{smallmatrix}$ tetrahedral

TABLE 19.4

Formula of acid	Name of acid	Formula of anion	Name of anion
$HClO$	Hypochlorous	ClO^-	Hypochlorite
$HClO_2$	Chlorous	ClO_2^-	Chlorite
$HClO_3$	Chloric	ClO_3^-	Chlorate
$HClO_4$	Perchloric	ClO_4^-	Perchlorate

These predictions fit the data for the oxo-acids of chlorine—hypochlorous acid ($m=0$) is a weak acid, while perchloric acid ($m=3$) is one of the strongest acids known. Sulphuric acid ($m=2$) is only moderately strong: although almost fully ionized in aqueous solution, it behaves as a weak acid when dissolved in acetic acid.

Oxo-acids and oxo-ions are also formed by some metals, notably the higher oxidation states of the transition metals. Table 19.5 gives some instances of this.

TABLE 19.5

Name of acid	Formula of anion	Name of anion
Permanganic acid (stable in solution only)	MnO_4^-	Permanganate [manganate(VII)]
Chromic acid	$Cr_2O_7^{2-}$ CrO_4^{2-}	Dichromate Chromate [chromate(VI)]

19.11 Non-aqueous systems

Acid-base reactions are readily carried out in solvents other than water, and the results are often interesting. For instance, acetic acid, CH_3COOH (m.p. 16·6°C) is a solvent which auto-ionizes like water, giving solvated protons and solvated anions. Compare the two processes:

$$CH_3COOH(l) \rightleftharpoons H^+(ac) + CH_3COO^-(ac)$$
$$H_2O(l) \rightleftharpoons H^+(aq) + OH^-(aq)$$

Evidence for this is provided by the fact that pure acetic acid, like water, has a slight electrical conductivity. Perchloric acid, $HClO_4$, is a strong electrolyte when dissolved in acetic acid, and so also is sodium acetate.[*]

$$HClO_4(l) + ac \longrightarrow H^+(ac) + ClO_4^-(ac)$$
$$CH_3COONa(c) + ac \longrightarrow$$
$$Na^+(ac) + CH_3COO^-(ac)$$

These solutions can be titrated in a manner exactly analogous to an aqueous titration. Methyl violet is a suitable indicator.

$$H^+(ac) + CH_3COO^-(ac) \longrightarrow CH_3COOH(l)$$
(spectator ions, $Na^+(ac)$, $ClO_4^-(ac)$)

In general, acids ionize less strongly in glacial acetic acid than they would in water. For instance, sulphuric acid can be classed as a strong acid in aqueous solution, but is relatively weak in acetic acid.

Another solvent which can be used in place of water is liquid ammonia, b.p. $-33°C$. Liquid ammonia auto-ionizes thus:

$$2 NH_3 \rightleftharpoons NH_4^+ + NH_2^-.$$

An ammonium salt is a typical acid in liquid ammonia, as the ammonium ion is effectively a solvated proton, $H^+ + NH_3$. A typical base would be a metal *amide*, for instance sodium amide, $NaNH_2(c)$.

Non-aqueous reactions are often carried out

---

[*] The state symbol (ac) here denotes ions *solvated* with acetic acid, in the same way as (aq) denotes ions *hydrated* with water molecules.

when the reactants are attacked by water, or fail to dissolve in it, and non-aqueous titrations are often used industrially.

19.12 The Brønsted-Lowry definition of acid–base behaviour

The fact that reactions analogous to aqueous acid-base neutralization can also take place in non-aqueous media such as acetic acid or liquid ammonia, means that a more general definition of acid and base is required. Brønsted in 1923 defined an acid as **any substance with a tendency to release a proton.** Conversely a species which tends to accept a proton is defined as a base. When an acid releases a proton, the **conjugate base** is said to be formed, for instance:

$$CH_3COOH \rightleftharpoons H^+ + CH_3COO^-$$

acid   conjugate base

An acid with a strong tendency to release protons will have a weak conjugate base. Conversely if a base has a strong tendency to accept protons, its conjugate acid must be weak. Thus, on the Brønsted definition acetate ions are relatively strongly basic since acetic acid itself is not a very strong acid.

It is possible to list acids in order of decreasing strength. Table 19.6 shows a list of half-equations arranged empirically in such an order. These are *not* complete chemical reactions but are analogous to electronic half-equations used to describe redox processes. The equation for an acid-base reaction is derived by combining two half-equations.

The position of equilibrium is determined by the relative strengths of the species involved. For instance perchloric acid has a stronger tendency than H_3O^+ ($= H^+(aq)$) to give up a proton, and H_2O has a stronger tendency to accept protons than ClO_4^-. Consequently the following equilibrium lies to the right:

$$HClO_4 + H_2O \rightleftharpoons H_3O^+ + ClO_4^-$$

strong   base   weaker acid   weaker base
acid   than ClO_4^- than H_2O

TABLE 19.6
SOME HALF-EQUATIONS FOR BRØNSTED ACIDS

	Acid	$\rightleftharpoons$ Conjugate base	+ proton	
decreasing strength of acid	HClO$_4$	$\rightleftharpoons$ ClO$_4^-$	$+ H^+$	increasing strength of base
	HNO$_3$	$\rightleftharpoons$ NO$_3^-$	$+ H^+$	
	H$_2$SO$_4$	$\rightleftharpoons$ HSO$_4^-$	$+ H^+$	
	H$_3$O^{+*}	$\rightleftharpoons$ H$_2$O	$+ H^+$	
	CH$_3$COOH	$\rightleftharpoons$ CH$_3$COO$^-$	$+ H^+$	
	NH$_4^+$	$\rightleftharpoons$ NH$_3$	$+ H^+$	
	H$_2$O	$\rightleftharpoons$ OH$^-$	$+ H^+$	
	NH$_3$	$\rightleftharpoons$ NH$_2^-$	$+ H^+$	

* It should be noted that we are writing H_3O^+ here where previously we have been writing $H^+(aq)$. $H^+(aq)$ is more correct, but H_3O^+ is more convenient in balancing equations when applying the Brønsted–Lowry definition. The student should be familiar with both methods.

This equation is nothing more than a slightly elaborate form of the equation usually written for the ionization of a strong acid when added to water:

$$HClO_4 + aq \rightleftharpoons H^+(aq) + ClO_4^-(aq)$$

The Brønsted theory accounts for what is known as the **levelling** action of water. The species $HClO_4$, HNO_3 and H_2SO_4, in Table 19.6 are all stronger acids than H_3O^+, and hence all react almost completely to give that ion. The same is true of many other acids, for instance the hydrogen halides (except HF). It makes little difference whether the equilibrium constant is 10^{10} or 10^{20}, since almost complete formation of H_3O^+ will be the result in every case (see section 19.1).

Water exerts a similar levelling action at the basic end of the scale. For instance if sodium amide is added to water, ammonia is produced:

$$NaNH_2(c) + H_2O(l) \longrightarrow NH_3(aq) + NaOH(aq)$$

Brønsted theory interprets this by saying that NH_2^- is a stronger base than OH^-, and that H_2O is a stronger acid than NH_3. The equation may be rewritten ionically thus:

$$NH_2^- + H_2O \longrightarrow NH_3 + OH^-;$$
$$\text{base}_1 \quad \text{acid}_2 \qquad \text{acid}_1 \quad \text{base}_2$$

(spectator ions, Na^+)

Note that species like H_2O and NH_3 can act as both acids and bases, since they can both donate and accept protons.

19.13 Hydrolysis of ions as an instance of acid–base behaviour

The Brønsted–Lowry theory is very successful in rationalizing many of the experimental effects observed with pH of aqueous solutions. One such effect is that commonly referred to as **hydrolysis** of salts. Hydrolysis literally means reaction with water, and two main types of effect are observed:

(i) Salts of weak acids dissolve in water to give solutions with alkaline pH. For instance sodium sulphide is quite strongly alkaline, and the solution smells noticeably of hydrogen sulphide:

$$S^{2-} + 2\,H_2O \rightleftharpoons 2\,OH^- + H_2S;$$
$$\text{base}_1 \quad \text{acid}_2 \qquad \text{base}_2 \quad \text{acid}_1$$

(spectator ions, Na^+).

Since H_2S is a weak acid, the sulphide ion is a relatively strong base, which can react with water by removing protons from it.

(ii) Solutions of ions of the weaker metals dissolve in water to give acidic solutions. This again is interpreted by postulating that the aqueous metal ions are themselves weak acids. It may seem strange at first that a metal ion can donate protons, but in fact these ions have a definite number of water molecules more or less firmly attached, and are in effect complex ions. For instance, $Al^{3+}(aq)$ consists mainly of the species $Al(H_2O)_6^{3+}$.

$$\overset{aq}{Al(H_2O)_6^{3+}} \rightleftharpoons [Al(H_2O)_5(OH)]^{2+} + H^+(aq)$$
$$\text{weak acid} \qquad \text{conjugate base} \qquad pH < 7$$

19.14 Choice of indicators for titrations

Hydrolysis of ions has a profound effect on the suitability of an indicator for a titration. The purpose of a titration is to carry out accurately a desired chemical reaction, and the indicator must be capable of detecting accurately the pH which exists at the endpoint. For instance if acetic acid is to be titrated with sodium hydroxide, the end-point will be reached when the pH is that of a solution of sodium acetate. Since acetate ions are hydrolysed, the end-point will occur at an alkaline pH. Phenolphthalein (pK = 9·6) is a suitable indicator for this titration. Conversely if a weak base is titrated with a strong acid the end-point will be acidic and methyl orange or methyl red will probably be chosen.

Figure 19.4 shows how the pH varies during different titrations. Table 19.7 summarizes the choice of indicator.

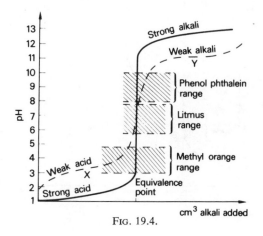

Fig. 19.4.

TABLE 19.7

	pH range of endpoint	Example
Strong acid + strong base	3—10 approx.	screened methyl orange clearest, and is unaffected by atmospheric CO_2
Strong acid + weak base	3—6 approx.	methyl orange or methyl red
Weak acid + strong base	8—10 approx.	phenolphthalein

When a polybasic acid is titrated, it is often possible to detect successive stages of the reaction, due to the different acidic strengths of the ions formed. For instance phosphoric acid can form a series of salts NaH_2PO_4 (pH of aqueous solution $\simeq$ 4·7), Na_2HPO_4 (pH $\sim$ 9·7), and Na_3PO_4(pH $\sim$ 13). An indicator such as methyl orange will change colour at the formation of NaH_2PO_4, while phenolphthalein will detect the formation of Na_2HPO_4. Figure 19.5 shows how pH varies when sodium hydroxide is added to a solution of phosphoric acid.

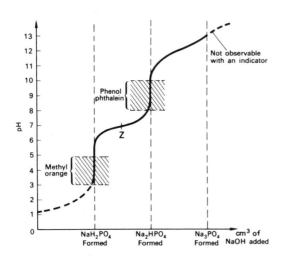

FIG. 19.5.

Carbonates can be titrated with acids if methyl orange is used as an indicator. Carbon dioxide is formed, which reacts to form a weakly acidic solution of pH approximately 4.

$$CO_2 + H_2O \rightleftharpoons H^+(aq) + HCO_3^-(aq)$$

When a weak acid is titrated with a weak base, the change in pH at the endpoint is not abrupt enough to allow a colour indicator to be used (Fig. 19.6). In this case titrations can sometimes be carried out using a recording pH meter.

Buffer solutions are solutions with compositions represented by the near-horizontal portions of the curves in Fig. 19.4 and 19.5. Examples of buffer solutions are:

(a) $CH_3COOH(aq) + CH_3COONa(aq)$;   e.g. point X in Fig. 19.4.

(b) $NH_3(aq) + NH_4Cl(aq)$; e.g. point Y in Fig. 19.4.

(c) $Na_2HPO_4(aq) + NaH_2PO_4(aq)$; e.g. point Z Fig. 19.5.

Another way of detecting end-points of acid–base reactions is by a **conductometric titration.** If a strong alkali and strong acid are titrated, for instance, the electrical conductivity will fall to a minimum at equivalence point. On either side of the endpoint there will be an excess of H^+(aq) or OH^-(aq) and the solution will have a greater conductivity. It is necessary to use an a.c. supply to avoid electrolysis occurring. Figure 19·6 shows the plot of current against cm³ of alkali added, for a typical acid–alkali titration. If either the acid or base is weak, or a precipitate is formed, a differently shaped graph will be obtained. Normally slightly curved lines, rather than straight lines, are obtained because of the increasing volume of the solution during the titration.

$$CO_2(g) + NaOH(c) \longrightarrow NaHCO_3(c);$$

$$\Delta H = -128 \text{ kJ}$$

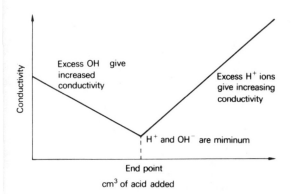

FIG. 19.6. Conductometric titration.

19.15 Limitations of the Brønsted-Lowry concept of acids

The Brønsted-Lowry theory is admirable for rationalizing acid-base reactions which occur in solution in protonic solvents such as water, ammonia and acetic acid, though it fails to include a whole series of reactions which, for convenience, are often considered to be acid-base neutralizations.

Consider for instance the reactions of **acidic oxides.** On the Brønsted–Lowry theory, one can hardly call an oxide such as $SO_3(c)$ an acid, since it does not possess any protons which it can donate. Nevertheless it reacts vigorously with water to give a solution of pH ≪ 7, sulphuric acid. Most non-metal oxides behave likewise.

Furthermore, these acidic oxides will react directly with alkalis in the absence of water. For instance, a tube containing calcium oxide or sodium hydroxide (frequently soda-lime, a mixture of $Ca(OH)_2$ and NaOH) will absorb carbon dioxide with the evolution of heat:

$$CO_2(g) + CaO(c) \longrightarrow CaCO_3(c);$$

$$\Delta H = -178 \text{ kJ}$$

An alternative concept of acid-base behaviour was put forward by Lewis, in 1938. Lewis defined a base as a substance capable of donating an electron-pair in the form of a donor bond. A Lewis acid is a substance which can accept an electron-pair by bond formation. Ammonia is a typical Lewis base, for it contains a lone-pair which can donate to a substance with vacant energy levels. (It is also a Brønsted base.) The simplest Lewis acid is the proton, for the proton readily accepts electron pairs, for example:

$$\begin{array}{ccc} H^+ & + & :NH_3 & \longrightarrow NH_4^+ \\ \text{proton} & & \text{Lewis base} \end{array}$$

| electron-pair acceptor | electron-pair donor | acid-base complex |

It follows that all Brønsted acids and bases are also Lewis acids and bases respectively. However, the term Lewis acid can be applied to substances which do not contain protons and are not therefore Brønsted acids. Another typical Lewis acid is boron(III) fluoride:

$$\begin{array}{ccc} F & H & FH \\ F:\ddot{B} & + & :\ddot{N}:H \longrightarrow F:\ddot{B}:\ddot{N}:H \\ \ddot{F} & \ddot{H} & \ddot{F}H \end{array}$$

| electron pair acceptor | electron pair donor | addition compound (adduct) |

It is usual to reserve the term *Lewis acid*, or *electron-acceptor*, for substances such as BF_3. When the word *acid* is used loosely, it can generally be taken to mean *Brønsted acid* or *proton donor*. Acidic oxides are Lewis acids, but they are not Brønsted acids.

Study Questions

1. (a) Calculate the concentration of $H^+(aq)$ in water at $0°$, $25°$ and $100°C$ if the ionic product, K_w, at these temperatures is 10^{-15}, 10^{-14} and 5×10^{-13} mol^2 dm^{-6} respectively.

(b) Calculate the pH at each temperature.

2. The pK_a values of acetic acid, monochloracetic acid, dichloracetic acid and trichloracetic acid are 4·8, 2·9, 1·3 and 0·7 respectively, at 25°C.

(a) Which is the strongest acid?

(b) Calculate the approximate pH of a molar solution of each acid, assuming the acids to be weak.

(c) Can you suggest a reason for the differences between the acids?

3. Calculate the pH values of the following solutions at 25°C. (At 25°C, $K_w = 10^{-14}$, HCl and NaOH are strong electrolytes, K_a for the monobasic formic acid $= 2 \times 10^{-4}$, K_b for ammonia $= 1·75 \times 10^{-5}$.)

(a) M HCl.   (b) 0·1 M HCl.
(c) 0·2 M HCl.   (d) 5 M HCl.
(e) M HCOOH.   (f) 0·1 M HCOOH.
(g) 0·2 M HCOOH.   (h) 5 M HCOOH.
(i) M NaOH.   (j) 0·1 M NaOH.
(k) 0·2 M NaOH.   (l) 5 M NaOH.
(m) M NH_3.   (n) 0·1 M NH_3.
(o) 0·2 M NH_3.   (p) 5 M NH_3.

4. (a) Calculate the pH values of the following at 25°C ($K_w = 10^{-14}$, HCl and sodium acetate are strong electrolytes, K_a for acetic acid $= 1·8 \times 10^{-5}$.)

   (i) A litre of M HCl.
   (ii) A litre of M acetic acid.
   (iii) A litre of water.
   (iv) A litre of a solution containing one mole of acetic acid and one mole of sodium acetate.

(b) What would the pH become in each case if 4 g of NaOH were added to each? (NaOH = 40)

(c) What would the pH become in each case if 3·65 g of HCl(g) were added to each? (HCl = 36·5)

(d) What name is given to such solutions as (iv)?

(e) Give an example of a solution like (iv), but with a pH greater than 7.

5. (a) Calculate the pH values of
   (i) 0·1 M H_2SO_4.
   (ii) 0·02 M H_2SO_4.

(Sulphuric acid is a dibasic acid, totally dissociated in solution)

(b) Calculate the pOH values, and hence the pH values of
   (iii) 0·1 M $Ba(OH)_2$.
   (iv) 0·02 M $Ba(OH)_2$.

6. (a) Suggest suitable indicators for the following acid-base titrations:
   (i) HCl + NaOH.
   (ii) Acetic acid + NaOH.
   (iii) HCl + ammonia.
   (iv) Acetic acid + ammonia.
   (v) Phosphorous acid (dibasic acid, pK_a values 1·8 and 6·2) + NaOH.

(b) Draw curves to show how the pH varies with the amount of base added to a given amount of acid in each case.

7. What reagents would be needed to make the following salts?

(a) Potassium bromide.
(b) Ammonium nitrate.
(c) Cobalt chloride.
(d) Calcium sulphate.
(e) Sodium hydrogen sulphate.
(f) Calcium hydrogen sulphite.

8. What salts will form when

(a) NaOH neutralizes sulphuric acid.
(b) Magnesium reacts with nitric acid.
(c) Lead nitrate reacts with sodium chloride.
(d) Sodium carbonate reacts with excess HCl(aq).
(e) Equimolar quantities of sodium carbonate and HCl(aq) react.

9. Will the following form acidic, neutral or basic aqueous solutions? Give reasons for your answers.

(a) Sodium acetate.
(b) Sodium chloride.
(c) Ammonium acetate.
(d) Ammonium chloride.
(e) Iron(III) chloride.
(f) Aluminium chloride.

10. Label the Brønsted acids and bases in the following equilibria:

(a) $HCl + H_2O \rightleftharpoons H_3O^+ + Cl^-$.
(b) $CH_3NH_2 + H_2O \rightleftharpoons CH_3NH_3^+ + OH^-$.
(c) $CH_3COONa + HNO_3 \rightleftharpoons CH_3COOH + NaNO_3$.
(d) $HSO_4^- + H_2O \rightleftharpoons H_3O^+ + SO_4^{2-}$.
(e) $N^{3-} + 3H_2O \rightleftharpoons NH_3 + 3OH^-$.

11. Classify the following substances as either electron donors or electron acceptors (Lewis bases and acids).

(a) Ammonia.
(b) Sulphur dioxide.
(c) Ether.
(d) Silicon tetrafluoride.
(e) Cr^{3+} ions.
(f) I^- ions.

12. The dissociation constant for the hydrolysis of a salt such as sodium acetate is given the symbol K_h.

$$CH_3COO^- + H_2O \rightleftharpoons CH_3COOH + OH^-$$

Find the relationship between K_h, K_w, and K_a (the dissociation constant of the acid).

13. Borax, $Na_2B_4O_7 \cdot 10H_2O$ can be titrated as a weak base. A solution containing 1·91 g is exactly neutralized by 20 cm^3 of 0·5 M HCl(aq).

(a) Suggest a suitable indicator for the titration.
(b) How many moles of borax react with how many moles of HCl?
(c) Write down an equation for this reaction.

†14. H_3PO_2 is a monobasic, H_3PO_3 a dibasic, and H_3PO_4 a tribasic acid. A hydrogen atom bound to phosphorus is not easily ionized, unlike a hydrogen atom bound to oxygen. On the basis of these data suggest possible structures for the three acids.

†15. Certain non-aqueous solvents ionize as follows:

$$2\,NH_3(l) \rightleftharpoons NH_4^+ + NH_2^-.$$
$$2\,HCl(l) \rightleftharpoons H^+ + HCl_2^-.$$
$$2\,BrF_3(l) \rightleftharpoons BrF_2^+ + BrF_4^-.$$

(All the ionic species are solvated).

(a) What are typical acids and bases for each solvent?
(b) Give an example of a neutralization reaction in liquid ammonia.
(c) Are ammonia and hydrogen chloride differentiating or levelling solvents for aqueous acids?
(d) To what extent is it fair to refer to acids and bases when bromine trifluoride is the solvent?

†16. When $SnBr_2F_{10}$ is titrated conductometrically against $KBrF_4$ in $BrF_3(l)$, it is found that the conductivity is at a minimum when 0·390 g of $KBrF_4$ have reacted with 0·469 g of $SnBr_2F_{10}$.

(a) What sort of reaction is taking place?
(b) How many moles of $KBrF_4$ react with how many moles of $SnBr_2F_{10}$?
(c) Write down an equation for the reaction.
(d) What is the nature of the substance $SnBr_2F_{10}$?

CHAPTER 20

The s-block metals and aluminium

THESE final chapters are not intended to fulfil the function of a reference book, but rather to provide a guide to general principles.

20.1 A note on the classification of metals

The term s-block refers to that block of elements in the periodic table following the noble gases where only the s-electron orbitals are being filled, namely Group IA (the alkali metals) and Group IIA (the alkaline earth metals). Although aluminium is not an s-block element, since its valence shell has the configuration $3s^2p^1$, its properties are quite closely related to the s-block elements adjacent to it in the periodic table.

We define a B-metal as a metal which follows a transition series in a given period.* By this definition, a number of similar metals are grouped together and they will be considered in Chapter 25. Aluminium forms the cation Al^{3+} with a noble-gas structure like the cations of s-block metals, and many of its properties are similar to the properties of beryllium. It is convenient, therefore, to treat it in this chapter, while bearing in mind that many of its properties are more akin to those of the B-metals, especially gallium, indium, and thallium.

---

* Some older textbooks use the A and B classification differently, viz. B for transition elements and A for "typical" elements.

TABLE 20.1

Principal quantum number of valence shell	Electronic structure of noble-gas core	Group			
		0 s^0	I s^1	II s^2	III s^2p^1
2	$1s^2$	He	Li	Be	B
3	$1s^2$; $2s^2p^6$	Ne	Na	Mg	Al
4	$1s^2$; $2s^2p^6$; $3s^2p^6$	Ar	K	Ca	
5	$1s^2$; $2s^2p^6$; $3s^2p^6d^{10}$; $4s^2p^6$	Kr	Rb	Sr	
6	$1s^2$; $2s^2p^6$; $3s^2p^6d^{10}$; $4s^2p^6d^{10}$; $5s^2p^6$	Xe	Cs	Ba	

TABLE 20.2

Group IA		Group IIA	
Lithium	Occurs mainly as complex silicates;	Beryllium	Occurs mainly as complex silicates; fairly rare
Sodium	Extensive underground deposits. Approx. 0·5 mol dm⁻³ of $Na^+(aq)$ in sea	Magnesium	Principally as sulphate, carbonate and silicates, and *Carnallite*, $MgCl_2$. KCl. 6 H_2O
Potassium	Occurs in association with anions such as Cl^-. Extractable as KCl	Calcium	Occurs as carbonate and sulphate
Rubidium	Very rare; traces occur in	Strontium	Occur as carbonate and
Caesium	association with potassium	Barium	sulphate
Francium	Radioactive; does not occur naturally on Earth	Radium	Radioactive; found in Th and U ores

Group IIIA — Aluminium — occurs mainly as complex silicates. *Bauxite*, Al_2O_3.2 H_2O, also common.

20.2 Occurrence of the s-block elements

None of the *s*-block elements occurs native (that is, as the free element) since all are far too reactive. The most electropositive elements are found in nature as cations combined with the most electronegative anions. The elements occur as shown in Table 20.2.

20.3 A general survey of properties and uses of s-block metals

s-block metals have the following general properties in common, though there are exceptions and some of the properties referred to are shared with other classes of elements:

(a) They are all metals, most of them being markedly reactive with non-metals.

(b) They have relatively low binding energies, due to the fact that there are only one (Group IA) or two (Group IIA) electrons per atom available for bond formation. Moreover, the binding *s*-electrons are present outside a noble gas core and are shielded from the direct attraction of the charge on the atomic nucleus. As consequences of this:

(i) They have melting and boiling points which are lower than those of the transition metals to their right. (See Figs. 1.6 and 1.7.)

(ii) they have relatively low densities, when compared with the metals which are to their right in the periodic table, as a result of their relatively large atomic radii and large atomic volumes (Fig. 1.10). Lithium is so light that it will even float on the oil in which it is customarily stored.

(iii) They have low ionization energies, the outermost *s*-electrons being lost with relative ease. This is the main factor which enables these metals to form positive ions readily. In aqueous solution:

$$M(c) \longrightarrow M^+(aq) + e^- \quad \text{(Group IA)}$$
$$M(c) \longrightarrow M^{2+}(aq) + 2e^- \quad \text{(Group IIA)}$$

(c) One of their most characteristic chemical properties is that of constant oxidation number. Ions such as Ca^+ and Na^{2+} never occur in chemical compounds. Transition metals and B-metals both show several oxidation states.

(d) They are all powerful reducing agents (electron suppliers) due to (b) above. Reducing power is measured by the half cell potential ε° (Chapter 11), and the data listed in Appendix 1 show that the metals of Group IA and Group IIA appear at the extreme "reducing" end of the list.

(e) With few exceptions their compounds show evidence of the presence of ions. This is in contrast with the transition metals, and with the B-metals, where ionic properties are often less well-defined or where the ions which form are generally complex.

(f) Among the s-block elements, the larger ions are not noticeably hydrolysed in aqueous solution. The smaller ions especially Be^{2+}(aq), Li^+(aq), and Mg^{2+}(aq), are hydrolysed and behave as weak acids (Chapter 19).

(g) Their hydroxides are soluble in water for the most part, exceptions being $Mg(OH)_2$ and $Be(OH)_2$ which are very sparingly soluble, and $Ca(OH)_2$ which is only slightly soluble. The term *alkali metal* derives from the fact that the metals react with water to form soluble hydroxides, alkalis.

(h) Apart from magnesium and beryllium, both of which have relatively high excitation energies, all the s-block elements give characteristic colours when they are introduced into a flame. This property can be used in analysis.

(i) The reactivity of these metals towards non-metals is very marked, though other factors may obscure it. Caesium and rubidium are so reactive that they inflame spontaneously if their surfaces are exposed to the air, and all the alkali metals burn if heated in air. Although magnesium reacts explosively with most non-metals when heated in a finely divided state, large pieces of the metal react only slowly with oxygen owing to the formation of a protective layer of surface oxide.

Uses. Magnesium is a useful structural metal on account of its lightness and is often alloyed with aluminium, for instance in airframe construction. Beryllium ought in theory to be a very attractive proposition as a structural metal, for it is less reactive than magnesium and is even lighter. Unfortunately, however, apart from its cost it possesses two serious snags: in the first place it is rather toxic, and second, it is very brittle except when very pure indeed. It has found uses in nuclear reactors, on account of its capacity to slow down neutrons.

Sodium nowadays finds some rather remarkable applications: formerly most sodium was consumed for chemical purposes, such as special organic reductions. In the laboratory it requires careful handling, owing to its violent reaction with water. It does, however, possess high electrical and thermal conductivity in common with the other Group IA metals, and this has led to its use as a heat transfer agent in nuclear reactors (where it is pumped around as a liquid), and as an electrical conductor. Recent trials have shown that plastic-covered sodium wire, which is very cheap to make by extrusion, is excellent for cables on account of its flexibility and good conductivity. The danger from fire due to chemical action is considerably less than that due to electrical short-circuiting.

Apart from sodium and magnesium (and to a lesser extent lithium which finds some structural uses in alloys) only limited quantities of the free s-block metals are manufactured and their uses are specialized. Compounds of these metals will be referred to under separate headings below.

TABLE 20.3

		I_A	II_A	III_A
increasing density, metallic and ionic radii, atomic ↓ volume	decreasing m.p., b.p., hardness, ionization ↓ energy	Li Na K Rb Cs Fr	Be Mg Ca Sr Ba Ra	Al (Sc) (Y) (La) (Ac)

increasing m.p., b.p.,
hardness, density
——————————————→
decreasing metallic and
ionic radii, atomic vol-
ume, thermal and electri-
cal conductivity

20.4 Principal trends in physical properties

Figures 20.1, 20.2, 20.3 show plots of melting points, boiling points, and atomic volumes of these metals. These plots should be compared with Figs. 1.6, 1.7 and 1.8 which show the same properties plotted in sequence for all the elements.

It is a characteristic of a Group of the periodic table that its elements show (a) a **similarity,** (b) a **gradation,** in properties.

(a) SIMILARITIES

The similarities within a Group can be traced back to their similar electronic structures. For the isolated atoms in their ground states, these structures are:

Group I_A—noble gas plus one outer *s*-electron.

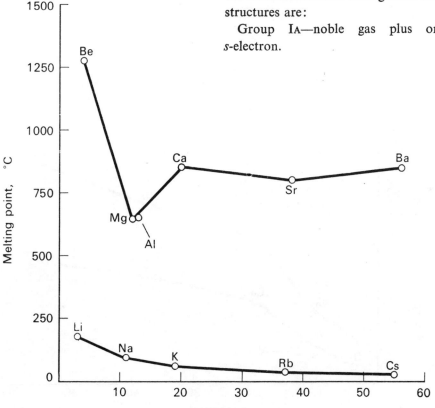

FIG. 20.1. The melting points of the *s*-block elements.

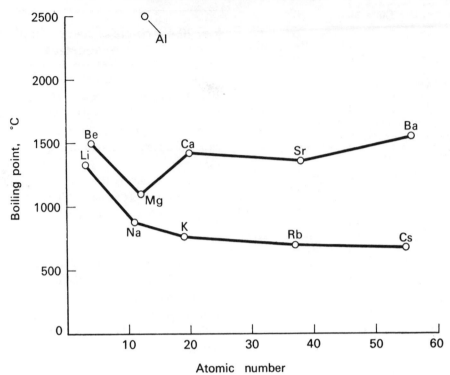

FIG. 20.2. The boiling points of the *s*-block elements.

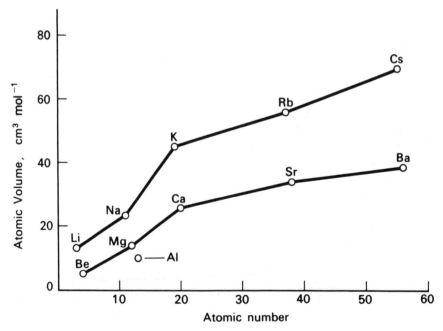

FIG. 20.3. The atomic volumes of the *s*-block elements.

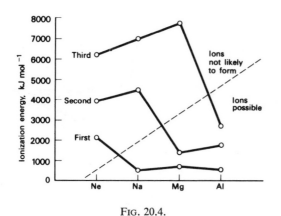

FIG. 20.4.

Similar behaviour is shown by other sets of elements at the beginning of a period. Figure 20.5 shows the first ionization energy of the alkali metals, and the sum of the first and second ionization energies of the Group IIA metals, plotted against atomic number.

(b) GRADATIONS

The gradual increase in size observed when moving down a given group can be attributed simply to the increasing size of the noble gas "core". Properties such as atomic volume are related closely to metallic radius, because structures are similar within a given group.

Figure 20.6 shows the estimated sizes of atoms and ions of *s*-block elements.

The Be^{2+} ion occurs only rarely, since its very small size causes it to exert an intense electrostatic field at its surface, and it polarizes other ions and forms bonds.

Table 20.4 gives data for the ionic conductances of some ions. One might expect the smallest ions to be more mobile, and to have higher

Group IIA—noble gas plus two outer *s*-electrons. The noble gas "core" plays no part at all in the chemistry of these elements, because the extra energy required to remove an electron from it is far too large. Figure 20.4 shows the energy required to remove one, two and three electrons from each of the elements Ne, Na and Mg. The diagram shows clearly that Na^{2+}, Na^{3+} and Mg^{3+} are unlikely to form.

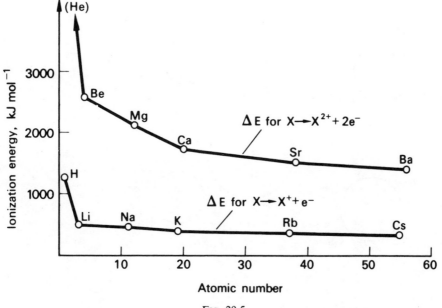

FIG. 20.5.

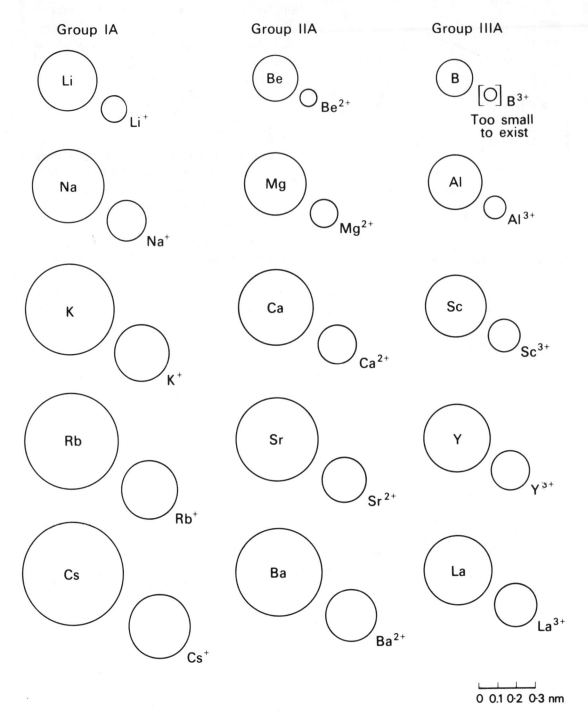

FIG. 20.6. The relative sizes of atoms and ions of *s*-block elements.

TABLE 20.4

IONIC CONDUCTANCES (Ω^{-1}) OF GROUP
IA METAL IONS

Li$^+$	Na$^+$	K$^+$	Rb$^+$	Cs$^+$
33·5	43·5	64·6	67·5	68·0

conductances, so that an ion like Li$^+$(aq) seems to be anomalous; however, the small size of lithium causes a large number of water molecules to be attracted to it. The ion is really Li(H$_2$O)$_x^+$, relatively large in size. This hypothesis explains other phenomena, such as the abnormal half-cell potential of lithium (section 20.5) and the higher degree of hydration of some of its crystalline salts.

The decrease in ionization energy down a Group (Fig. 20.5) is a result of the increasing distance of the outer electron(s) from the centre of positive charge provided by the core. This

is the main factor causing the higher reactivity of the heavier metals.

Larger atoms exert weaker attractions on one another, because electrostatic forces decrease with increasing internuclear distance. The heavier metals within a Group are therefore softer and more volatile.

Heat of sublimation provides a measure of interatomic forces, and this is lower for the heavier elements within a Group (Fig. 20.7).

The metals of Group IIA are harder and less volatile than their Group IA neighbours, showing that their interatomic forces are greater.

This observation is consistent with our notions about the nature of the metallic bond: a metal is regarded as a fairly closely packed assembly of positive ions "welded" together by a mobile electron "gas" (Chapter 5). In the case of the alkali metals there is only one electron per atom available for metallic bonding, whereas in Group IIA there are two electrons per atom available. Moreover, the doubly-charged ions M^{2+} of Group IIA are smaller than their

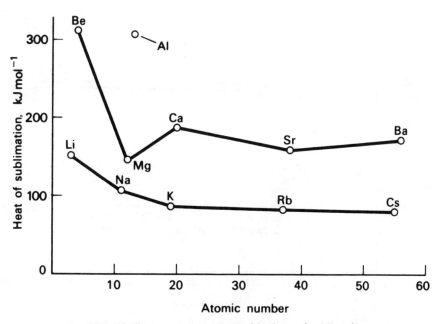

FIG. 20.7. Binding energy, as measured by heat of sublimation.

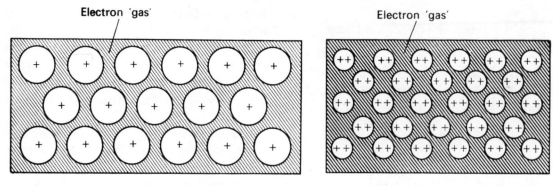

FIG. 20.8.

left-hand neighbours in Group IA, M^+, because of the increased nuclear charge in Group IIA (Fig. 20.8).

20.5 Formation of positive ions by s-block elements

The s-block elements form positive ions due to the ready loss of their s-electrons. One of the important characteristics of s-block metals is that only *one* charge is encountered for a given element. Group IA elements form compounds which dissolve to give the ions $M^+(aq)$ (Fig. 20.9), while in Group IIA the characteristic ion is $M^{2+}(aq)$. Thus s-block elements have a tendency to lose their outermost s-electrons completely. The relative energies needed to form aqueous ions are related to the ionization energies, but depend on other factors as well. Figure 20.10 shows by means of an energy cycle the factors which determine the formation of aqueous ions.

The strongly negative $\varepsilon°$ values of these ele-

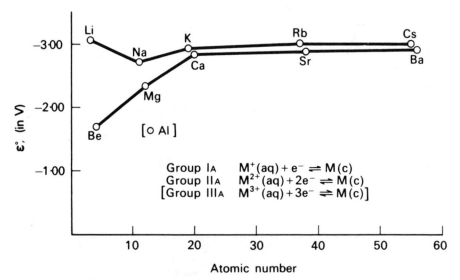

FIG. 20.9. The half-cell potentials of the s-block elements.

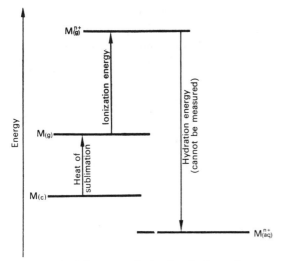

FIG. 20.10. Energy cycle showing the formation of an aqueous ion.

TABLE 20.5

Element	Reactivity	Element	Reactivity
Li	Violent with acids. Moderate with water (cf. barium).	Be	Dissolves rapidly in acids, no reaction with water.
Na	Violent with water and acids. Does not usually catch fire in water in small amounts.	Mg	Slow reaction in water, fast in acids.
K	Violently in water and acids. Usually catches fire.	Ca	Dissolves steadily in cold water, violently in acids.
Rb Cs	Even more violent reaction than potassium.	Sr Ba	About like lithium.

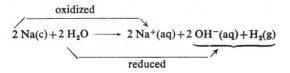

ments mean that all react vigorously with dilute acids, $H^+(aq)$. Those with very negative values react vigorously with water also, Table 20.5.

20.6 Halides of the *s*-block elements

The *s*-block elements all form simple halides MX (Group IA), or MX_2 (Group IIA), where X is a halogen. These can all be obtained in the anhydrous state, though in cases where the metal ion is small and highly charged the solid which crystallizes from aqueous solution contains water of crystallization, for instance, $CaCl_2 . 6 H_2O$ (Ca^{2+} attracts water molecules), and $BaCl_2 . 2 H_2O$ (Ba^{2+} is larger and has less attraction for water molecules, though more attraction than an alkali metal ion, M^+).

The *s*-block halides can be regarded as fairly "ideal" ionic compounds—their physical and chemical properties can be interpreted with

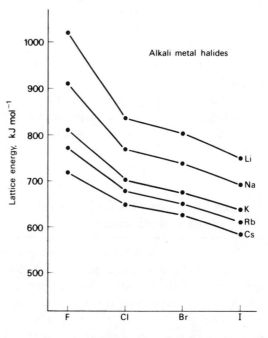

FIG. 20.11. The lattice energies of Group IA halides.

reasonable accuracy on the assumption that ions are present in the solid lattice. There is plenty of evidence for the presence of ions in the molten state: on electrolysis the metal appears at the anode, and the halogen at the cathode. Electrolysis of the fused chloride is the method normally chosen for the extraction of the Group IA metals, calcium and magnesium.

The trends in boiling point and lattice energy of the Group IA halides are summarized in Figs. 6.12 and 20.11. Lattice energy may be defined as the energy required to convert one mole of the ionic lattice into gaseous ions: lattices with smaller ions have higher lattice energies.

The halides of s-block elements can all be made by the reaction between the metal and the halogen, though this reaction is rarely carried out except as a laboratory exercise. The energetics of this process were considered in Chapter 9, where it was shown that such a reaction is exothermic (Born–Haber cycle, section 9.7).

Most of the alkali metal halides have 6 : 6 structures, though CsCl, CsBr, and CsI crystallize out with 8 : 8 coordination. The anhydrous Group IIA halides show a variety of structures depending upon the relative sizes of the ions.

The chlorides, and to a lesser extent bromides and iodides, of sodium and magnesium occur in considerable amounts in sea-water, from which they may be obtained by crystallization.

20.7 Oxides and hydroxides of s-block elements

All the s-block elements burn vigorously in air, reactivity being greatest in the bottom left-hand corner of the periodic table (caesium and francium). In addition to the expected oxides M_2O (Group IA) and MO (Group IIA), the more reactive members of the groups form peroxides and even superoxides. Peroxides have

been shown to contain the ion $[O-O]^{2-}$, and superoxides the ion $[O-O]^-$. On descending the periodic table the tendency is for the element to combine with an increasing amount of oxygen:

$$2\,Li(c) + \tfrac{1}{2}\,O_2(g) \longrightarrow Li_2O(c)$$
lithium oxide, white solid

$$2\,Na(c) + O_2(g) \longrightarrow Na_2O_2(c)$$
sodium peroxide, white solid

$$Ba(c) + O_2(g) \longrightarrow BaO_2(c)$$
barium peroxide, white solid

$$K(c) + O_2(g) \longrightarrow KO_2(c)$$
potassium superoxide, yellow solid

This observation is a manifestation of a fairly general rule concerning ionic compounds, namely that the stability of lattices MX, where X is a *large* anion, increases as the size of M increases.

In this case O_2^{2-} is clearly larger than O^{2-}, and it is the larger metal ions at the foot of the periodic table which form stable ionic peroxides. The same effect may be noted for carbonates (section 20.12), nitrates, and such compounds as polyhalides, e.g. KI_3.

The normal oxides contain the oxide ion, O^{2-}, which is an extremely strong base. Hence they react with water to form the hydroxide by proton exchange:

$$\overset{\text{gains proton}}{O^{2-} + H_2O \rightleftharpoons OH^- + OH^-}$$
loses proton

If water is added to calcium oxide (quicklime) CaO, for instance, the above equation can be expressed in full thus:

$$CaO(c) + H_2O(l) \longrightarrow Ca(OH)_2(c);$$
$$\Delta H = -65 \text{ kJ at } 25°C$$

Magnesium and beryllium oxides do not react vigorously with water. If water is added to the product of burning magnesium in oxygen, a weakly alkaline solution (pH about 10) in equilibrium with much undissolved solid is obtained.

Barium peroxide has been used commercially in the manufacture of hydrogen peroxide, since the peroxide ion reacts with water or dilute acid:

$$\overbrace{O_2^{2-} + 2\,H_2O}^{\text{gains protons}} \underset{\text{loses protons}}{\rightleftharpoons} H_2O_2 + 2\,OH^-$$

The solubilities of the *s*-block element hydroxides in water are plotted in Fig.20.12. The two main factors which lead to insolubility of hydroxides are (i) higher cation charge, and (ii) smaller cation size. It is often said that hydroxides like KOH and NaOH are strong bases, while Ca(OH)$_2$ and Mg(OH)$_2$ are weak. The real *base* in aqueous solution is however the ion OH$^-$, and

the "weakness" of an alkali like calcium hydroxide is due to its relative insolubility in water.

The Group IA hydroxides will only decompose on very strong heating, but the Group IIA hydroxides are thermally unstable, giving the normal oxide and water vapour:

$$Ca(OH)_2(c) \longrightarrow CaO(c) + H_2O(g);$$
$$\Delta H = +109 \text{ kJ at } 25°C$$

TABLE 20.6

Compound	$K_s = [M^{2+}][OH^-]^2$ at 20°C	ΔH for M(OH)$_2$(c) → MO(c) + H$_2$O(g) at 25°C
Be(OH)$_2$	Be^{2+} not formed	+ 54 kJ
Mg(OH)$_2$	$1\cdot2\times10^{-11}$	+ 81 kJ
Ca(OH)$_2$	$1\cdot2\times10^{-5}$	+109 kJ
Sr(OH)$_2$	$3\cdot2\times10^{-4}$	+127 kJ
Ba(OH)$_2$	$1\cdot2\times10^{-2}$	+146 kJ

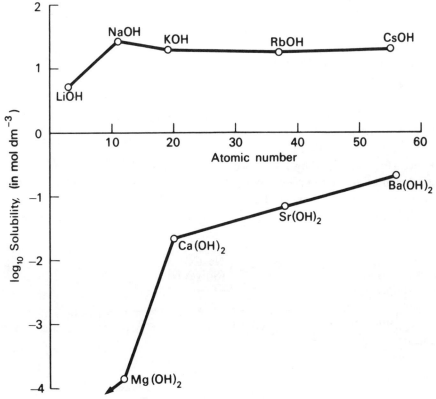

FIG. 20.12. Solubilities of hydroxides.

Table 20.6 gives some data for the solubility and thermal stability of the Group IIA hydroxides.

Beryllium hydroxide is exceptional in this Group, since it can behave as an acid as well as a base; such a substance is said to be **amphoteric.** If an excess of $OH^-(aq)$ is added to beryllium hydroxide it will dissolve forming the complex, tetrahydroxoberyllate(II):

$$Be(OH)_2(c) + 2OH^-(aq) \longrightarrow Be(OH)_4^{2-}(aq)$$

Other amphoteric hydroxides are those of tin, lead, aluminium and chromium; with all these metals there is a strong tendency for the metal atom to occur in the anion in compounds.

20.8 Manufacture and uses of alkalis

Alkalis are used for a variety of purposes in the chemical industry and in the laboratory; among the most common are sodium hydroxide (caustic soda), and calcium hydroxide (slaked lime). Smaller quantities of potassium hydroxide (caustic potash) and other soluble hydroxides are required for special purposes. Calcium oxide, quicklime, is itself used extensively for the neutralization of acids; the use of lime on acidic soils is common agricultural practice.

The flowsheet (Fig. 20.13) shows the two main industrial routes to sodium hydroxide.

Reaction (1) occurs on account of the insolubility of calcium carbonate:

$$CO_3^{2-}(aq) + Ca(OH)_2(c, aq)$$
from   "milk of lime"
Solvay   (suspension of
process   solid in water)
$$\longrightarrow 2\ OH^-(aq) + CaCO_3(c);$$
             precipitate
spectator ion $Na^+(aq)$ (1)

The route from sodium chloride (2) commences with a concentrated solution of brine, obtained by extraction of underground deposits of sodium chloride by pumping hot water. Various cells are in use for the electrolysis, but all depend upon the following *overall* cathode reaction:

$$H^+(aq) + e^- \longrightarrow \tfrac{1}{2} H_2(g).$$

The result of discharging $H^+(aq)$ ions is that Na^+ and OH^- ions are left behind in the cathode region.

The chief technical problem in the design of cells is avoiding interference from the reaction which occurs at the anode:

$$Cl^-(aq) \longrightarrow \tfrac{1}{2} Cl_2(g).$$

Steps must be taken to prevent the chlorine diffusing across the cell and being hydrolysed by the OH^- ions;

$$Cl_2(aq) + 2\ OH^-(aq)$$
$$\longrightarrow Cl^-(aq) + ClO^-(aq).$$

One very ingenious design which overcomes this problem is the *Kellner–Solvay cell* (Fig. 20.14). This uses a mercury cathode, which is

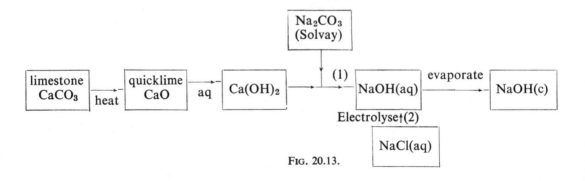

FIG. 20.13.

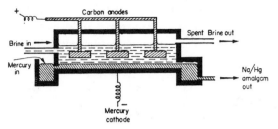

FIG. 20.14. The Kellner–Solvay cell.

remarkable in that it allows *sodium* to be discharged in preference to hydrogen. The sodium dissolves in the cathode to form a dilute amalgam, and the half-cell potential for this process is much less negative than for the discharge of pure sodium. Furthermore, the energy of activation (Chapter 17) required to form hydrogen molecules at a mercury surface is very large.

The liquid sodium amalgam is run off, and reacts with water in an iron vessel giving sodium hydroxide solution which is concentrated by evaporation.

20.9 Reactions of alkalis

The reactions of aqueous alkalis are essentially reactions of the hydroxide ion.

(i) Acids and acidic oxides give salts. Excess of a polybasic acid will give rise to one or more acid salts.

(ii) Insoluble hydroxides are precipitated when the ions are mixed.

$$M^{n+}(aq) + n\ OH^-(aq) \longrightarrow M(OH)_n(c).$$

(iii) Amphoteric hydroxides dissolve in excess alkali to give a soluble hydroxo-complex:

$$Be(OH)_2 + 2\ OH^-(aq)$$
$$\longrightarrow Be(OH)_4^{2-}(aq);$$
also Al, Sn, Pb, Cr, etc.

(iv) Salts of weak bases react to liberate the free base:

$$NH_4^+(aq) + OH^-(aq)$$
$$\longrightarrow NH_3(aq,\ g) + H_2O.$$

$$C_6H_5NH_3^+(aq) + OH^-(aq)$$

soluble salt $\longrightarrow C_6H_5NH_2(1) + H_2O.$
(e.g.: aniline
hydrochloride) aniline, (oil
immiscible
with water)

(v) Many non-metals are attacked; sometimes a disproportionation reaction occurs when the element, in zero oxidation state, gives one compound containing the element in a negative oxidation state, and another compound in a positive oxidation state. One example is the action of alkali on chlorine described in the previous section (bromine and iodine behave similarly). Another example is the action of alkali on white phosphorus giving phosphine:

$$P_4(c) + 3\ OH^-(aq) + 3\ H_2O$$
$$\longrightarrow 3\ H_2PO_2^-(aq) + PH_3(g)$$

(vi) Many organic substances are attacked by the OH^- ion which inserts itself in the molecule and displaces a negatively charged ion. Examples are the hydrolyses of esters, alkyl halides and acyl halides.

(vii) Many reactions are catalysed by $OH^-(aq)$, for instance, the polymerization of aldehydes.

20.10 Sulphides of *s*-block elements

All the *s*-block elements combine with sulphur by direct reaction, though the reactions are dangerous and should not be attempted in the laboratory. Group IIA metals combine explosively on warming with powdered sulphur forming MS; Group IA metals form M_2S.

The Group IA sulphides are soluble in water, and the Group IIA sulphides sparingly soluble. They are ionic and the sulphide ion is a weak base (proton acceptor, Chapter 19) which is hydrolysed by water to produce an alkaline solution, which smells of hydrogen sulphide:

$$\overbrace{S^{2-} + 2H_2O}^{\text{accepts protons}} \underset{\text{loses protons}}{\rightleftharpoons} H_2S + 2\,OH^-$$

The Group IIA sulphides perform this reaction on warming.

20.11 Other binary compounds with non-metals

Most of the s-block elements will combine directly with other non-metals, such as phosphorus, silicon and selenium. Nitrogen is a very unreactive element, though it combines fairly readily with lithium, forming Li_3N, and magnesium, forming Mg_3N_2. These compounds behave as if they contain the nitride ion, N^{3-}, which is hydrolysed by water forming ammonia; the reaction is analogous to the hydrolysis of sulphide ion given above. The elements also react directly with hydrogen if heat is supplied, forming ionic hydrides (sometimes called *saline*, or salt-like, hydrides), which contain the ion H^- (Chapter 18). Stable carbides are formed by calcium, strontium and barium:

$$CaO(c) + 3\,C(\text{graphite})$$
$$\longrightarrow CaC_2(c) + CO(g);$$
$$\Delta H = +460 \text{ kJ.}$$

20.12 Carbonates and bicarbonates

Two series of salts characteristic of the s-block elements are the *carbonates*, salts of carbonic acid, H_2CO_3, in which the metal cation is in

association with the anion CO_3^{2-}, and *bicarbonates* (hydrogen carbonates), in which the metal cation is in combination with the ion HCO_3^-:

$$O=C{<}^{O^-}_{O^-} \qquad O=C{<}^{O^-}_{O-H} \qquad O=C{<}^{O-H}_{O-H}$$

*carbonate ion (planar triangle)	bicarbonate ion	carbonic acid

$$\Updownarrow$$
$$CO_2 + H_2O$$

The bicarbonate ion is itself a weak acid with a tendency to release protons. Solid bicarbonates are formed only by the more electropositive Group I metals, Na, K, Rb, Cs and Fr. On heating they decompose readily (below 100°C) giving carbon dioxide and steam:

$$2\,NaHCO_3(c)$$
$$\longrightarrow \underset{\text{stable to heat}}{Na_2CO_3(c)} + CO_2(g) + H_2O(g)$$

Solid bicarbonates of the alkaline earth metals cannot be obtained though their ions can exist together in solution. Calcium and magnesium bicarbonates are responsible for *temporary hardness* of water. On boiling the solution the bicarbonate ion decomposes, and the insoluble carbonate is removed from the solution by precipitation:

$$Ca^{2+}(aq) + 2\,HCO_3^-(aq)$$
$$\longrightarrow CO_2(g) + H_2O(l) + CaCO_3(c).$$

The stability of carbonates themselves towards heat is greatest for metal whose ions have large size and low charge (cf. section 20.7). It may be imagined that the carbonate ion decomposes more easily to give $CO_2(g)$ when it is polarized by the cation. The general reactions

---

* The negative charges are in fact equally shared among the three oxygen atoms.

are illustrated by the following equations:

$$Li_2CO_3(c) \longrightarrow Li_2O \ c) + CO_2(g)$$

(Other alkali metal carbonates only decompose with great difficulty.)

$$MgCO_3(c) \longrightarrow MgO(c) + CO_2(g)$$

(All Group IIA carbonates can be decomposed by Bunsen heat, but $BaCO_3$ only with great difficulty.)

The carbonate ion is a fairly strong base, the conjugate base of the weakly acidic HCO_3^- ion. Solutions of the carbonate ion behave as alkalis, due to the following hydrolysis reaction:

$$\overset{\text{loses protons}}{\underset{\text{gains protons}}{CO_3^{2-}(aq) + 2H_2O \longrightarrow H_2CO_3(aq) + 2OH^-(aq)}}$$

The carbonates of the alkaline earth Group are all insoluble in water, and occur in the Earth's crust. Although carbon-containing compounds are comparatively rare in the Earth's crust, deposits containing $CaCO_3$ (chalk, limestone, marble and Iceland spar), and $MgCO_3$ (such as dolomite, $MgCO_3 \cdot CaCO_3$) are extensive. These are derived originally from microscopic organisms.

The most important carbonate industrially is sodium carbonate, which is a cheap alkali used for glass making, soap manufacture, and a large number of industrial processes. Many million tons are produced annually by the Solvay process, developed by the Belgian chemist Ernest Solvay in 1872.

___

20.13 The Solvay process

In this process brine (sodium chloride solution) is saturated with ammonia. Carbon dioxide, obtained by heating limestone, is then blown through the brine under pressure (Fig. 20.15). The carbon dioxide reacts with the

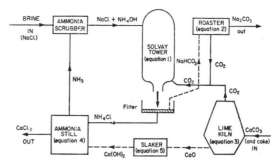

FIG. 20.15. The Solvay process.

water according to the equilibrium:

$$CO_2 + H_2O \rightleftharpoons HCO_3^- + H^+(aq)$$

The ammonia displaces this equilibrium to the right by combining with hydrogen ions, and the solubility product of $NaHCO_3(c)$ is exceeded, causing most of the bicarbonate ions to be precipitated from solution. The sodium bicarbonate is washed, dried and decomposed into sodium carbonate by heating.

The calcium oxide from the limestone is converted into calcium chloride by reaction with the ammonium chloride solution; this regenerates ammonia which can be recycled together with the carbon dioxide.

The principal reaction is carried out in high towers, which have to be kept cool in order to precipitate as much sodium bicarbonate as possible. The final product can be made into washing soda, $Na_2CO_3 \cdot 10 H_2O$; these crystals are efflorescent. Millions of tons of sodium carbonate are produced annually for use as a cheap alkali. The overall equation for the process can be summarized:

$$CaCO_3 + 2NaCl \longrightarrow Na_2CO_3 + CaCl_2.$$

Most of the calcium chloride produced is wasted, though a certain amount is required for use as a drying agent, and a certain amount for conversion into calcium metal by electrolysis. The process is a very economical one, however and affords a good illustration of how to make the maximum use of materials.

The Solvay process cannot be adapted to make potassium carbonate, because the essential first stage, the precipitation of potassium bicarbonate, cannot occur because of its high solubility in water.

The Solvay process can be summarized by five equations (Fig. 20.15):

(1) $CO_2 + H_2O + Na^+ \longrightarrow NaHCO_3 + H^+$.

(2) $2\,NaHCO_3 \longrightarrow Na_2CO_3 + CO_2 + H_2O$.

(3) $CaCO_3 \longrightarrow CaO + CO_2$.

(4) $2\,NH_4Cl + Ca(OH)_2$
   $\longrightarrow CaCl_2 + 2\,NH_3 + 2\,H_2O$.

(5) $CaO + H_2O \longrightarrow Ca(OH)_2$.

20.14 Nitrates of s-block elements

Nitrates are salts of the strong acid, nitric acid, HNO_3. The nitrate ion, NO_3^-, is stable in aqueous solution and is not hydrolysed.

nitrate ion
planar triangle
(The bonds are in fact
equivalent cf. carbonate)

nitric acid
a strong acid
(contrast carbonic acid)

Nitrates, like carbonates, are decomposed by heating. The nitrates of Group IA elements decompose losing oxygen only, forming the nitrite (the ending -ite indicates less oxygen):

$NaNO_3(c) \longrightarrow NaNO_2(c) + \frac{1}{2}O_2(g)$;

white   sodium nitrite   glowing
crystals   stable to heat   splint test

$\Delta H = +107$ kJ at 25°C.

In Group IIA the nitrites themselves are unstable to heat, and the overall reaction is the evolution of oxygen and nitrogen(IV) oxide, leaving the metal oxide residue. This latter reaction is typical of most metal nitrates:

$Ba(NO_3)_2(c) \longrightarrow BaO(c) + 2\,NO_2(g) + \frac{1}{2}O_2(g)$;

white   white   brown   glowing
crystals   solid   fumes   splint test

$\Delta H = +500$ kJ at 25°C.

Nitrates, either made synthetically or derived from sources like Chile saltpetre, $NaNO_3$, are an important source of nitrogen for plants.

All common metal nitrates are soluble in water. Neither the anion nor the cation is hydrolysed appreciably for the more electropositive metals, so the solutions have pH values near to 7 at room temperature. Since the ions $Li^+(aq)$, $Be^{2+}(aq)$, and $Mg^{2+}(aq)$ are appreciably hydrolysed by water, nitrates of these metals will be appreciably acidic in water. For instance, magnesium nitrate has a pH of about 6.

$$Mg(H_2O)_x^{2+} + aq \rightleftharpoons \left[Mg^{(H_2O)_{x-1}}_{(OH)}\right]^+$$

$[= Mg^{2+}(aq)]$ hydrolysed aqueous ion

$+ H^+(aq)$
pH = 6.

The phenomenon of hydrolysis of metal ions is closely linked to the formation of insoluble hydroxides. The ions which are readily hydrolysed form precipitates if further $OH^-(aq)$ is added, and this is a further stage in the hydrolysis process. For instance, if sodium hydroxide solution is added to a solution of magnesium nitrate, a precipitate of magnesium hydroxide will result. The overall equation is

$2\,NaOH(aq) + Mg(NO_3)_2(aq)$
   $\longrightarrow Mg(OH)_2(c) + 2\,NaNO_3(aq)$

though this may be written ionically as a continuation of the hydrolysis equation above:

$$\left[Mg^{(H_2O)_{x-1}}_{(OH)}\right]^+ + OH^-(aq)$$
   $$\longrightarrow Mg(OH)_2(c) + (x-1)H_2O$$

(spectator ions, Na^+ and NO_3^-).

20.15 Sulphates, phosphates and silicates of *s*-block elements

These are considered under a common heading since they are derived from the oxo-acids of the second-row non-metals, and have certain features in common. It was shown in Chapter 19 that the acids increase in strength across the row. The weaker the oxo-acid, the greater is its tendency to "polymerize", as Table 20.7 shows.

Acid salts can exist for all except the perchlorates. The main *s*-block elements which occur in nature as silicates are lithium, beryllium and occasionally magnesium. The phosphate ion is found in nature as calcium phosphate, and the sulphate ion occurs as magnesium sulphate (*Epsom salt*, $MgSO_4.7H_2O$), gypsum, $CaSO_4.2H_2O$, and barytes, $BaSO_4$. (Note the decreasing amount of water of crystallization in these salts, cf. section 20.6.)

Perchlorates do not occur in nature and their properties will not be considered further here.

Sulphates, phosphates and silicates are markedly stable to heat. This is in contrast to the salts containing the first-row anions, nitrate and carbonate. Very strong heat indeed will cause sulphates to give off a mixture of $SO_2(g)$ and $SO_3(g)$ but such temperatures are not usually achieved in the laboratory.

Calcium silicate is one of the principal components of cement. Cement is made by heating powdered *marl*, a natural mixture of limestone, and clay, in an oil-fired furnace. On adding water to the product, a rigid macromolecular structure is obtained, and heat is evolved in the hydration reaction. Concrete consists of cement mixed with gravel or ballast.

Calcium sulphate, which occurs as *gypsum*, is of importance as a source of sulphur dioxide in the manufacture of sulphuric acid:

$$2\,CaSO_4(c) + C(coke)$$
$$\longrightarrow 2\,CaO(c) + 2\,SO_2(g) + CO_2(g)$$

Alumina, Al_2O_3, and silica, SiO_2, are added to a powdered mixture of gypsum and coke, and calcium silicate and aluminate form a by-product which is a useful cement clinker;

$$CaO + SiO_2 \longrightarrow CaSiO_3.$$

Barium sulphate is one of the most important salts of barium; it is highly insoluble and unreactive. The heavy barium nuclei make it opaque to X-rays, and a suspension of it is administered to patients requiring X-ray examination of the digestive tract. It is used as a white pigment, and occasionally as a paper filler.

An important distinction between Group IA and Group IIA is observed in this series of salts: almost all the Group IA salts are soluble in water, whereas nearly all the Group IIA salts are insoluble. Thus the sulphates, phosphates and silicates of Ca, Sr Ba and Ra are insoluble ($CaSO_4$ sparingly soluble in water); beryllium and magnesium sulphates are however soluble.

TABLE 20.7

acids	$Si(OH)_4$ very weak acid	$O{=}P(OH)_3$ fairly weak acid	$O_2S(OH)_2$ fairly strong acid	$O_3Cl(OH)$
anions formed	SiO_3^{2-} and many complex "polymer" species	PO_4^{3-}, HPO_4^{2-}, $H_2PO_4^-$ and various "polymer" species, e.g. $P_2O_7^{4-}$	SO_4^{2-}, HSO_4^- and a few "polymers" such as $S_2O_7^{2-}$	ClO_4^- only

20.16 Aluminium and its relationship to *s*-block elements

Aluminium shows a number of similarities to the *s*-block metals; it shows a strong resemblance to the element diagonally above it, beryllium. This is often referred to as a **diagonal relationship.** The reason for the similarity can be seen by examining Figs. 20.3, 20.6 and 20.7, where the trends in atomic and ionic radius, interatomic forces, ionization energy, etc., are seen to cancel out. The metals become more electropositive down a Group, but less electropositive from left to right:

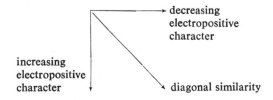

The aluminium atom has the ground state electronic structure $1s^2$; $2s^2p^6$; $3s^2p^1$. The underlined electrons are present in the valence shell and are available for bond formation. Figures 1.8 to 1.11 show how the physical properties of aluminium fit in with periodic trends.

Aluminium is one of the commonest elements in nature. Being a relatively "weak" metal, it is not found as chloride or nitrate, or even as carbonate or sulphate, but it occurs widely as complex aluminosilicates such as those present in clay. Such minerals are too stable chemically to be an economic source for extraction of the metal, but deposits of the hydrated oxide, bauxite, $Al_2O_3xH_2O$, also occur and this can be used to obtain the metal. Aluminium oxide is an extremely stable substance (free energy of formation $\Delta G° = -1580$ kJ mol^{-1}), since the ions Al^{3+} and O^{2-} are both quite small.

Despite the high energy of ionization needed to remove the three outer electrons to form

the ion Al^{3+}, this is the stable ion in aqueous solution since considerable energy is regained as hydration energy. The reducing power, as measured by the half-cell potential $\varepsilon°$, is lower than for most *s*-block elements as far as aqueous solutions are concerned (Fig. 20.9).

$$Al(c) \rightleftharpoons Al^{3+}(aq) + 3e^-; \quad \epsilon° = -1·66 \text{ V}$$

Like beryllium, aluminium shows a tendency to form molecular rather than ionic compounds. The ion $Al^{3+}(aq)$ is quite strongly hydrolysed, and behaves as an acid in aqueous solution.

$$[Al(H_2O)_6]^{3+} \rightleftharpoons H^+(aq) + [Al^{(OH)}_{(H_2O)_5}]^{2+}.$$

Aluminium hydroxide, $Al(OH)_3$, is formed as a white gelatinous precipitate when alkali is added to a solution of $Al^{3+}(aq)$ ions. Like beryllium hydroxide, it is amphoteric:

$$Al(OH)_3(c) + OH^-(aq) \longrightarrow Al(OH)_4^-(aq)$$
white ppt   excess   tetrahydroxo-
             alkali   aluminate

Addition of more alkali gives species such as $Al(OH)_5^{2-}$ and $Al(OH)_6^{3-}$

Aluminium is widely used as a light structural metal, either alone or alloyed with other metals such as magnesium. Duralumin is an alloy containing about 4% copper, which has superior strength and corrosion resistance. A hard oxide coat on its surface protects it from corrosion by the atmosphere. The formation of oxide makes aluminium difficult to weld. Typical uses of aluminium are aircraft construction, roofing, liquid containers, domestic utensils and packaging. Its electrical conductivity, though inferior to that of copper, is good enough to be useful when light-weight cables are required.

20.17 Compounds of aluminium

Aluminium reacts directly with most nonmetals, including nitrogen, to form compounds in which it has an oxidation number of $+3$, for instance $Al_2O_3(c)$, $Al_2Cl_6(c)$, $AlN(c)$, and $Al_2S_3(c)$. It burns vigorously in air when finely divided to form a hard refractory solid, m.p. 2,050°C. This oxide occurs in various crystalline forms, such as *corundum* which is used in the manufacture of artificial gemstones. The oxide is also used for making refractory crucibles on account of its good stability towards heat.

Aluminium burns readily in chlorine with the formation of a volatile white smoke, Al_2Cl_6, which sublimes at 180°C. This is a molecular substance. Its configuration may be written

$$
\begin{array}{ccc}
Cl & Cl & Cl \\
 & Al & Al \\
Cl & Cl & Cl
\end{array}
$$

On heating to high temperatures this molecule dissociates into $AlCl_3(g)$ molecules. Aluminium bromide and iodide have similar structures, but the fluoride crystallizes as an ionic lattice.

Aluminium dissolves in dilute hydrochloric acid to form an ionic solution of aluminium chloride, but the anhydrous compound cannot be obtained by evaporation owing to hydrolysis of $Al^{3+}(aq)$. $AlCl_3$ itself reacts vigorously with water. In either case aluminium hydroxide is the result:

$$Al(c) + 3\,HCl(aq) \longrightarrow AlCl_3(aq) + \tfrac{3}{2}\,H_2(g)$$

$$AlCl_3(aq) + 3H_2O \xrightarrow{\text{evaporate}} Al(OH)_3(c) + 3HCl(g).$$

Aluminium reacts explosively with sulphur, and this reaction should not be attempted. Al_2S_3 is attacked slowly by water, rapidly by $H^+(aq)$, producing gaseous hydrogen sulphide.

Aluminium forms normal salts with most acids: their formulae can be calculated by balancing the charges on the ions. For instance Al^{3+} and SO_4^{2-} form aluminium sulphate, $Al_2(SO_4)_3$, which crystallizes out as the hydrate $Al_2(SO_4)_3 . 18\,H_2O$. This can be made by dissolving aluminium, or aluminium hydroxide, in dilute sulphuric acid. When dilute nitric acid is used, the metal itself is rendered passive, though the hydroxide reacts to give aluminium nitrate.

Aluminium sulphate is less stable to heat than most sulphates:

$$Al_2(SO_4)_3(c) \longrightarrow Al_2O_3(c) + 3\,SO_3(g);$$
$$\Delta G^\circ = +408 \text{ kJ at } 25°C.$$

The very negative free energy of formation of Al_2O_3 undoubtedly has a part to play here.

An important salt is the double salt *potash alum*, of empirical formula $KAl(SO_4)_2.12\,H_2O$. This is less soluble in water than aluminium sulphate itself, and is often used preferentially for applications such as the above. It is used as a *mordant* in dyeing: aluminium hydroxide is precipitated and adsorbed on to the fibres of a fabric, and the dye bonds to the hydroxide more easily than to the fabric alone. On a small scale it is used in styptic pencils for stopping bleeding. Here the high charge on Al^{3+} causes the colloidal particles in blood to coagulate (Chapter 16).

20.18 Summary

(1) PHYSICAL PROPERTIES

Soft, relatively light, easily fusible metals.

GROUP IIA: Harder, denser, less fusible than Group IA neighbours.
ALUMINIUM: Similar to beryllium.

(2) CHEMICAL PROPERTIES

IONIZATION: Only one oxidation state; M^+ (Group I) and M^{2+}(Group II) formed, with noble-gas structures; ion formation occurs readily.

REACTIVITY:

(a) All burn in air, and combine directly with most non-metals.

(b) Heavier members of both Groups react vigorously with water; Be and Al hardly react at all, and Mg very slowly;

(c) They are vigorous reducing agents, due to ready formation of positive ions.

COMPOUNDS

(a) Only the smallest ions (Be^{2+} and Al^{3+}) have a marked tendency to form covalent bonds.

(b) The largest ions form the most stable compounds with large anions. Hence carbonates, nitrates, peroxides and tri-iodides for instance are most stable in Group I and at the bottom of a given Group.

(c) The largest ions form the most soluble hydroxides, and hence the most basic oxides.

(3) TRENDS IN THE s-BLOCK

(a) Ionic radii increase, and hence ionization energies decrease, down a Group.

(b) Ionic radii decrease from left to right.

Trends (a) and (b) tend to cancel along a diagonal, making Li and Mg similar, also Be and Al; (diagonal relationship, Chapter 24).

(c) Stabilities and solubilities of compounds usually follow well-marked trends within a Group (see above).

Although quite a wide range of properties is encountered over the entire range of the s-block, extremes being represented by beryllium and caesium, the block of elements forms a closely related set. Group trends are readily observed, and within a Group the elements appear to be closely related. In later Groups, for instance Group V (Chapter 25), the range of chemical behaviour is often wider, and these straightforward relationships are often obscured.

Study Questions

1. (a) Why do ionic radii (i) increase down a Group? (ii) decrease across a period?
   (b) Why do ionization energies decrease down a Group, but tend to increase across a period?

2. What are the formulae of the simplest binary compounds formed by the following pairs of elements?
   (a) Li and Cl. (b) Li and O. (c) Li and N.
   (d) Ca and Br. (e) Ca and S. (f) Ca and P.
   (g) Al and I.   (h) Al and O. (i) Al and N.

3. In the text it is noted that lithium floats on the oil in which it is stored. (a) Why is it stored in oil? (b) Which other s-block elements would you expect to be stored in this way?

4. Calcium forms a fluoride CaF_2. (a) Why does CaF_3 not form with an excess of fluorine? (b) Can you suggest why CaF does not form?

5. The following chlorides are all readily soluble in water. Arrange the solutions in order of increasing pH: NaCl, LiCl, $BeCl_2$, $MgCl_2$ and $AlCl_3$.

6. The properties of elements in the same Group of the periodic table show both similarities and a gradation. Show how this applies to (a) the sulphates of the Group IIA metals (b) the reactivity of the Group IA metals with water.

7. Classify the following as acid-base or redox reactions. Label each species to show how the transfer of electrons or protons takes place.
   (a) $2 Na + Cl_2 \rightleftharpoons 2 NaCl$.
   (b) $BaO_2 + 2 H_2O \rightleftharpoons Ba(OH)_2 + H_2O_2$.
   (c) $Sr + 2 H_2O \rightleftharpoons Sr(OH)_2 + H_2$.
   (d) $Ca_3P_2 + 6 H_2O \rightleftharpoons 2 PH_3 + 3 Ca(OH)_2$.
   (e) $2 NaBr + Cl_2 \rightleftharpoons 2 NaCl + Br_2$.

8. Write equations for the following reactions of NaOH(aq):
   (i) When it is added to a solution of a zinc salt (Zn^{2+}), a white precipitate forms which dissolves in excess alkali.
   (ii) On adding it to a white crystalline solid X, ammonia is evolved.
   (iii) On adding it hot to iodine, a solution containing both I^- and IO_3^- forms.
   (iv) On refluxing with ethyl acetate, ethyl alcohol is formed.

9. Why are Group IIA salts generally much less soluble than Group IA salts? The solubilities of the Group IIA sulphates are as follows: $BeSO_4$, 440; $MgSO_4$, 260; $CaSO_4$, 2·0; $SrSO_4$, 0·11; $BaSO_4$ $2·5 \times 10^{-3}$ g dm^{-3}.

(i) Calculate the solubility of each salt in moles per litre.
(ii) Give the solubility products $[M^{2+}]$ $[SO_4^{2-}]$ of each sulphate.
(iii) What trend do you observe? How does this agree with the data for the hydroxides? (Fig. 20.12.)

10. Why is magnesium used
(i) to extract uranium from UF_4?
(ii) in aircraft alloys?
(iii) in fireworks?

11. What have NaF, MgO and AlN in common? What variations would you expect in the properties of these compounds?

12. The lattice energies of LiF, NaF, KF, RbF and CsF are 1000, 890, 795, 760 and 725 kJ mol^{-1}.

(a) How can you account for the gradually diminishing values?
(b) On this basis, which of the Group IA fluorides would you expect to be least soluble in water?
(c) What other factor is important in determining these solubilities? (Draw an energy level diagram.)

13. Account for the following observations:

(a) Molten aluminium bromide is a poor conductor of electricity.
(b) An aqueous solution of the same substance is a good conductor.
(c) This aqueous solution is acidic.
(d) In the gas phase, the molecular weight of aluminium bromide is approximately 530.

14. (a) List as many similarities in the chemistries of Al and Be as you can.
(b) It is believed that, in strong alkali, aluminium forms the complex ions $Al(OH)_4^-$, $Al(OH)_5^{2-}$ and $Al(OH)_6^{3-}$. Yet, in strong alkali, BeO forms only $Be(OH)_4^-$, but no $Be(OH)_5^{2-}$ nor $Be(OH)_6^{3-}$. Why is this?

15. Write equations for the following reactions:

(a) Calcium oxide reacts with carbon to give calcium carbide, which reacts with water to give acetylene.
(b) Aluminium combines directly with carbon at 1600°C to give Al_4C_3. On hydrolysis, the carbide gives an inflammable gas of m.wt. 16.
(c) A carbide of magnesium, Mg_2C_3, gives allylene, $CH_3-C\equiv CH$ on hydrolysis.

16. (a) 8·8 g of an element reacted with excess dilute hydrochloric acid to give 2250 cm^3 of hydrogen at s.t.p. Assuming that the element forms ions with only one, two or three positive charges, what possible values are there for the atomic weight of the element?
(b) The element was then ignited; it burned with a red flame to give a white solid that dissolved in water to give a basic solution. What sort of element is it?
(c) When CO_2 was passed into this solution, a white precipitate formed. Which element is it?

17. The elements francium and radium are radioactive and therefore their chemistries (especially that of Fr) have not been investigated in great detail.

(a) How would you expect Fr to react with (i) air (ii) water and (iii) chlorine? What method would you use to try and obtain a sample of metallic francium from one of its compounds?
(b) How would radium react with (i) air, (ii) water, (iii) chlorine?
(c) How soluble would you expect the hydroxide and the sulphate of radium to be? What would the formulae of these compounds be?
(d) Use the diagrams in the chapter to predict values of the boiling points, the electrode potentials, the atomic volumes and the ionization energies of these elements.

18. Selenium is a non-metal in the same group as sulphur, Group VIB. Predict what will happen when powdered magnesium and powdered selenium are heated together. What will be the formula of the product? How might you expect the product to react with water?

19. Which ion in Group IIA would you expect to have the highest ionic mobility? Refer to Fig. 20.9 and Table 20.4.

20. Which of the following compounds will be paramagnetic: K_2O, K_2O_2, KO_2, K_3N? Name all these compounds.

Commercial samples of Na_2O_2 are often coloured yellow because of NaO_2 impurities. The addition of water to such samples causes a little oxygen to be evolved and the salt becomes colourless. When a trace of MnO_2 is added, a lot of oxygen is evolved. How do you account for these observations?

21. The hydration energies of the Group IA ions have been estimated to be: Li^+ 515, Na^+ 405, K^+ 322, Rb^+ 293, and Cs^+ 263 kJ mol^{-1}.

(a) What is meant by the hydration energy?
(b) Can you give a reason for the trend that is observed?
(c) How would the hydration energies of the Group IIA ions compare with these values?

22. Make a list of similarities and differences between lithium and magnesium. Explain these relationships as far as possible on the basis of physical principles.

CHAPTER 21

The halogens

A TYPICAL NON-METAL GROUP

In this chapter an important family of non-metals, the halogens, will be surveyed. Many of their properties are characteristic of non-metals in general. The Group as a whole, Group VIIʙ of the periodic table, affords another good example of the kind of trends observed for the s-block metals (Chapter 20), namely a *similarity* and a *gradation* in properties.

21.1 Occurrence and extraction of the halogens

Table 21.1 lists the halogens, with their ground state electronic structures.

The halogens are far too reactive to occur in nature as the free elements, and all are therefore found in combination with metals, generally as

simple halides. This is due to the strong tendency of the halogens to gain an electron and form a negative ion, either in aqueous solution or in a crystal lattice. The tendency to form the ion X^- is weakest for astatine, and is relatively weak for iodine—iodine occurs in nature largely as sodium iodate, $NaIO_3$. Astatine is radioactive, and has not been found in nature except in very low concentrations. Fluorine occurs in cryolite, Na_3AlF_6, and in *fluorite*, CaF_2.* Chlorine occurs mainly as chloride ion in sea water, though deposits of rock salt, *halite*, $NaCl$, are common. The chief source of bromine is the

---

* Fluorite is used as a flux, and it is this property which gives rise to the name fluorine (Latin, *fluo*, flow). The term fluorescence comes from the same source, as fluorite has been known for many years to exhibit this property.

TABLE 21.1

Name	Principal quantum number of valence shell	Electronic structure in ground state (abbreviated form)	Electronic structure (full form)
Fluorine	2	2, 7	$1s^2$; $2s^2p^5$
Chlorine	3	2, 8, 7	$1s^2$; $2s^2p^6$; $3s^2p^5$
Bromine	4	2, 8, 18, 7	$1s^2$; $2s^2p^6$; $3s^2p^6d^{10}$; $4s^2p^5$
Iodine	5	2, 8, 18, 18, 7	$1s^2$; $2s^2p^6$; $3s^2p^6d^{10}$; $4s^2p^6d^{10}$; $5s^2p^5$
Astatine	6	2, 8, 18, 32, 18, 7	$1s^2$; $2s^2p^6$; $3s^2p^6d^{10}$; $4s^2p^6d^{10}f^{14}$; $5s^2p^6d^{10}$; $6s^2p^5$

bromide ion in sea water. Iodine occurs in low concentrations in sea water, but is absorbed by certain seaweeds. Seaweed ash contains considerable quantities of iodide ion. Table 21.2 shows the relative abundance of the halogens in nature.

TABLE 21.2

Element	% by weight in Earth's crust	molarity in sea water (average)
Fluorine	0·08	10^{-4}M
Chlorine	0·19	0·56M
Bromine	0·01	8×10^{-4}M
Iodine	10^{-4}	v. small
Astatine	negligible	negligible

The extraction of the free elements from their simple ions is essentially a problem of electron removal, that is, oxidation. The extraction method chosen depends on the gradation of properties down the group. Figure 21.1 shows that electron removal is going to be most difficult for fluorine, slightly easier for chlorine, and so on. In fact the half-cell potential for the reaction $F_2(g) \rightleftharpoons 2F^-(aq)+2e^-$ is more positive than for almost any other known couple, making it impossible to extract fluorine with a chemical oxidizing agent. If electrolysis is used instead, there is no limit to the potential which can be applied to effect the oxidation. In practice chlorine and fluorine are both generally extracted by electrolysis of a fused salt or suitable solution (see below). With bromine and iodine, chemical oxidation is feasible and generally more convenient. A good oxidizing agent is chlorine, which has the merit of being cheap, since its relative abundance is much greater. The extraction of bromine from sea-water depends upon the reaction:

$$Cl_2(g)+2\,Br^-(aq) \longrightarrow Br_2(aq)+2\,Cl^-(aq);$$
$$\Delta\epsilon° = 1·36-1·07\ V$$
$$= +0·29\ V$$
$$\Delta G = -28\ kJ.$$

Extraction of fluorine. Fluorine is extracted by electrolysing a solution of potassium fluoride in anhydrous hydrogen fluoride, at a fairly high temperature (between 100° and 270°, depending upon the electrolyte composition). The chief

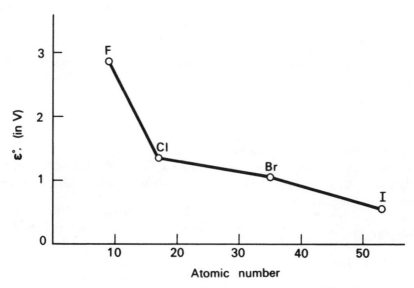

FIG. 21.1. Half-cell potentials for the halogens, $X_2 \rightleftharpoons 2X^-+2e^-$.

problem is to prevent the fluorine from attacking the cell or the anode. All metals are attacked by fluorine, but a protective coat of fluoride often forms, and special steel is used in modern cells, together with a special non-graphitic carbon anode.

Extraction of chlorine. An early process (invented by Deacon in 1868) for obtaining chlorine was the oxidation of hydrogen chloride by oxygen in the presence of a catalyst of $Cu^{2+}(aq)$.

$$4\,HCl(g) + O_2(g) \xrightarrow{450°C} 2\,H_2O(g) + 2\,Cl_2(g);$$

In the laboratory, chlorine is made by oxidizing concentrated HCl(aq) with a convenient oxidizing agent such as potassium permanganate (no heat required) or manganese(IV) oxide (gentle heat needed). The latter oxidizing agent was once of industrial importance, but the process using it is now obsolete.

In the United Kingdom, the production of chlorine is closely tied up with the electrolysis of brine for making sodium hydroxide, and this is the main source of the gas.

Extraction of bromine. The usual commercial method for bromine extraction is to evaporate sea water; chlorides, mainly sodium chloride, crystallize out leaving the "mother liquor" relatively rich in bromide ion. Chlorine gas is then bubbled through, and bromine obtained from the solution by distillation. A less common commercial method, but one commonly carried out as a laboratory reaction, is to heat a bromide with concentrated sulphuric acid and manganese(IV) oxide. The reaction can be regarded as two distinct processes:

(i) $KBr(c) + H_2SO_4(l)$

$\longrightarrow KHSO_4(c) + HBr(g).$

Notice that the acid salt is formed, and that

the "driving force" of the reaction is the formation of a gaseous product.

(ii) $MnO_2 + 4\,H^+ + 2\,Br^-$

$\longrightarrow Mn^{2+} + Br_2 + 2\,H_2O.$

The overall equation becomes:

$$MnO_2 + 2\,KBr + 3\,H_2SO_4$$
$$\longrightarrow MnSO_4 + 2\,KHSO_4 + Br_2 + 2\,H_2O.$$

Extraction of iodine. The majority of the world's iodine comes from the small percentage of sodium iodate present in *caliche (Chile saltpetre)*. Iodate ions, $IO_3^-(aq)$ are reduced with sodium bisulphite, which provides $HSO_3^-(aq)$. The reaction occurs in two stages:

(i) $3\,HSO_3^- + IO_3^- \longrightarrow I^- + 3\,HSO_4^-$

(ii) $IO_3^- + 5\,I^- + 6\,H^+ \longrightarrow 3\,I_2 + 3\,H_2O.$

Reaction (ii) involves two different oxidation states of iodine becoming one (the reverse of disproportionation).

Smaller amounts of iodine are obtained from certain natural brines containing $I^-(aq)$, or from seaweed ash.

Astatine. Astatine does not occur in nature in appreciable amounts. It is made by bombarding bismuth with α-particles, and little is known of its chemistry and physical properties. Its behaviour appears to resemble that of a non-metal more than that of a metal. For instance, it is fairly volatile, and can be dissolved in benzene.

21.2 General survey of reactions and uses of the halogens

The reactions of the halogens are dominated by the strong tendency for the half-reaction:

$$X_2 + 2e^- \longrightarrow 2X^-.$$

Fluorine reacts with all metals, even the noble metals (platinum, gold etc.), and with most non-metals. Chlorine attacks nearly all metals though it is less reactive with non-metals. Bromine and iodine show progressively less reactivity towards metals.

Fluorine gas finds few uses, and it is generally converted into a substance such as SbF_3 which is a milder fluorinating agent. A range of organic compounds containing fluorine in place of hydrogen has been developed in recent years, and a number of such compounds are of commercial importance, for instance the refrigerant gas *freon* (containing CCl_2F_2, $CClF_3$, etc.) and the polymer PTFE (polytetrafluoroethylene), $(C_2F_4)_x$ (section 24.6).

Most of the uses of chlorine depend upon its oxidizing power. It is used as a disinfectant, and as a wood-pulp bleach in the manufacture of paper. A large amount of chlorine is used in the organic chemical industry, for instance in the making of dyes, plastics such as polyvinyl chloride, insecticides such as DDT, and explosives.

Bromine is used chiefly in the manufacture of silver bromide for photography, and in drug manufacture (potassium bromide is a widely used sedative).

Iodine is an essential element to health. A deficiency of iodine leads to abnormal swelling of the thyroid gland in the neck—a condition known as *goitre* which is now prevented by controlled addition of iodide to water supplies which require it. Small quantities of iodine are consumed in a miscellany of different ways such as for antiseptics, organic catalysts, and in the manufacture of "polaroid".

21.3 Trends in physical properties of the halogens

The halogen elements all form diatomic molecules X_2, due to the sharing of one electron pair from the uppermost energy level. The atoms in X_2 therefore all have completed octets, and can be regarded as being isoelectronic with their noble-gas neighbours. Figure 21.2 shows how the sizes of atoms and ions, and the bond lengths, increase down the group.

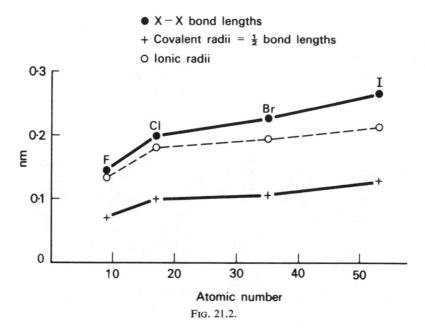

● X – X bond lengths
+ Covalent radii = ½ bond lengths
O Ionic radii

FIG. 21.2.

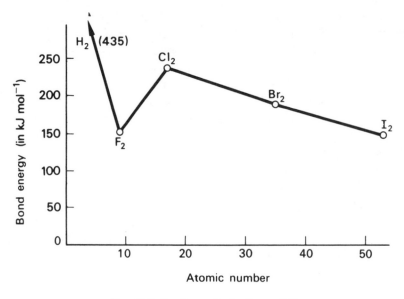

FIG. 21.3. Bond energies in Group VIIB.

Figure 21.3 shows a plot of bond dissociation energy against atomic number of X. Fluorine has an unexpectedly weak bond, and this is part of the reason for fluorine's great reactivity. In contrast the bonds formed between fluorine atoms and other atoms are generally fairly strong. Table 21.3 shows how the strengths of bonds H–X and C–X vary, in comparison with X–X.

TABLE 21.3

BOND ENERGIES (kJ mol^{-1})

	X−X	H−X	C−X
X = fluorine	155	565	438
X = chlorine	242	430	330
X = bromine	192	363	276
X = iodine	150	297	238

The melting points and boiling points of the halogens both increase with molecular weight. Iodine is a solid at room temperature (sublimes at 183°C) and bromine is a liquid.

Refer back to Fig. 3.2 and you will see that the halogens all have high ionization energies, the lowest being iodine (excluding astatine). Hence we would not expect the halogens to form positive ions, X^+, in chemical compounds.*

The tendency for the atoms to form negative ions, X^-, on the other hand, is very marked. The halogens all possess high values of **electron affinity**. The electron affinity of an atom X is equal to the ionization energy of the negative ion X^-, that is:

$$X^-(g) \longrightarrow X(g); \quad \text{electron affinity} = \Delta E.$$

Electronegativity. The electronegativity of an element may be loosely defined as "the power of an atom in a molecule to attract electrons". Various attempts have been made to assign numerical values to electronegativity coefficients but none is entirely satisfactory. One such attempt, by Mulliken (Fig. 21.4), is based on

---

* Compounds of iodine containing the element in an oxidation state of +1, for instance ICl, frequently exist, but they are not ionic.

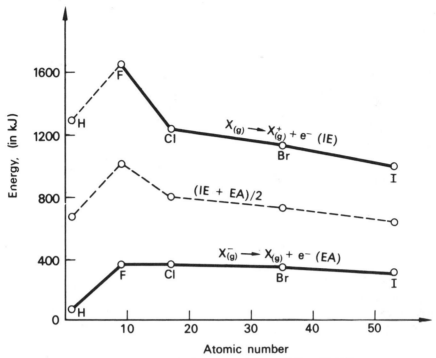

FIG. 21.4. Mulliken's electronegativity = (I.E. + E.A)/2.

the relationship

$$\text{electronegativity} \propto \frac{\text{ionization energy} + \text{electron affinity}}{2}$$

In this book the concept will only be used qualitatively.

The halogen molecules are non-polar and consequently are not very soluble in water, except where chemical reaction plays a part. Fluorine attacks water and "dissolves" in it in the same sense that sodium "dissolves" in water —a variety of products including ozone and the fluoride ion are formed—while chlorine is moderately soluble due to slow chemical reaction:

$$Cl_2(g) + H_2O(l) \longrightarrow HCl(aq) + HOCl(aq).$$

Bromine is sparingly soluble in water giving a red solution. Bromine water kept in the light on the laboratory shelf slowly decolorizes due to hydrolysis. Iodine is practically insoluble in water. All the halogens dissolve readily in non-polar solvents such as carbon tetrachloride.

21.4 The hydrogen halides

Certain of the properties of the hydrogen halides have already been dealt with under different headings in the book. They are summarized here together with the chapter references:

(i) Hydrogen fluoride has a high boiling point (19°C) due to hydrogen bond formation (section 18.7), but the remainder are gaseous substances consisting of molecules HX.

(ii) They are all acids (proton donors) which react with water as follows:

$$HX + aq \rightleftharpoons H^+(aq) + X^-(aq).$$

(iii) They are all highly soluble in water, forming maximum boiling azeotropes (section 14.15); This again indicates bond formation with water.

(iv) Their bond energies decrease down the group (section 21.3 above). For this reason, hydrogen iodide is fairly easily dissociated on heating, hydrogen bromide only with difficulty, and hydrogen chloride and fluoride hardly at all.

(v) They are all colourless gases which fume in moist air. This property of fuming is due to reaction between HX and water vapour to form a "fog" of HX(aq).

General methods of making hydrogen halides. Industrially, large amounts of hydrogen chloride, for conversion into hydrochloric acid, are made by burning hydrogen and chlorine together. This method is not suitable for hydrogen fluoride as hydrogen and fluorine explode on contact, even at $-200°C$ in the dark.

In the laboratory any hydrogen halide can be made by hydrolysis of a suitable halide of a metal or non-metal. Taking hydrogen chloride as an example, most non-metal chlorides fume in moist air due to the formation of hydrogen chloride:

$$BCl_3(g) + 3\,H_2O \longrightarrow H_3BO_3(aq) + 3HCl(aq, g)$$

$$PCl_5(c) + \ H_2O \longrightarrow POCl_3(l) \ + 2\,HCl(aq, g)$$

Some of the weaker metals have chlorides which resemble non-metal chlorides in their readiness to react with water: aluminium chloride is an example (section 20.17). Other metal halides come to an equilibrium when dissolved in water, without any violent reaction: the only effect observed is an acidic pH value due to hydrolysis. Note that halides of carbon, such as CCl_4, do not react with water (section 24.6).

A hydrogen halide is produced whenever an organic hydroxy-compound is attacked by a suitable non-metal halide. A common test for organic OH-groups is the reaction of the compound with phosphorus(V) chloride:

$$ROH + PCl_5(c) \longrightarrow RCl + HCl(g) + POCl_3.$$

Frequently a hydrogen halide can be produced by displacement with a stronger, less volatile acid. Concentrated sulphuric acid is suitable for making HF and HCl, and this is a common method for obtaining these compounds. Notice that the *acid* salt is formed in this displacement reaction.

$$NaCl(c) + H_2SO_4(l)$$
$$\longrightarrow NaHSO_4(c) + HCl(g).$$

$$CaF_2(c) + 2\ H_2SO_4(l)$$
$$\longrightarrow Ca(HSO_4)_2(c) + 2\,HF(g).$$

Sulphuric acid is not a suitable reagent for making HBr and HI, however, since these are oxidized. The reaction between sulphuric acid and the other halides can be used as a test for the halide ion (Table 21.4).

TABLE 21.4

Halide	Reaction with concentrated sulphuric acid
Fluoride	Bubbles of HF evolved which stick to the glass test-tube (HF attacks glass)
Chloride	Colourless fuming gas, HCl, evolved
Bromide	Brown acidic gas evolved (some HBr, some bromine vapour, and some reduction products such as SO_2)
Iodide	Violet vapour, I_2, evolved, and reduction products such as SO_2, H_2S and sulphur.

Hydrogen iodide is a powerful reducing agent, being so readily oxidized to iodine, and it is often used for reducing organic compounds.

Hydrochloric acid finds extensive industrial use where an acid is required, and it is used for cleaning metal surfaces. Hydrofluoric acid is a relatively weak acid, but possesses the unusual property of attacking glass:

$$6\,HF(aq) + \underset{\text{in glass}}{SiO_2}$$

$$\longrightarrow 2\,H^+(aq) + \underset{\text{water soluble}}{SiF_6^{2-}(aq)} + 2\,H_2O.$$

All the aqueous hydrogen halides show typical "acidic" properties, reacting with metals to form hydrogen, with carbonates to form carbon dioxide, and with oxides and hydroxides to form salts (halides). The anhydrous substances or their solutions in dry benzene do not show these acidic properties; for instance anhydrous hydrogen chloride dissolved in benzene will not attack zinc or calcium carbonate. This shows that the "acid" in aqueous solution is $H^+(aq)$ and not HX.

21.5 Interhalogen compounds

A compound consisting of two elements, both halogens, is called an **interhalogen.** Table 21.5 shows the formulae of some interhalogens. Only one or two are of practical importance, and the best known is probably iodine monochloride, ICl(l). The physical properties depend on the molecular weights and polarities of the molecules. The shape of these molecules is of great interest: it will be left to the reader to deduce from arguments of electron-pair re-

TABLE 21.5

SOME INTERHALOGENS

Linear	T-shaped	Square pyramid	Pentagonal bipyramid
ClF	ClF$_3$		
BrF	BrF$_3$	BrF$_5$	
BrCl	IF$_3$	IF$_5$	IF$_7$
ICl	ICl$_3$		

pulsions why these molecules have the shapes they do (refer back to section 6.2).

Iodine has a strong tendency to aggregate other halogen atoms around itself, and in addition to molecular compounds, **polyhalide** ions are formed in aqueous solution. The best known of these is the ion I_3^-(aq), formed when iodine is dissolved in a solution already containing iodide ions:

$$I_2 + I^-(aq) \rightleftharpoons I_3^-(aq);$$
$$K = 725\ \text{mol}^{-1}\ \text{dm}^3\ \text{at } 25°C$$

This complex ion is relatively easily decomposed, and a solution of tri-iodide ions behaves in effect like a solution of free iodine. Another example of a polyhalide ion is the ion ICl_4^-(aq). Salts of polyhalides are stable when the metal cation is large (cf. section 20.12). Among the alkali metals for instance, KI_3, RbI_3 and CsI_3 are formed though not LiI_3 and NaI_3.

21.6 Oxides and oxo-acids

Fluorine forms a series of oxides at low temperatures, but F_2O(g) is the only one stable at room temperature. They are more correctly regarded as oxygen fluorides, since oxygen is less electronegative than fluorine.

Chlorine forms Cl_2O(g), ClO_2(g), Cl_2O_6(l), and Cl_2O_7(l). Cl_2O_7, chlorine(VII) oxide (chlorine heptoxide) is made by powerfully dehydrating perchloric acid with phosphorus(V) oxide, and it reacts vigorously with water to re-form the acid:

$$2\,HClO_4(l) + \tfrac{1}{2}P_4O_{10}(c)$$
$$\longrightarrow Cl_2O_7(l) + 2\,HPO_3(l)$$
$$Cl_2O_7(l) + H_2O \longrightarrow 2\,HClO_4(l).$$

The oxides of chlorine are in general unstable and explosive.

Bromine forms a series of oxides, all of which are unstable.

Iodine forms various oxides, among them iodine(V) oxide (iodine pentoxide), $I_2O_5(c)$. This is made by heating iodic acid crystals above 100°C. At about 300° it dissociates into its elements, but it is more stable thermally than most halogen oxides:

$$2\,HIO_3(c) \xrightarrow{100°} I_2O_5(c) + H_2O(g)$$

$$I_2O_5(c) \xrightarrow{300°} I_2(g) + \tfrac{5}{2}O_2(g)$$

I_2O_5 is an oxidizing agent, and can be used in the quantitative determination of carbon monoxide:

$$I_2O_5(c) + 5\,CO(g) \xrightarrow{70°C} I_2(g) + 5\,CO_2(g).$$

Although there does not appear to be a clearly discernible pattern of behaviour among the halogen oxides, the oxo-acids follow a distinct pattern. Fluorine is too electronegative to assume a positive oxidation number, so it does not form true oxo-acids, but among the remaining halogens oxidation numbers of $+1$, $+3$, $+5$ and $+7$ are observed. Table 21.6 shows the names and formulae of the halogen oxo-ions.

TABLE 21.6

Oxidation number	Nomenclature	Fluorine	Chlorine	Bromine	Iodine
$+1$	hypo-ite	—	ClO^-	BrO^-	IO^-
$+3$	-ite	—	ClO_2^-	—	IO_2^-
$+5$	-ate	—	ClO_3^-	BrO_3^-	IO_3^-
$+7$	per- -ate	—	ClO_4^-	—	IO_4^-

Bromine oxo-acids are in general rather unstable, and no one so far has succeeded in making bromites and perbromates.

Hypochlorites and hypochlorous acid. Sodium hypochlorite is the product when cold aqueous sodium hydroxide is reacted with chlorine. This can be regarded as hydrolysis of the element, which disproportionates into the oxidation states $+1$ and -1:

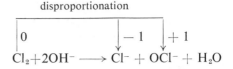

A solution of sodium hypochlorite, mixed with sodium chloride, is made by electrolysing brine and allowing the products, OH^- and Cl_2, to react together. In an electrolytic cell for making sodium hydroxide it is necessary to devise some means of keeping the chlorine separate from the sodium hydroxide.

It is not possible to obtain a pure hypochlorite. Any attempt to recover a solid product by evaporation leads to decomposition and loss of oxygen and chlorine oxides. Aqueous sodium hypochlorite is used as a household bleach, sold commercially under names such as "Domestos". Bleaching powder is a non-stoichiometric solid formed by absorbing chlorine in solid calcium hydroxide; it can be regarded as $xCa(OH)_2.yCa(OCl)_2.zCaCl_2$. The bleaching action of the hypochlorite ion is due to its being a powerful oxidizing agent which can oxidize a dye to a colourless product.

Hypochlorous acid can only exist in dilute solution, and the action of dilute acid on hypochlorite ion causes decomposition in which chlorine is evolved:

$$2\,H^+ + OCl^- + Cl^- \longrightarrow Cl_2 + H_2O.$$

Chlorates. On heating a dilute solution of a hypochlorite it disproportionates giving the oxidation states $+5$ and -1:

$$\underset{3\,ClO^-(aq)}{\overset{+1}{}} \quad \text{disproportionation} \quad \underset{2\,Cl^-(aq)}{\overset{-1}{\longrightarrow}} + \underset{\underset{\text{chlorate}}{ClO_3^-(aq)}}{\overset{+5}{}}$$

The same overall reaction occurs on reacting chlorine with *hot* alkali:

$$3\,Cl_2 + 6\,OH^- \longrightarrow 5\,Cl^- + ClO_3^- + 3\,H_2O.$$

Sodium and potassium chlorates are white solids, soluble in water, and quite powerful oxidizing agents. Potassium chlorate is a weed-killer.

Perchlorates. Heating potassium chlorate to just above its melting point leads to further disproportionation, forming potassium perchlorate:

$$\overset{+5\quad\quad +7\quad\;\; -1}{4\,KClO_3 \longrightarrow 3\,KClO_4 + KCl}$$

If potassium perchlorate is distilled with concentrated sulphuric acid, anhydrous perchloric acid can be distilled. This is a very powerful oxidizing agent, which is liable to react explosively with organic matter such as dust.

$$H_2SO_4 + KClO_4 \longrightarrow HClO_4 + KHSO_4$$
distil under reduced pressure

Iodic acid and iodates. Iodine oxoacids and oxo-ions are relatively stable, and deposits of sodium iodate occur naturally as a constituent of *caliche* (Chile saltpetre, mainly $NaNO_3$). Iodic acid is the product when iodine is oxidized with a powerful oxidizing agent such as concentrated nitric acid. The equation may be derived as follows:

(i) $I_2 + 6\,H_2O \longrightarrow 2\,IO_3^- + 6\,H^+ + 5e^-$

(ii) $2\,HNO_3 + 6H^+ + 6e^- \longrightarrow 2NO + 4H_2O$

Take five times (ii) added to six times (i):

$$6I_2 + 10HNO_3 \longrightarrow 6H^+ + 12IO_3^- + 10NO + 2H_2O$$

Iodic acid, and the iodate ion, are fairly powerful oxidizing agents. A common reaction is that between iodate and iodide:

$$\overset{+5\quad -1\quad\quad 0}{HIO_3 + 5\,HI \longrightarrow 3\,I_2 + 3\,H_2O}$$

21.7 Titrations involving iodine

A solution of iodine will oxidize a wide variety of reducing agents quantitatively, and a standard solution can be used for titration purposes. Iodine is almost insoluble in water, but it dissolves readily in potassium iodide solution, forming the tri-iodide ion (section 21.5). For practical purposes such a solution can be regarded as containing free iodine, since the I_3^- readily decomposes.

Relatively weak reducing agents are oxidized by iodine solution, and a common example used in practice is the thiosulphate ion, $S_2O_3^{2-}(aq)$. Sodium thiosulphate, $Na_2S_2O_3.5H_2O(c)$, can be obtained pure and is used to standardize a solution of iodine. The reaction is:

$$\overset{\text{oxidized}}{2\,S_2O_3^{2-} + \quad I_2 \quad \longrightarrow S_4O_6^{2-} + 2\,I^-}$$
reduced

If a strong oxidizing agent is treated in neutral or acid solution with a large excess of iodide ion, the latter acts as a reducing agent and the oxidant is quantitatively reduced. An equivalent amount of iodine is liberated and this is then titrated with a standard solution of a reducing agent, usually sodium thiosulphate. Below are three examples of equations which can be employed in titrations. Study Question 6 gives practice in balancing a few more.

$$H_2O_2 + 2\,H^+ + 2\,I^- \longrightarrow 2\,H_2O + I_2$$
$$2\,HNO_2 + 2\,H^+ + 2\,I^-$$
$$\longrightarrow 2\,NO + I_2 + 2\,H_2O$$
$$Br_2 + 2\,I^- \longrightarrow 2\,Br^- + I_2$$

Iodine is quite strongly brown-coloured in potassium iodide solution, and to some extent acts as its own indicator. The colour can be strongly accentuated by adding starch solution indicator, which reacts with iodine to form a dark blue substance of indefinite composition.

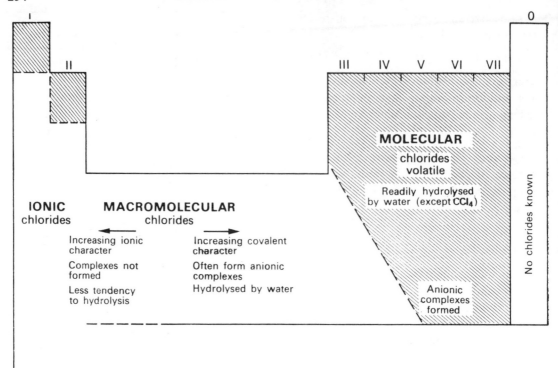

Fig. 21.5.

21.8 A general survey of chlorides

In order to understand the properties of halides of the elements, we will first consider the chlorides. Figure 21.5 shows diagrammatically how the properties of chlorides depend upon the position of the element in the periodic table. Note the following points:

(a) Chlorides of the s-block metals, except for $BeCl_2$ and $MgCl_2$, are more or less ideally ionic: they are soluble in water and their ions are hardly hydrolysed at all (pH of aqueous solutions near 7). $MgCl_2$(c) and $BeCl_2$(c) do not fit the ionic model so well, and indeed beryllium chloride has a structure analogous to Al_2Cl_6, which is the result of directional bonds between atoms rather than that due to the packing

together of oppositely charged ions:

$$
\begin{array}{ccccccc}
& Cl & & Cl & & Cl & & Cl \\
& \diagdown & \swarrow & & \swarrow & & \swarrow & \diagdown \\
& Be & & Be & & Be & \\
\diagdown & & \diagup & \diagdown & \diagup & \diagdown & \diagup & \diagdown \\
& Cl & & Cl & & Cl & & Cl
\end{array}
$$

(b) Chlorides of the non-metals are molecular. Apart from CCl_4 they generally react with water forming hydrogen chloride.

(c) Chlorides of the B-metals and transition metals have structures which suggest the formation of electron-pair bonds between atoms. Those of higher oxidation state are usually molecular. Most are soluble in water, but silver chloride, AgCl, mercury(I) chloride, Hg_2Cl_2, and lead(II) chloride are rather insoluble. Transition metals and B-metals have a tendency to form anionic complexes, for instance:

$PbCl_4^{2-}(aq)$ tetrachloroplumbate(II)

$CuCl_4^{2-}(aq)$ tetrachlorocuprate(II)

$BiCl_4^-(aq)$ tetrachlorobismuthate(III).

In naming a complex ion, the ending -ate is used to indicate an anion, followed by the oxidation number of the metal. The oxidation number of the metal is easily derived by imagining the ion to dissociate into its constituents:

$$PbCl_4^{2-} \longrightarrow Pb^{2+} + 4\,Cl^-$$
$$\text{lead(II)}$$

$$PbCl_6^{2-} \longrightarrow Pb^{4+} + 6\,Cl^-$$
$$\text{lead(IV)}$$

Apart from a few chlorides such as those mentioned above, most metal chlorides are soluble in water. If a solution of a metal ion in water gives a precipitate on adding dilute hydrochloric acid, the following metal ions can be suspected: silver(I), lead(II) or mercury(I).

The chloride ion is colourless, and therefore if a solution of a chloride is coloured, the colour is that due to the metal ion present; only transition metals give coloured ions in solution.

(d) Apart from one or two exceptions, chlorides of transition metals and B-metals hydrolyse with water giving HCl(g). It is not generally possible to prepare an anhydrous chloride by evaporating an aqueous solution to dryness. Zinc chloride for instance gives a basic chloride,

and aluminium chloride hydrolyses completely to hydroxide:

$$\overset{\text{evaporate}}{ZnCl_2(aq) + H_2O \longrightarrow Zn(OH)Cl(c) + HCl(g)}$$
$$AlCl_3(aq) + 3\,H_2O \longrightarrow Al(OH)_3(c) + 3\,HCl(g)$$

The replacement of chlorine atoms by hydroxy-groups is not a phenomenon peculiar to inorganic chemistry. Organic chlorides hydrolyse in an exactly analogous way:

$$CH_3COCl + H_2O \longrightarrow CH_3COOH + HCl$$
$$\text{acetyl chloride} \qquad\qquad \text{acetic acid}$$

$$C_4H_9Cl + H_2O \longrightarrow C_4H_9OH + HCl$$
$$\text{butyl chloride} \qquad\qquad \text{butyl alcohol}$$

Anhydrous chlorides of the metals which hydrolyse can be prepared by using a method which avoids the use of water. The following methods are available:

(i) Pass a stream of *dry* chlorine gas over the heated metal. Where a metal can exist in more than one oxidation state, this reaction produces a higher state because chlorine is an oxidizing agent. Aluminium chloride, $Al_2Cl_6(c)$, and iron(III) chloride, $Fe_2Cl_6(c)$, can be made in this way (Figure 21.6).

(ii) Pass a stream of *dry* hydrogen chloride gas over the heated metal. This produces a chloride of lower oxidation state, because hydrogen chloride is not such a powerful oxidizing agent. For instance iron filings give mainly iron(II) chloride, $FeCl_2(c)$.

(iii) Heat the hydrated chloride in a stream of dry hydrogen chloride gas. This reaction works by driving the equilibrium composition over to the right:

$$Zn(OH)Cl + HCl \rightleftharpoons ZnCl_2 + H_2O$$
$$\text{passed into}\quad\text{residue}\quad\text{driven from}$$
$$\text{reaction}\qquad\qquad\text{reaction}$$

TABLE 21.7

Giant structures		Molecular structures	
Formula	m.p.	Formula	m.p.
$FeCl_2$	670°C	Fe_2Cl_6	319°C
$PbCl_2$	498°C	$PbCl_4$	−15°C
$SnCl_2$	247°C	$SnCl_4$	−33°C

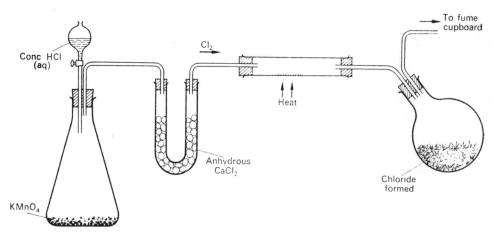

FIG. 21.6. Preparation of the volatile chloride of a metal, e.g. Al_2Cl_6.

21.9 Reactions of the chloride ion

Certain reactions of the chloride ion are quite commonly met with in the laboratory, and they are summarized here for convenience:

(1) $Cl^-(aq) + Ag^+(aq) \rightarrow AgCl(c)$. This reaction forms the basis of a common test for chloride ions; silver nitrate is added, and a white precipitate of silver chloride appears. In order to make the test specific for chloride, the reaction is carried out in acid solution (dilute nitric acid is added first), and the resultant precipitate is confirmed by adding an excess of dilute ammonia solution. $AgBr(c)$ (cream coloured) and $AgI(c)$ (yellow) both form precipitates under the same conditions but these will not dissolve in dilute ammonia solution.

(2) Chloride ions are oxidized to chlorine by any powerful chemical oxidant (ε° must be greater than $+1\cdot36$ V), or at the anode in electrolysis. Suitable chemical oxidants include manganese(IV) oxide, MnO_2, and potassium permanganate, $KMnO_4$.

21.10 Fluorides compared with chlorides

In many respects fluorides resemble chlorides in their properties. The major point of difference is that fluorides generally behave as more ideally ionic compounds. Their structures are those to be expected from the packing together of oppositely charged ions, whereas chlorides frequently have layer structures due to directional bond formation. Moreover, the solubility of fluorides in water is generally what would be expected from ionic substances. Silver fluoride for instance, is soluble in water.

Fluorine has the property of bringing out the maximum oxidation state in the elements with which it is combined due to a combination of exceptionally high oxidizing power and small size. In fact it is often difficult to prepare a fluoride of low oxidation number. Among the non-metals, compare the formula of SF_6 with SCl_4, and IF_7 with ICl_3. Among the metals, note that uranium forms UF_6 (used in isotope separation) but not UCl_6.

21.11 Bromides and iodides compared with chlorides

We have already seen that whereas fluorine tends to form ideally ionic compounds with metals, chlorine has a tendency to form compounds which depart from the ionic model (layer structures, insolubility in water). The same trend is observed in even more marked degree with bromides and iodides; thus many iodides are insoluble in water. Complex ions readily form, and many iodides dissolve in an excess of iodide ion, for instance:

$$Hg^{2+}(aq) + 2\,I^-(aq) \longrightarrow HgI_2(c)$$
$$\text{red solid}$$

$$HgI_2(c) + 2\,I^-(aq) \longrightarrow HgI_4^{2-}(aq)$$
$$\text{excess} \qquad \text{colourless solution}$$

The formation of complex ions is just another manifestation of the tendency of iodine to form bonds rather than simple ions.

Bromides are intermediate in properties between chlorides and iodides. The trend down the group is well summarized by considering the halides of silver (Table 21.8).

TABLE 21.8

Halide	Structure	Solubility in water product $mol^2\ dm^{-6}$	Solubility in ammonia(aq)
AgF	Simple cubic (6 : 6)	soluble	soluble
AgCl	Simple cubic (6 : 6)	$K_s \simeq 10^{-10}$	soluble
AgBr	Simple cubic (6 : 6)	$K_s \simeq 10^{-12}$	soluble only in conc NH_3(aq)
AgI	Tetrahedral (4 : 4)	$K_s \simeq 10^{-16}$	insoluble even in conc NH_3 (aq)

Figures 21.7 and 21.8 are summary charts giving some of the reactions of chlorine and iodine and their important compounds.

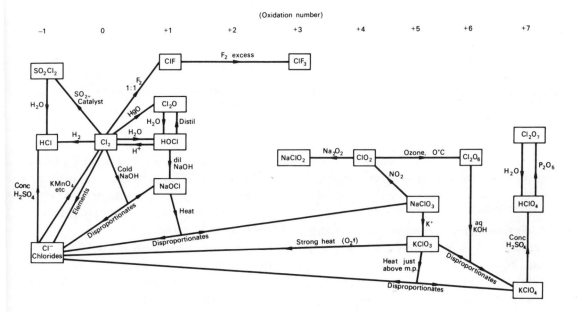

FIG. 21.7. Reactions of chlorine.

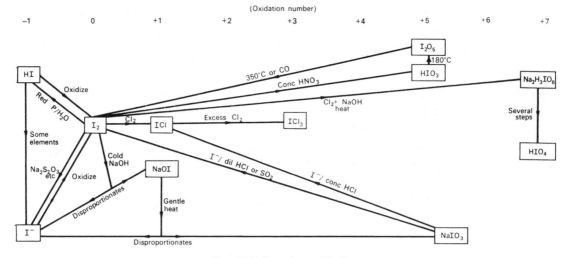

Fig. 21.8. Reactions of iodine.

21.12 Summary

(1) PHYSICAL PROPERTIES

X_2 molecules; increasing intermolecular-forces down Group, hence $Br_2(l)$ and $I_2(c)$, at room temperature.

(2) CHEMICAL PROPERTIES

IONIZATION:

(a) X^- (noble-gas structure) readily formed

(b) Some oxo-ions XO^-, XO_2^-, XO_3^- and XO_4^- formed (not fluorine).

REACTIVITY:

(a) They combine vigorously with metals.

(b) The elements are oxidizing agents.

COMPOUNDS:

(a) Ionic lattices containing X^- readily formed with electropositive metals, e.g. in the s-block.

(b) Covalent lattices, or complex ions, readily formed with weaker metals, e.g. transition metals and B-metals.

(c) Molecular compounds formed with the non-metals, including hydrogen.

(3) TRENDS DOWN THE GROUP

(a) Ionic radii increase down the Group: the larger ions are readily polarized.

(b) Ionization energies decrease down the group (cf. trends in s-block, section 20.18).

(c) Order of oxidizing power is $F_2 > Cl_2$ $> Br_2 > I_2$ (relatively weak).

(d) Order of reducing power is $I^- > Br^- >$ $> Cl^- > F^-$ (last two almost negligible).

(e) Increasing tendency away from non-metallic behaviour down the Group. (cf. trends in Group V, Chapter 23).

The halogens, like the alkali metal and alkaline earth metal Groups, form a closely related family, though the range of chemical behaviour is wider. Fluorine, the first member of the series, departs from the regular Group trends in some respects. Note in particular:

(a) The acid HF is relatively weak.

(b) Fluorine does not form oxo-ions.

(c) Fluorine is exceptionally reactive, due primarily to the unexpectedly small F–F bond dissociation energy.

Study Questions

1. Use Table 21.3 to calculate the energy change that accompanies each of the following reactions. (C–H $= 415$ kJ mol^{-1}; X is any halogen.)

(a) $X_2 + 2\,HI \longrightarrow 2\,HX + I_2$.

(b) $X_2 + 2\,CH_4 \longrightarrow 2\,CH_3X + 2\,HX$.

2. Describe what happens when dilute hydrochloric acid is boiled.

3. (a) How will the following react with water?

      (i) NaCl(c).

      (ii) FeCl$_3$(c).

      (iii) SiCl$_4$(l).

      (iv) ICl(l).

      (v) CCl$_4$(l).

(b) NCl$_3$ reacts very slowly with water while PCl$_3$ reacts very rapidly. Can you account for this difference?

4. Write equations for the preparation of hydrogen bromide from:

(a) The hydrolysis of silicon tetrabromide.

(b) The burning of hydrogen in bromine at 300°C.

(c) The reaction of phosphorus tribromide with ethyl alcohol.

(d) Distilling potassium bromide with orthophosphoric acid, (H$_3$PO$_4$).

5. Predict the shapes of the following molecules and ions: (a) PBr$_3$; (b) SF$_4$; (c) XeF$_4$; (d) I$_3^-$; (e) ClO$_3^-$; (f) ClO$_4^-$.

6. Balance the following equations. Show the nature of the oxidation-reduction process in each case.

(a) $Ce^{4+} + I^- \longrightarrow Ce^{3+} + I_2$.

(b) $BrO_3^- + I^- + H^+ \longrightarrow Br^- + I_2 + H_2O$.

(c) $MnO_4^- + I^- + H^+ \longrightarrow Mn^{2+} + I_2 + H_2O$.

(d) $Cr_2O_7^{2-} + I^- + H^+ \longrightarrow Cr^{3+} + I_2 + H_2O$.

(e) $OCl^- + I^- + H^+ \longrightarrow Cl^- + I_2 + H_2O$.

(f) $Cu^{2+} + I^- \longrightarrow CuI + I_2$.

(g) $I_2 + KIO_3 + HCl \longrightarrow KCl + ICl + H_2O$.

(h) $BrO_3^- + N_2H_4 \longrightarrow Br^- + N_2 + H_2O$.

7. Chromium forms chlorides in oxidation states 2 and 3. Give equations for the reactions that will probably occur when chromium reacts with (a) HCl(g), (b) Cl$_2$(g).

8. When the sodium nitrate has been extracted from caliche, an iodate solution containing 10 g of iodine per dm^3 remains. How many (a) moles, (b) grams, of sodium bisulphite must be added to the solution to reduce exactly the iodate to iodine?

9. Classify the following halides as molecular, ionic or giant structures. Which of them will be hydrolysed by water?

(a) BCl$_3$, (b) AlBr$_3$, (c) NaCl, (d) BaI$_2$, (e) TiCl$_2$, (f) TiCl$_4$, (g) CHI$_3$, (h) IBr.

10. (a) What would you expect a sample of astatine to be like at room temperature?

(b) Will HAt be a strong or a weak acid?

(c) How stable would you expect At$_3^-$ to be?

(d) How soluble would AgAt be?

(e) How would conc H$_2$SO$_4$ react with an astatide?

(f) What other physical and chemical properties would you expect astatine to show?

11. (a) Give an account of the ways in which the chemistry of fluorine differs from the other halogens.

(b) What physical factors are responsible for the differences?

12. (a) Excess iodide, I$^-$(aq), in weak acid, was added to 25 cm^3 of 0·1 M iodate solution, IO$_3^-$ (aq). The iodine that was liberated was titrated against 0·5 M sodium thiosulphate solution: 30 cm^3 of the latter was required.

(b) The same iodate solution was titrated into 25 cm^3 of 0·1 M KI solution in very strong hydrochloric acid, in the presence of a little CCl$_4$. Initially, the layer of CCl$_4$ became violet, but after 12·5 dm^3 of the iodate solution had been added, it became colourless.

Explain these reactions and give equations for them.

13. Show how the reactions of the halogens can be interpreted in terms of their half-cell potentials.

CHAPTER 22

Oxygen and sulphur

SHOWING DIFFERENCES BETWEEN FIRST AND SECOND ROW ELEMENTS

Oxygen and sulphur are the two lightest elements of Group VIB, and this chapter is concerned mainly with the relationship between them. The remaining elements, selenium, tellurium and polonium, are relatively rare and they will not be considered in detail.

A number of differences are observed between oxygen, the "parent" element of the Group, and the remainder. It is in fact a characteristic of any Group of the periodic table that the first member often differs from the rest. Although this chapter stresses the differences between oxygen and sulphur, Group VIB is not exceptional in this respect.

22.1 Occurrence and extraction of oxygen and sulphur

Table 22.1 lists the Group VIB elements, with their ground state electronic structures. Oxygen and sulphur are encountered native —oxygen as a constituent of the atmosphere, and sulphur in certain rocks—whereas the halogens are only found in chemical combination.

Oxygen is present in the atmosphere as diatomic molecules $O_2(g)$, forming approximately one-fifth of the atmosphere by weight. In the upper atmosphere, a certain amount of atomic oxygen is formed due to the action of ultraviolet energy and cosmic rays causing dissociation:

$$O_2(g) + h\nu \longrightarrow O(g) + O(g);$$
$$\Delta E = +497 \text{ kJ}.$$

TABLE 22.1

Name	Principal quantum number of valence shell	Electronic structure in ground state (abbreviated form)	Electronic structure (full form)
Oxygen	2	2, 6	$1s^2$; $2s^2p^4$
Sulphur	3	2, 8, 6	$1s^2$; $2s^2p^6$; $3s^2p^4$
Selenium	4	2, 8, 18, 6	$1s^2$; $2s^2p^6$; $3s^2p^6d^{10}$; $4s^2p^4$
Tellurium	5	2, 8, 18, 18, 6	$1s^2$; $2s^2p^6$; $3s^2p^6d^{10}$; $4s^2p^6d^{10}$; $5s^2p^4$
Polonium	6	2, 8, 18, 32, 18, 6	$1s^2$; $2s^2p^6$; $3s^2p^6d^{10}$; $4s^2p^6d^{10}f^{14}$; $5s^2p^6d^{10}$; $6s^2p^4$

Free oxygen atoms can combine with ordinary oxygen leading to triatomic molecules of *ozone*, $O_3(g)$:

$$O_2(g) + O(g) \longrightarrow O_3(g);$$

Oxygen exists in chemical combination in two main forms:

(i) as oxides;

(ii) as oxo-anions.

Examples of compounds in which oxygen is present in the oxo-anions are nitrate, NO_3^-, carbonate, CO_3^{2-}, sulphate, SO_4^{2-}, and ortho-phosphate, PO_4^{3-}.

Oxygen is extracted from the atmosphere on a large scale by the fractional distillation of liquid air. Air is liquefied by making use of the Joule–Thomson effect (cooling by expansion of the gas), and water vapour and carbon dioxide are removed by solidification. Oxygen boils at a higher temperature than nitrogen, and the two gases form a near-ideal mixture. The principles of the separation are discussed in section 14.13.

Sulphur is fairly widespread in nature, constituting about 0·1 per cent of the Earth's crust. It occurs mainly as sulphides of metals, but also as sulphates and the native element. The chief sources of free sulphur are the following:

(a) *Frasch process*. Sulphur is extracted from underground sulphur-bearing rock in Texas by sinking a shaft about a foot in diameter containing three concentric tubes. Compressed air and superheated water (160°C and about 10 atm) are forced down, and this melts the sulphur which comes up as a foam mixed with air and water. Earthy material is left behind and sulphur of high purity (about 99·5%) is obtained (Fig. 22.1).

(b) *From petroleum*. Most deposits of crude petroleum contain sulphur, and some are economical for extraction when other sources of

the element are scarce. The element is usually obtained as hydrogen sulphide, $H_2S(g)$, which can be converted to sulphur by burning with limited air:

$$2\,H_2S + O_2 \longrightarrow 2\,H_2O + 2\,S.$$

The chief use of sulphur is for the conversion into sulphur dioxide and thence to sulphuric acid, and so other methods of extraction are

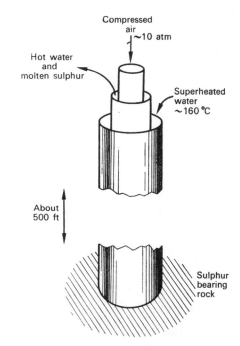

FIG. 22.1. The Frasch process.

employed to obtain sulphur dioxide directly. Two important methods are:

(a) *From iron pyrites, and other sulphide ores*. Iron pyrites, FeS_2, is a common sulphide ore, and when it is heated in air (roasted) it is converted into $Fe_2O_3(c)$ and sulphur dioxide. Other sulphide ores are often roasted as a means of obtaining the metal, and in these cases sulphur dioxide is a useful by-product (section 13.6).

(b) *From sulphates.* In the United Kingdom, *anhydrite*, $CaSO_4$, is an important raw material, and this is converted to sulphur dioxide for sulphuric acid manufacture by reduction with carbon.

22.2 Reactions and uses of oxygen and sulphur

Both elements combine directly with most metals. Only the noble metals are totally immune to attack by oxygen. Frequently a protective coating of oxide prevents further reaction from taking place. This is true of aluminium for instance.

The relative stabilities of oxides and sulphides play an important part in the extraction of metals from their ores, and this subject was discussed in some detail in Chapter 13.

The reactions of other elements with oxygen and sulphur differ largely because of the different structures and physical states of the two elements. Sulphur has a marked tendency to **catenate** (it forms chains and S_8 rings) whereas oxygen forms diatomic molecules. The structure of sulphur was discussed in section 5.8, and its allotropy further discussed in sections 14.4 and 14.5.

Both elements will combine with most nonmetals, though not always by direct reaction. All non-metals will burn in oxygen except the halogens, noble gases, nitrogen and selenium.

Even nitrogen will combine directly with oxygen above 2000°C in an endothermic reaction, giving nitrogen(II) oxide (section 23.8).

In general, oxygen combines more readily with electropositive elements than does sulphur but sulphur combines more easily with the weaker metals such as copper, silver and lead. In nature the weaker metals tend to occur as sulphides rather than as oxides, for example Ag_2S, HgS, PbS and ZnS.

Considerable quantities of oxygen are consumed in the use of oxy-acetylene welding equipment and in metallurgical processes. A large amount is used in the petrochemical industry in the manufacture of ethylene oxide, and the gas is used extensively in industry for a wide variety of other oxidizing reactions. Smaller amounts of highly purified oxygen are required for medical use.

Sulphur is mainly used for conversion into sulphur dioxide, and thence to sulphuric acid (section 22.7). Smaller amounts are used for vulcanizing (hardening) rubber; the tangled molecules of rubber become cross-linked with –S–S– bridges.

22.3 Physical trends down Group VIB

Figure 22.2 shows how the sizes of atoms and ions vary down Group VIB. All the ions X^{2-} are iso-electronic with the nearest noble gas.

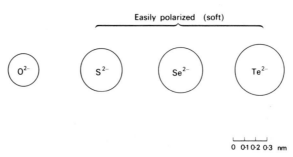

Easily polarized (soft)

O^{2-} S^{2-} Se^{2-} Te^{2-}

0 0·1 0·2 0·3 nm

FIG. 22.2.

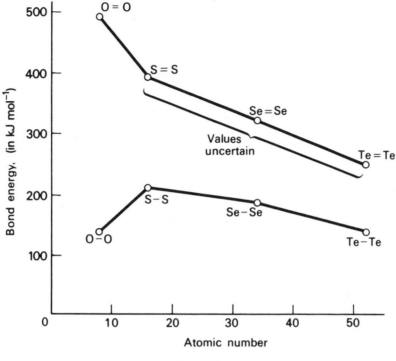

FIG. 22.3.

Mere consideration of size however is not adequate in explaining the marked differences between oxides and sulphides. The ion O^{2-} is far less easily distorted—less *polarizable*—than the remaining ions of the group. This is sometimes expressed another way by saying that O^{2-} is a *hard* ion, while the remainder are soft ions.

Figure 22.3 is a plot of bond energies for the bonds X–X and X=X. It is seen from this that only oxygen has a preference for double bonds $O{=}O^{*}$.

The melting and boiling points of the Group VIB elements show marked differences which reflect the different structures of the elements. It will be left to the reader to plot these out and to compare them with other Groups.

---

* The double bond in O_2 is unusual in that two of the bonding electrons are unpaired.

Like the halogens, the Group VIB elements all have quite high ionization energies, and consequently positive ion formation is not common. Polonium is, however, a metal, and tellurium does show some metallic properties, these being the elements of the group with the lowest ionization energies.

Those forms of sulphur which contain S_8 molecules are soluble in non-polar solvents like benzene and carbon disulphide. "Polymeric" forms of sulphur, such as plastic and amorphous sulphur, do not dissolve readily in any solvent.

22.4 Compounds with hydrogen

The Group VIB elements all form simple hydrides, H_2X. Water is exceptional in the following respects:

(a) it has a higher melting point and boiling point;

(b) it is a weaker acid.

These effects are attributed to hydrogen bonding (section 18.7). Hydrogen bond formation involving sulphur atoms is rare, and is always very weak. Hence $H_2S(g)$ is relatively insoluble in water (a saturated solution at room temperature and pressure is only about 0.1 M).

Hydrogen and sulphur combine together to form *polysulphides*, H_2S_x, where x can be as large as five. This is another manifestation of the property of catenation—sulphur atoms can link to form chains. Oxygen has a weaker tendency to catenate, and the highest known oxide of hydrogen is hydrogen peroxide, H_2O_2.

22.5 Hydrogen peroxide

Hydrogen peroxide, H_2O_2 (m.p. $-0.5°C$, b.p. $158°C$), is a colourless viscous liquid when pure. It decomposes violently if heated, the reaction $H_2O_2 \rightarrow H_2O + \frac{1}{2}O_2$ being exothermic. It has a bent structure

Redox reactions of hydrogen peroxide. Hydrogen peroxide is capable of acting both as an oxidizing and as a reducing agent.

As an oxidizing agent:

$$H_2O_2(aq) + 2H^+(aq) + 2e^- \longrightarrow 2H_2O \quad (1)$$

As a reducing agent:

$$H_2O_2(aq) \longrightarrow O_2(g) + 2H^+(aq) + 2e^- \quad (2)$$

The reducing reactions are generally favoured by alkaline solution (addition of OH^- removes

H^+), while oxidizing actions generally proceed in an acidic medium (addition of H^+ favours equation (1)). The following equations illustrate both these processes:

(a) Reducing reactions:

(i)

(ii)

(b) Oxidizing reactions:

(i) $4 H_2O_2 + PbS \longrightarrow 4 H_2O + PbSO_4$

   $\quad\quad\quad\quad$ black $\quad\quad\quad\quad\quad\quad\quad$ white

(ii) $2Fe^{2+} + H_2O_2 + 2H^+ \longrightarrow 2Fe^{3+} + 2H_2O$

(iii) $2 I^- + H_2O_2 + 2 H^+ \longrightarrow I_2 + 2 H_2O$.

Uses of hydrogen peroxide. Many dyes are bleached by oxidation with hydrogen peroxide; textiles and wood pulp are bleached in this way. Since the only by-product is water, elaborate washing after bleaching is unnecessary.

Preparation of hydrogen peroxide. Hydrogen peroxide can be made by the action of $H^+(aq)$ on a metal peroxide. Sodium or barium peroxide (made by burning the metals in oxygen) and dilute sulphuric acid have been employed:

$$Na_2O_2 + 2 H^+ \longrightarrow H_2O_2 + 2 Na^+.$$

A *peroxide* is an oxide which contains structural units $[O—O]^{2-}$.*

---

* Older systems of nomenclature sometimes incorrectly refer to oxides as peroxides when they contain a metal in a high oxidation state. Lead(IV) oxide, PbO_2, for instance, was known as lead peroxide.

A *peroxo-salt* is one containing an —O—O— linkage in the anion. The peroxo-disulphate ion provides an example. It is prepared by the anodic oxidation of sulphuric acid under conditions where sulphate ions are discharged preferentially (60% solution, high current density, low temperature).

$$2 \begin{bmatrix} & O & \\ & \uparrow & \\ O \leftarrow & S & \rightarrow O \\ & \downarrow & \\ & O & \end{bmatrix}^{2-} \longrightarrow \begin{bmatrix} & O & & O & \\ & \uparrow & & \uparrow & \\ O \leftarrow & S & —O—O— & S & \rightarrow O \\ & \downarrow & & \downarrow & \\ & O & & O & \end{bmatrix}^{2-}$$
$$+ 2e^-$$

Action of dilute mineral acid on a peroxo-acid or its salt produces hydrogen peroxide.

Volume concentration of hydrogen peroxide. The concentration of hydrogen peroxide is often expressed as "10-volume", or "20-volume", etc. An *n*-volume solution is defined as one of such concentration that 1 volume of solution gives *n* volumes of oxygen at s.t.p.

$$2 H_2O_2 \longrightarrow 2 H_2O + O_2.$$

2 moles of $H_2O_2 \rightarrow 1$ mole of $O_2 = 22\,400$ cm^3 of O_2 at s.t.p.

∴ 2 M H_2O_2(aq) = a 22·4-volume solution = 68 g dm^{-3}.

Similarly, a 10-volume solution of H_2O_2 is $\left(2 \times \dfrac{10}{22 \cdot 4}\right)$ M.

22.6 Hydrogen sulphide

Hydrogen sulphide is a colourless gas, highly poisonous, with a strong smell resembling rotten eggs. Sulphur, unlike oxygen, does not react readily with hydrogen, though some hydrogen sulphide is produced if hydrogen is bubbled through boiling sulphur. The following equations show that oxygen has a greater affinity for hydrogen than sulphur has.

$$H_2(g) + \tfrac{1}{2} O_2(g) \longrightarrow H_2O(g);$$
$$\Delta H = -240 \text{ kJ at } 25°C$$

$$H_2(g) + S(\text{rhombic}) \longrightarrow H_2S(g);$$
$$\Delta H = -19 \cdot 6 \text{ kJ at } 25°C.$$

The gas is usually obtained by the action of dilute acid on the sulphide of a metal.

$$FeS + 2 H^+(aq) \longrightarrow H_2S(g) + Fe^{2+}(aq).$$

For making the pure gas, antimony(III) sulphide and hydrochloric acid are very suitable reagents.

Hydrogen sulphide burns in excess oxygen forming water and sulphur dioxide; in limited air, some free sulphur is produced.

Hydrogen sulphide is a weak acid:

$$H_2S(aq) \rightleftharpoons H^+(aq) + HS^-(aq);$$
$$K_a = 10^{-7} \text{ mol dm}^{-3} \text{ at } 25°C.$$

It reacts as a dibasic acid, though the concentration of free ions S^{2-}(aq) in a solution of hydrogen sulphide is very low indeed.

Hydrogen sulphide is a powerful reducing agent, and sulphur is formed:

$$S + 2 H^+ + 2e^- \rightleftharpoons H_2S;$$
$$\epsilon° = +0 \cdot 141 \text{ V}$$

Examples of species which hydrogen sulphide will reduce are iron(III) ions to iron(II), chlorine to chloride ion, sulphuric acid to sulphur dioxide and further to sulphur, and permanganate to manganese(II). Equations may be readily balanced using the half-cell reactions.

Worked Example. Obtain an equation for the reduction of iron(III) chloride to iron(II) chloride by hydrogen sulphide.

$$2 Fe^{3+}(aq) + 2e^- \longrightarrow 2 Fe^{2+}(aq)$$

Subtracting the half-cell equation for hydrogen sulphide, we have,

$$2Fe^{3+} + H_2S \longrightarrow 2Fe^{2+} + 2H^+ + S.$$

Adding the *spectator ions*, $6Cl^-$, the full equation becomes:

$$2\,FeCl_3 + H_2S \longrightarrow 2\,FeCl_2 + 2\,HCl + S.$$

Hydrogen sulphide is a familiar laboratory reagent, as it used in the identification of metal ions by the formation of insoluble sulphides of distinctive properties. The properties of some sulphides are considered in Section 22.13.

OXIDES AND OXO-ACIDS OF SULPHUR

Sulphur forms a range of oxides, among which sulphur dioxide, SO_2 (m.p. $-73°C$, b.p. $-10°C$) and sulphur trioxide, SO_3 (m.p. $17°C$, b.p. $45°C$) are the most important. A large number of oxo-acids also exist, the most important being sulphuric acid, H_2SO_4.

Pure sulphurous acid cannot be isolated: in this respect it resembles carbonic acid, nitrous acid and hypochlorous acid.

Sulphur dioxide is not inflammable, though it will combine with oxygen in the presence of a catalyst. This reaction is put to good use in the **contact process** for manufacture of sulphuric acid:

$$SO_2(g) + \tfrac{1}{2}O_2(g) \longrightarrow SO_3(g);$$
$$\text{converted to } H_2SO_4$$
$$\Delta H = -98 \text{ kJ at } 25°C.$$

It will only support combustion in the case of metals whose free energy of oxide formation exceeds that of sulphur dioxide itself. Aluminium and magnesium, for instance, will burn in the gas liberating sulphur and forming $Al_2O_3(c)$ and $MgO(c)$ respectively.

Sulphur dioxide is reduced to sulphur by hydrogen sulphide, and hence sulphur is precipitated when $H_2S(g)$ is bubbled into a solution of a sulphite or hydrogen sulphite.

$$SO_2(g) + 2\,H_2S(g)$$
$$\longrightarrow 3\,S(\text{amorphous}) + 2\,H_2O(g).$$

This is one of the rare reactions in which sulphur dioxide is acting as an oxidizing agent: in general it is a fairly powerful reducing agent:

$$SO_4^{2-}(aq) + 4\,H^+(aq) + 2e^-$$
$$\rightleftharpoons SO_2(aq) + 2\,H_2O;$$
$$\epsilon° = +0·17 \text{ V}$$

(cf. hydrogen sulphide above).

22.7 Sulphur dioxide and sulphurous acid

Sulphur dioxide, SO_2, is a colourless gas with a sharp, choking smell, which liquefies readily under pressure (cf. boiling point above). It is produced industrially either by burning sulphur in air, or as a by-product of *roasting* of a sulphide ore:

$$S + O_2 \longrightarrow SO_2;$$
$$\Delta H = -297 \text{ kJ at } 25°C$$

$$4\,FeS_2 + 11O_2 \longrightarrow 8\,SO_2 + 2\,Fe_2O_3;$$
iron
pyrites $\qquad \Delta H = -3300 \text{ kJ at } 25°C.$

Sulphur dioxide is also made in the United Kingdom by a process in which anhydrite, $CaSO_4(c)$, is reduced with carbon.

Sulphur dioxide is readily soluble in water (one volume of water dissolves 45 of the gas at $15°C$), and the solution contains the dibasic acid, sulphurous acid, $H_2SO_3(aq)$:

$$SO_2(g) + H_2O \rightleftharpoons H_2SO_3(aq)$$
$$\rightleftharpoons H^+(aq) + HSO_3^-(aq); \; pk_1 = 1·8 \text{ at } 25°C.$$
$$\Updownarrow$$
$$H^+ + SO_3^{2-}; \; pk_2 = 6·2 \text{ at } 25°C.$$

The action of sulphur dioxide on aqueous solutions of oxidizing agents is similar to that of hydrogen sulphide, except that sulphur is not precipitated. For instance:

$$2\,MnO_4^- + 5\,SO_2 + 2\,H_2O$$
$$\longrightarrow 5\,SO_4^{2-} + 2\,Mn^{2+} + 4\,H^+$$

Adding the spectator ions for potassium permanganate, this equation becomes:

$$2\,KMnO_4 + 5\,SO_2 + 2\,H_2O$$
$$\longrightarrow K_2SO_4 + 2\,MnSO_4 + 2\,H_2SO_4.$$

Sulphur dioxide is used industrially as a bleaching agent, a fungicide, and as a refrigerant gas, in addition to its use as an intermediate in the manufacture of sulphuric acid. In the laboratory it is usually obtained from a syphon. Alternatively small quantities can be made by the action of $H^+(aq)$ on a sulphite, or by reducing concentrated sulphuric acid with a metal such as copper.

The sulphur dioxide molecule is bent, with the sulphur atom in the middle, due to the lone-pair on the sulphur atom. The sulphite ion is pyramidal for the same reason.

22.8 Sulphur trioxide

$SO_3(g)$ is the product of passing a mixture of sulphur dioxide and oxygen over a heated catalyst, vanadium pentoxide, V_2O_5. Sulphur tri-oxide has a very strong affinity for water, and consequently forms dense white fumes in moist air:

$$SO_3(g) + H_2O(g) \longrightarrow H_2SO_4(1);$$
$$\Delta H = -175 \text{ kJ at } 25°C.$$

On cooling to room temperature, a white solid is formed which contains different crystalline modifications, which are polymeric forms of the monomer SO_3 (Fig. 22.4).

Sulphur trioxide is formed as white fumes when sulphates of some metals are strongly heated, although sulphates are fairly resistant to thermal decomposition, for instance,

$$\underset{\text{white}}{CuSO_4(c)} \longrightarrow \underset{\text{black}}{CuO(c)} + SO_3(g)$$

22.9 Sulphuric acid

Sulphuric acid is a very important industrial chemical, and indeed it has been remarked that a country's consumption of it is a fair indication of that country's economic prosperity, for very many processes require its use at some stage or another. It is used in the manufacture of fertilizers such as ammonium sulphate and calcium superphosphate, and in making explosives, detergents, accumulators, varnishes and artificial fibres. A large amount is used in the preparation of other acids by displacement, and for cleaning metal surfaces.

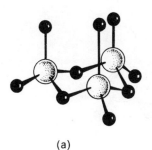
(a)

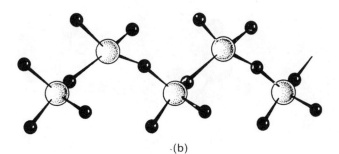
·(b)

FIG. 22.4. Forms of sulphur trioxide.

It is a heavy, oily liquid (density $1·834$ g cm^{-3} at $18°C$, b.p. $338°C$ with decomposition), made principally by two processes (a) the **chamber process,** which has been in operation for about two hundred years in some form, and which gives an impure acid of moderate concentration, and (b) the **contact process,** which can lead to acid of 100% purity.

(a) *The chamber process.* The chemistry of this process is complex but it involves the oxidation of sulphur dioxide by oxygen in the presence of $NO(g)$ and $NO_2(g)$. The oxides of nitrogen can be regarded as *homogeneous catalysts*, which operate by intermediate compound formation. Acid of about 78% purity is produced, and this is generally converted into fertilizer.

(b) *The contact process.* Sulphur trioxide is hydrated, by absorbing it in fairly concentrated sulphuric acid, previously prepared. If sulphur trioxide were added directly to water, the evolution of heat would be too great and a fine spray of acid would be formed. Anhydrous acid (100%) is readily made by the contact process, as is *oleum*, otherwise known as pyrosulphuric acid:

$$H_2SO_4(1)+SO_3(g) \longrightarrow H_2S_2O_7(1)$$

(ortho)sulphuric   pyrosulphuric
acid   acid

The structures of sulphuric acid and the sulphate ion are based on the tetrahedron (Fig. 22.5). Sulphuric acid has chemical properties as follows:

(a) *As an acid.* It is dibasic, and ionizes in dilute aqueous solution thus:

$$H_2SO_4+aq \rightleftharpoons H^+(aq)+HSO_4^-(aq)$$
$$HSO_4^-(aq) \rightleftharpoons H^+(aq)+SO_4^{2-}(aq)$$

The ions $SO_4^{2-}(aq)$ and $HSO_4^-(aq)$ are not oxidizing agents, and therefore dilute sulphuric acid liberates hydrogen when added to metals with negative half-cell potentials.

The concentrated acid is also an oxidizing agent, and this effect often obscures its simple acidic reactions. For instance metals react to evolve sulphur dioxide, but give no hydrogen.

(b) *As an oxidizing agent.* Concentrated sulphuric acid is readily reduced by many substances, giving sulphur dioxide. For example consider the reaction between copper and concentrated sulphuric acid:

half-reaction: $2e^-+2\,H^++H_2SO_4$
$\longrightarrow SO_2+2\,H_2O$ (reduction)

half-reaction: $\;H_2O+Cu$
$\longrightarrow CuO+2\,H^++2e^-$(oxidation)

$H_2SO_4+Cu \longrightarrow SO_2+H_2O+CuO$ (incomplete)

The equation is incomplete, because we know that CuO is a basic oxide:

$$CuO+H_2SO_4 \longrightarrow CuSO_4+H_2O$$

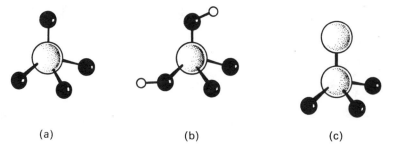

    (a)   (b)   (c)

FIG. 22.5. Structures of (a) the sulphate ion; (b) sulphuric acid; (c) the thiosulphate ion.

The final equation is therefore:

$$Cu(c) + 2 H_2SO_4(l)$$
$$\longrightarrow SO_2(g) + CuSO_4 + 2 H_2O.$$

Concentrated sulphuric acid will oxidize a number of other substances apart from metals, for instance carbon ($\rightarrow CO_2$), sulphur ($\rightarrow SO_2$), bromide ion ($\rightarrow$ bromine) and iodide ion ($\rightarrow$ iodine).

(c) *As a dehydrating agent*. Dehydration differs from drying: drying is simply the physical removal of absorbed water from a substance, while dehydration is the chemical removal of the elements hydrogen and oxygen in the ratio corresponding to H_2O. Sulphuric acid will dehydrate a number of organic substances, notably carbohydrates (of general formula $C_x(H_2O)_y$) which form carbon.

(d) *As a sulphonating agent*. Concentrated sulphuric acid and oleum are used extensively in the organic chemical industry for replacing hydrogen atoms with the sulphonic acid group $-SO_2OH$. For instance benzene forms benzene-sulphonic acid:

$$C_6H_6 + H_2SO_4 \longrightarrow C_6H_5SO_2OH + H_2O.$$

Sulphonation is an important reaction in the manufacture of detergents, which are generally salts of an alkylsulphonic acid.

22.10 Other oxo–ions of sulphur

A large number of oxo-ions of sulphur exist, and there would be little point in cataloguing all their properties. Certain ones are of special interest, if only because they frequently occur in chemical processes.

Sodium thiosulphate $Na_2S_2O_3 . 5 H_2O(c)$ is an important salt though its parent acid thio-sulphuric acid, $H_2S_2O_3$, does not exist. The prefix *thio* means "containing sulphur in place of oxygen", and the thiosulphate ion is structurally analogous to sulphate (Fig. 22.5).

It is prepared by boiling sodium sulphite with sulphur:

$$Na_2SO_3(aq) + S(rhombic) \longrightarrow Na_2S_2O_3(aq).$$

Sodium thiosulphate is used in volumetric analysis for quantitatively reducing solutions of iodine. It forms the *tetrathionate* ion, $S_4O_6^{2-}$.

Half-cell equation: $S_4O_6^{2-} + 2e^- \rightleftharpoons 2 S_2O_3^{2-}$

Half-cell equation: $I_2 + 2e^- \rightleftharpoons 2 I^-$

Hence one mole of sodium thiosulphate will reduce half a mole of iodine in the complete equation:

$$I_2 + 2 S_2O_3^{2-} \rightleftharpoons S_4O_6^{2-} + 2 I^-.$$

One curious property of sodium thiosulphate is its remarkable tendency to supersaturate in aqueous solution. If the solid crystals are gently warmed, they "melt", or dissolve in their own water of crystallization.

This solution can be cooled to room temperature without nucleation taking place to allow crystals to form. Seeding with a tiny crystal, or scratching the inside of the test-tube, causes rapid crystallization to form the solid again, with noticeable evolution of heat.

22.11 Acid–base character of oxides

Almost all the elements form compounds with oxygen (the only exceptions are the lighter noble gases) and these oxides may be classified in a variety of different ways: according to acid–base character, according to structure, according to bond type, and so on. One of the most useful classifications is according to **acid–base character.** The three main categories are:

(i) *Acidic oxides;*
(ii) *Basic oxides;*
(iii) *Oxides which do not display acid–base properties at all.*

This third category is very small and unimportant. Practically all oxides show *some* tendency to undergo acid–base reactions, though the rate of reaction may be slow. At high temperatures for instance, carbon monoxide reacts as an acidic oxide with sodium hydroxide, forming sodium formate.

A number of oxides exist which do not fit precisely into the above categories. For instance *amphoteric oxides* exist, which fall into categories (i) and (ii) simultaneously. Some oxides have a stoichiometric composition corresponding to the presence of two oxidation numbers in the same oxide. For instance red lead, $Pb_3O_4(c)$, can be regarded as being a *compound oxide* made up of $2\,PbO + PbO_2$. Its systematic name is therefore lead(II,II,IV) oxide and it overlaps categories (i) and (ii) above.

The acid–base character of an oxide depends upon two main factors (Fig. 22.6).

(a) The position of the element in the periodic table. Elements on the right (non-metals) form acidic oxides while those on the left form basic oxides.

(b) The oxidation state of the element. The higher the oxidation state, the more strongly acidic the oxide. This is especially true of some metals where a complete range can often be observed, e.g.

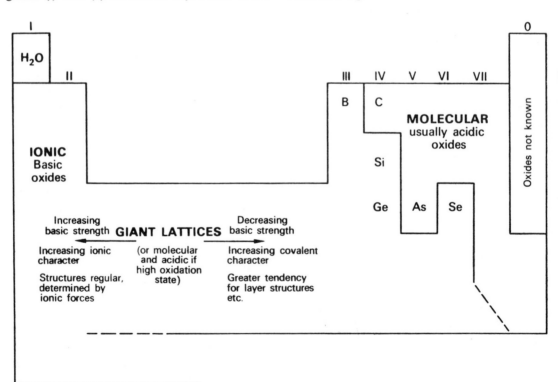

FIG. 22.6. The properties of oxides.

chromium(II) oxide, CrO, basic;
chromium(III) oxide, Cr_2O_3, amphoteric;
chromium(VI) oxide, CrO_3, acidic.

ACIDIC OXIDES

Most non-metals form acidic oxides, that is, oxides which will dissolve in alkali to form salts. Most acidic oxides react with water to form $H^+(aq)$, but this is not the case if the oxide is macromolecular. Thus $B_2O_3(c)$ and $SiO_2(c)$ are classed as acidic oxides since they react with alkalis, though they do not react with water.

BASIC OXIDES

Basic oxides may be sub-divided into two classes:

(a) those which are insoluble in, and not attacked by, water;
(b) those which react with water producing $OH^-(aq)$.

Oxides in category (a) are generally macromolecular structures in which the bonding is essentially covalent. Their structures are frequently layered, and do not correspond to those expected for the packing together of ionic particles. Oxides in category (b) are generally simple ionic lattices containing the

O^{2-} ion, though some elements of Group IA and IIA form *peroxides* (containing the $[O—O]^{2-}$ ion) and occasionally *superoxides* (containing the $[O—O]^-$ ion). These were considered in Chapter 20.

AMPHOTERIC OXIDES

These are oxides of metals which have the property of dissolving in acids as well as in alkalis, due to the ability of the metal to form complex ions of the type $M(OH)_x^{n-}$:

$$Al_2O_3(c) + 6\,H^+(aq) \longrightarrow 2\,Al^{3+}(aq) + 3\,H_2O.$$
$$Al_2O_3(c) + 2\,OH^-(aq) + 3\,H_2O \longrightarrow 2\,Al(OH)_4^-.$$

Other examples of amphoteric oxides are ZnO, SnO, SnO_2, PbO, PbO_2, and Cr_2O_3.

"NEUTRAL" OXIDES

A few oxides exist which seem to show no acid–base properties. Examples are nitrogen(I) oxide, N_2O, nitrogen(II) oxide, NO, and oxygen difluoride, OF_2.

Carbon monoxide is sometimes classed as a neutral oxide, but it will react with alkali at high temperatures, and this reaction is used industrially:

$$CO(g) + OH^-(aq) \longrightarrow H.COO^-(aq)$$
$$\text{formate ion}$$

TABLE 22.2

A SELECTION OF ACIDIC OXIDES
OF NON-METALS

IIIB	IVB	VB	VIB	VIIB
B_2O_3	CO_2	N_2O_3 N_2O_4 N_2O_5	—	—
	SiO_2	P_4O_6	SO_2 SO_3	Cl_2O_7
		P_4O_{10}		

22.12 Structure of oxides

An alternative way of classifying the oxides of elements is by their structure. This form of classification is in some ways similar to the acid–base classification given above, for structure does have an important bearing upon chemical properties. The main categories which may be distinguished are as follows:

(a) *Essentially ionic structures.* These oxides have regular structures of the form expected from the packing together of oppositely charged ions, usually O^{2-}. Peroxides and superoxides come into this category. These essentially ionic structures are usually soluble in water, though in cases where the lattice energy is high the solubility may be small, (as for instance with MgO). Ionic oxides occur on the left-hand side of the periodic table (Fig. 22.6).

(b) *Essentially covalent, macromolecular structures.* Where an oxide has a distorted structure, or a layer structure, it may be concluded that simple ionic forces are not operating but that directional bonds exist between the atoms. Such oxides are common among the transition metals and the B-metals. They are insoluble in water, and often only react with alkali on prolonged boiling. The acidic oxides B_2O_3 and SiO_2 come into this category, though most macromolecular oxides are basic or amphoteric. Many covalent macromolecules of this type are markedly non-stoichiometric.

(c) *Molecular structures.* Molecular oxides are usually acidic; even if they are not gaseous at room temperature, they are nevertheless volatile. They may be oxides either of non-metals or of metals in high oxidation states. Examples are $P_4O_6(c)$ and $Mn_2O_7(l)$.

22.13 Sulphides

It was observed in section 22.3 that the sulphide ion is "softer" than the oxide ion. In other words, the former is more easily distorted (polarized) by the proximity of adjacent positive ions. Most metal ions, except those of the *s*-block metals, are also "soft" ions, and when one attempts to form a compound between two soft ions, what actually happens is that bonds form instead. For this reason, although sulphides may be classified structurally in a similar way to oxides, the ionic category is very small (limited to the metals of Group IA in effect) while the covalent category is very large.

A typical ionic sulphide is sodium sulphide: it is soluble in water and it can be electrolysed when molten. Its physical properties (melting and boiling points, structure) are those expected

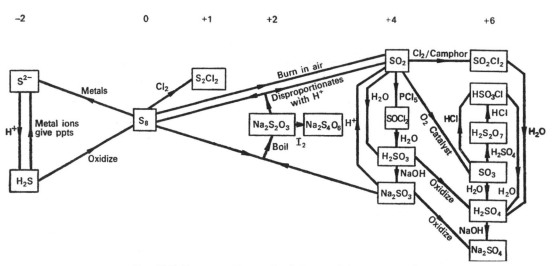

FIG. 22.7. Some reactions of sulphur and its compounds.

for simple ions. Most other metal sulphides are almost totally insoluble in water, and they form layer structures or distorted structures typical of bond formation. They are often highly coloured, the colours being different from those of the hydrated metal ions. The colours are due to energy absorbed ($\Delta E = h\nu$) when **charge transfer** occurs. For example mercury(II) sulphide is black because the energy of transfer of charge from S^{2-} to Hg^{2+} is relatively low, and corresponds to the visible region of the electromagnetic spectrum. In contrast, barium sulphide, BaS, is colourless because the charge transfer energy corresponds to absorption in the ultra-violet. The more polarizable the ions, the lower will be the charge transfer energy as a rule, and the greater the likelihood of the compound being coloured.

Most non-metals form sulphides, but only a few have achieved any importance practically. Carbon disulphide, CS_2, is used as a solvent and P_4S_3 finds uses in the match industry. Figure 22.7 summarizes some of the reactions of sulphur.

22.14 Summary

The trends down Group VI are similar to those down the halogen Group (see Study Question 15). The differences between oxygen and sulphur highlight an important feature of the periodic table, namely the differences between first-row and second-row elements.

(a) The relative increase in atomic radius (and hence decrease in ionization energy) is most marked between the first two members of a Group; hence the first-row elements are markedly more electronegative. For instance, N, O and F form *much* stronger hydrogen bonds than P, S and Cl.

(b) The valence shell of first-row elements is limited to eight electrons (Chapter 3) whereas second-row elements have $3d$ levels available for bond formation. Thus:

(i) SF_6 (but OF_2); PCl_5 (but NCl_3); SiF_6^{2-} (but no CF_6^{2-}); AlF_6^{3-} (but BeF_4^{-}).

(ii) $SiCl_4$ hydrolyses but CCl_4 is kinetically stable.

(c) Second-row elements have less tendency to form multiple bonds. Thus:

(i) $O\!=\!C\!=\!O$ is a gas, but SiO_2 is a polymeric solid, with single Si—O bonds.

(ii) Compare S_8 with O_2.

(iii) Compare the various allotropic forms of phosphorus with N_2.

Study Questions

1. (a) Use a data book to plot the melting and boiling points of the Group VIB elements against their atomic numbers.

    (b) How do the plots compare with those obtained for Group IA?

2. (a) The boiling points of CH_4, NH_3, H_2O and HF are $-160°$, $-33°$, $+100°$ and $+19°C$ respectively. How can you account for the maximum at water?

    (b) H_2O_2 boils at 158°C: What can you say about the structure of this substance?

3. Suggest how the molecular formulae of (a) H_2S, (b) SO_2 could be proved experimentally.

4. (a) How many moles of sulphuric acid are there in a litre of pure H_2SO_4 at room temperature?

    (b) Concentrated sulphuric acid reacts with carbohydrates to give carbon. What would you expect it to give with (i) formic acid, HCOOH, (ii) oxalic acid, $(COOH)_2$?

(c) Why does concentrated sulphuric acid decolorize blue copper sulphate crystals?

5. (a) Calculate the molarity of a 5-volume solution of hydrogen peroxide.
   (b) The decomposition of hydrogen peroxide is a redox process. Label the equation to show the nature of the oxidation and reduction.

6. Which oxide in the following pairs will be the more acidic?

(a) CaO and CO_2.   (b) Cr_2O_3 and CrO_3.
(c) MnO and Mn_2O_7.   (d) N_2O and NO_2.
(e) SeO_2 and TeO.

7. (a) Classify the following oxides in terms of structure:
      (i) CaO, (ii) Fe_2O_3, (iii) SiO_2, (iv) Cl_2O_7.
   (b) What relationship is there between the structure of oxides and their acid–base character?

8. Write equations for the reactions of:

(a) ZnO with (i) conc HCl, (ii) conc NaOH(aq).
(b) SO_2 with (i) Fe^{3+}(aq), (ii) $Cr_2O_7^{2-}$ (both in acid).

9. Balance the following equations, and show clearly the nature of the oxidation-reduction process in each case:

(a) $MnO_4^- + H_2O_2 + H^+ \longrightarrow Mn^{2+} + H_2O + O_2$.
(b) $Fe(CN)_6^{4-} + H_2O_2 + H^+ \longrightarrow Fe(CN)_6^{3-} + H_2O$.
(c) $Fe(CN)_6^{3-} + H_2O_2 + OH^-$
      $\longrightarrow Fe(CN)_6^{4-} + H_2O + O_2$.
(d) $Cl_2 + H_2O_2 \longrightarrow HCl + O_2$.
(e) $(NH_4)_2S_2O_8 + H_2O \longrightarrow NH_4HSO_4 + H_2O_2$.
(f) $PbS + H_2O_2 \longrightarrow PbSO_4 + H_2O$.

10. When SO_2 and chlorine are passed together over a camphor catalyst, a volatile liquid A can be obtained. When treated with water, the liquid slowly hydrolyses to a mixture of hydrochloric and sulphuric acids.

(a) 1·35 g of A were hydrolysed and the solution treated with excess silver nitrate solution. This precipitated all the chlorine as AgCl (2·87 g). How many moles is this?

(b) The same solution that was obtained in (a) was then treated with barium nitrate solution, which precipitated all the sulphur as $BaSO_4$ (2·33 g). How many moles is this?
(c) Work out the relative numbers of g-atoms of Cl, S and O to obtain the empirical formula of A.
(d) Suggest a structural formula for A, given that the molecular formula is the same as the empirical formula.
(e) Write down the equation for the reaction of A with water.

11. When SO_2 is passed over PCl_5, a mixture of two liquids, B and C are obtained: these can be separated by fractional distillation.

(a) 2·38 g of B hydrolysed violently with water to give 5·74 g of AgCl after treatment with silver nitrate solution. How many moles is this?
(b) Corrected to s.t.p., 224 cm^3 of B's vapour weighed 1·19 g. What is the molecular weight of B?
(c) Suggest a molecular formula for B.
(d) Suggest a structural formula for B.
(e) What is C? Give the equation for the reaction of SO_2 with PCl_5.

12. A solution of $Na_2S_2O_3$ was prepared from Na_2SO_3 and radioactive sulphur. When the sodium thiosulphate was hydrolysed with dilute HCl to give SO_2 and sulphur, it was found that none of the SO_2 contained radioactive sulphur, but that all the radioactivity remained with the precipitated elemental sulphur. Can you suggest an explanation for this?

13. In Chapter 21, it was noted that compounds of the tri-iodide ion, I_3^-, were thermally more stable when the cation was also large. Comment on the relative thermal stabilities you would expect for the following compounds of s-block elements:

(a) nitrates; (b) sulphates; (c) oxides and peroxides.

14. Summarize the ways in which the properties of sulphur *differ* from those of oxygen. Find examples of analogous behaviour in Groups III, IV and V.

†15. Use your knowledge of the trends in Groups V and VII to predict the properties of Se, Te and Po.

CHAPTER 23

The Group V elements

THE TRENDS DOWN A GROUP

In this chapter, as in Chapter 21, a complete family of elements will be examined. The gradation in properties is here even more strongly marked; the first member, nitrogen, is a typical non-metal, while the last member bismuth has the characteristic properties of a metal.

23.1 Occurrence and extraction of Group V elements

Table 23.1 lists the elements of Group VB, together with their ground state electronic structures.

The first member, nitrogen, occurs native where it constitutes approximately four-fifths of the Earth's atmosphere, and in chemical combination mainly as nitrates. Nitrogen is an essential element to living matter, being an important constituent of proteins and nucleic acids. Phosphorus is also necessary for life, while the remaining elements, especially arsenic, are toxic to human beings.

Phosphorus occurs in phosphate ores such as *apatite*, $3\,Ca_3(PO_4)_2.CaF_2$. The remaining elements of the Group behave essentially as weak metals, and accordingly occur chiefly as the sulphides; this is in accord with the trend noted in Chapter 13. Bismuth, being more electropositive than the other elements, occurs

TABLE 23.1

Name	Principal quantum number of valence shell	Ground state electronic structure (abbreviated form)	Electronic structure (long form)
N	2	2, 5	$1s^2;\quad 2s^2p^3$
P	3	2, 8, 5	$1s^2;\quad 2s^2p^6;\quad 3s^2p^3$
As	4	2, 8, 18, 5	$1s^2;\quad 2s^2p^6;\quad 3s^2p^6d^{10};\quad 4s^2p^3$
Sb	5	2, 8, 18, 18, 5	$1s^2;\quad 2s^2p^6;\quad 3s^2p^6d^{10};\quad 4s^2p^6d^{10};$ $5s^2p^3$
Bi	6	2, 8, 18, 32, 18, 5	$1s^2;\quad 2s^2p^6;\quad 3s^2p^6d^{10};\quad 4s^2p^6d^{10}f^{14};$ $5s^2p^6d^{10};\quad 6s^2p^3$

IPC—T

also with the more electronegative anion O^{2-} (as Bi_2O_3) but the chief ores of these elements contain $FeAsS(c)$, $As_2S_3(c)$, $Sb_2S_3(c)$, and $Bi_2S_3(c)$. Trace quantities of arsenic occur in many metal ores and their removal (essential on account of the toxicity) presents many problems.

Nitrogen is extracted on a large scale by the fractionation of liquid air (section 14.13). Large amounts are consumed in the manufacture of ammonia and nitric acid, and many metallurgical processes use it as an inert atmosphere where oxidation must be prevented.

Phosphorus is normally obtained as the red allotrope (Chapter 5) for the production of the mixture used on the sides of safety match boxes, or as the white allotrope for conversion to phosphoric acid and phosphates. It is extracted by heating, in an electric furnace, a mixture of silica, coke and a phosphate ore such as *apatite*. The process may be regarded as consisting of two essential reactions:

(1) slag formation: $Ca_3(PO_4)_2(c) + 3 SiO_2(c)$
$$\xrightarrow{2000°C} 3 CaSiO_3(l) + \tfrac{1}{2}P_4O_{10}$$

(2) reduction: $P_4O_{10} + 10 C(c)$
$$\longrightarrow P_4(g) + 10 CO(g)$$

Note that in stage (1) the more volatile acidic oxide, phosphorus(V) oxide, is displaced by the less volatile silicon(IV) oxide, and that in stage (2), $P_4(g)$ *molecules* are produced. Upon cooling, white phosphorus (the volatile, molecular allotrope, section 5.7) is the product, and this is precipitated electrostatically. Conversion from white phosphorus to red requires a catalyst, usually iodine, when done on a small laboratory scale, but for larger scale industrial conversion no catalyst is required, and white phosphorus is simply heated to about 270°C for 4 or 5 days until the transition is complete.

Demand for arsenic and its compounds is nowadays very low, on account of its highly toxic nature; in former years arsenic was used

medicinally, and arsenic(III) sulphide as a yellow pigment. Demand for metallic antimony and its compounds is far greater. Soft metals such as tin and lead are hardened by alloying with antimony, and type-metal alloys often contain antimony in varying proportions according to hardness required. Antimony is not appreciably attacked by sulphuric acid, and it is therefore used in the manufacture of accumulator grids. The metal is obtained by roasting the sulphide to give the oxide, followed by carbon reduction. This is the usual method for all but the most electropositive metals (Chapter 13):

$$Sb_2S_3(c) + 4\tfrac{1}{2}O_2 = Sb_2O_3(c) + 3 SO_2(g);$$
$$Sb_2O_3(c) + 3 C(s) = 2 Sb(l) + 3 CO(g).$$

The same method is applied to the extraction of bismuth metal, which is important in the manufacture of low melting alloys such as Wood's metal (section 14.21). Some bismuth occurs native and is extracted by melting.

23.2 General survey of reactions and trends in Group V

There is a strong gradation from non-metallic to metallic properties down the Group. It has been noted in other chapters that the first member of a Group often differs in properties from later members, and this is certainly true of nitrogen which is the only gaseous element in Group V at room temperature; this property can be attributed to its strong tendency to form triple bonds $N{\equiv}N$ in preference to single bonds $N—N$. Diatomic molecules are therefore stable for nitrogen but not for other members of the group. Phosphorus, arsenic and antimony all form molecules X_4 in the vapour state, in which single bonds $X—X$ are present. Figure 23.1 is a plot comparing the relative bond energies of these elements. Bismuth and

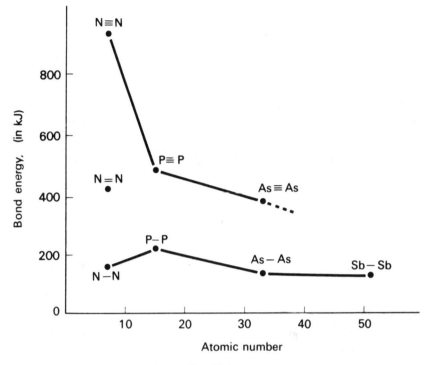

FIG. 23.1.

antimony both conduct electricity in the solid state, the conductivity decreasing when temperature is increased. They are therefore both metals. The N≡N bond in nitrogen is exceptionally strong (heat of dissociation 940 kJ mol^{-1}) and this makes it rather unreactive. It does not react with many metals directly, though some of the s-block metals react when heated in the gas.

$$3 \, Mg(c) + N_2(g) = Mg_3N_2(c);$$

magnesium
nitride

$$\Delta H = -430 \text{ kJ}$$

$$3 \, Li(c) + \tfrac{1}{2} N_2(g) = Li_3N(c);$$

lithium
nitride

$$\Delta H = -197 \text{ kJ at } 25°C.$$

Phosphorus, although less electronegative than nitrogen, combines more readily with metals to form phosphides as it is more reactive. The other elements of the Group also combine with metals, though in the case of antimony and bismuth the process is probably more correctly regarded as alloy formation.

The structures of the Group V elements present an interesting pattern. Nitrogen is very simple, forming diatomic molecules; the other elements form diatomic molecules at high temperature in the vapour. Phosphorus, arsenic and antimony all form tetrahedral molecules X_4 (Fig. 5.13), and this molecular structure persists in the solid state when the vapour is condensed.

In addition to these forms, all the elements form giant lattices with a gradual trend from covalent to metallic bonding and structure. At one end of the scale, violet phosphorus and red phosphorus form layered structures with a co-ordination number of three (Fig. 5.13). With

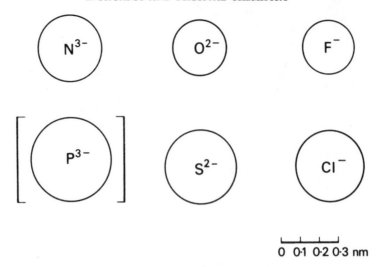

FIG. 23.2. The sizes of simple negative ions. The ion P^{3-} cannot exist, owing to polarization.

arsenic and antimony the layers are becoming more closely packed—these are intermediate covalent-metallic structures—and with bismuth the packing of atoms, and the bonding, is truly metallic.

23.3 Ion formation in Group V

The ground state electronic structures of the atoms of Group V elements are all *noble gas minus three electrons*. An examination of the energy levels suggests that an ion X^{3-} ought to be formed in favourable cases: in fact only nitrogen appears to form this ion, and even then only in certain nitrides of electropositive metals. Two factors weigh against the existence of stable ions of this type:

(i) It is difficult to form an ion of high negative charge, because each additional electron added to the atom is repelled by the ones already there.

(ii) The more negative charges an ion possesses, the *larger* it becomes (Figure 23.2) and a large ion is easily polarized (that is, it is "soft", cf. sulphide ion in Chapter 22). Moreover, a very large ion cannot fit easily into a lattice with smaller ions. For this reason, true ions P^{3-}, As^{3-}, etc., do not appear to exist. Compounds such as Li_3P, lithium phosphide, may *appear* to contain P^{3-} ion, but their structure is not in accord with the ionic model.

All the elements in this Group can exist in oxidation states of $+3$ or $+5$; in general negatively-charged oxo-ions are formed. Only the metals antimony and bismuth can form positive ions.

Table 23.2 summarizes the main oxo-ions of $+5$ oxidation state formed by Group V

TABLE 23.2

Element	Name of anion	Formula of anion
Nitrogen	Nitrate	NO_3^-
Phosphorus	Metaphosphate	PO_3^-
	orthophosphate	PO_4^{3-}
Arsenic	Arsenate	AsO_4^{3-}
Antimony	Antimonate	SbO_3^-
Bismuth	Bismuthate	BiO_3^-

elements; all such compounds have the ending *-ate*.

Bismuth is able to form a cation $Bi^{3+}(aq)$, though this is readily hydrolysed to the oxocation $BiO^+(aq)$:

$$Bi^{3+}(aq) + H_2O \rightleftharpoons BiO^+(aq) + 2 H^+(aq)$$

$Sb^{3+}(aq)$ is even more strongly hydrolysed to $SbO^+(aq)$. The two metals may be regarded as characteristic examples of B-metals, and the following factors determine the nature of the ions formed:

(i) The heavy B-metals, such as bismuth, show a tendency to have an oxidation state equal to the *Group number minus two*. This effect is known as the **inert pair effect** (Chapter 25).

(ii) The B-metals show a stronger tendency than s-block metals to form complex ions; such complexes are frequently formed in such a way as to neutralize some of the charge originally present. The ions $BiO^+(aq)$ and $SbO^+(aq)$ are of lower charge than the simple species from which they are derived, and ions such as $BiCl_4^-(aq)$ are also examples. B-metal cations are more easily polarized—they are softer—than s-block cations and for this reason a large number of B-metal compounds are essentially covalent in their bonding.

23.4 Group V hydrides

All the Group V elements probably form hydrides XH_3, though the existence of BiH_3 has been questioned—certainly if it exists it is exceedingly unstable. The lighter elements also form hydrides X_2H_4, and nitrogen forms hydrazoic acid, HN_3. These compounds are summarized in Table 23.3.

Hydrolysis of nitrides gives ammonia, and hydrolysis of phosphides gives phosphine. This is an example of the general reaction mentioned

TABLE 23.3

Element	Name of hydride XH_3	Formula of XH_3 (state at room temp)	Other hydrides
Nitrogen	Ammonia	$NH_3(g)$	Hydrazine, $N_2H_4(l)$
Phosphorus	Phosphine	$PH_3(g)$	Hydrazoic acid, HN_3 Diphosphine, $P_2H_4(l)$
Arsenic	Arsine	$AsH_3(g)$	
Antimony	Stibine	$SbH_3(g)$	
Bismuth	(Bismuthine)	$(BiH_3$, very unstable)	

decreasing thermal stability

in section 18.6: where an "X-ide" is hydrolysed, a hydride of X is the result.

$$Mg_3N_2 + 6 H_2O \longrightarrow 2 NH_3 + 3 Mg(OH)_2$$
$$Ca_3P_2 + 6 H_2O \longrightarrow 2 PH_3 + 3 Ca(OH)_2$$

This method is of limited practical application.

23.5 Ammonia

Considerable amounts of nitrogen are converted into ammonia by the Haber synthesis, $N_2 + 3 H_2 \rightleftharpoons 2NH_3$. This reaction has been used in this book to illustrate equilibrium constant and le Chatelier's principle, and the reader should refer to the relevant chapter (sections 10.3 and 10.4). The precise experimental conditions vary from one chemical plant to another, but all processes use (a) a high pressure (between 250 and 1000 atm), (b) the lowest temperature compatible with a reasonable rate of reaction (generally around 500°C), and (c) a catalyst based on iron and aluminium oxide (the precise details of catalysts are generally closely guarded industrial secrets).

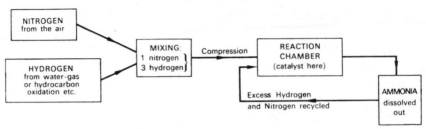

FIG. 23.3. The Haber synthesis.

The flow-sheet (Fig. 23.3) shows the essential details of the process. Despite the fact that the yield of a given equilibrium process may be only about one-tenth, almost total conversion is finally achieved since the gases are recycled after dissolving out the ammonia formed.

Ammonia is gaseous at room temperature but readily liquefied (b.p. $-33°C$). It is highly soluble in water (at room temperature and pressure a solution of 35% by weight, density 0.880 g cm^{-3}, called "880 ammonia" can be obtained). The high solubility and relatively low volatility of ammonia are attributed to hydrogen bond formation (section 18.7). The ammonia molecule is pyramidal, with one lone pair, and its structure and electronic configuration are shown in Fig. 23.4. In aqueous solution a small amount of ionization takes place, causing ammonia to behave as a weak base:

$$NH_3(g) + H_2O(l) \rightleftharpoons NH_3 \ldots H_2O(aq)$$
$$\text{hydrogen-bonded}$$
$$\rightleftharpoons NH_4^+(aq) + OH^-(aq);$$
$$K = 1.8 \times 10^{-5} \text{ mol dm}^{-3} \text{ at } 25°C.$$

Liquid ammonia is itself a liquid rather like water, even to the extent of autoionizing (section 19.11). It is described as an ionizing solvent because many ionic compounds dissolve in it, and reactions analogous to aqueous reactions can be carried out in it.

Ammonia will burn in air with difficulty, forming mainly nitrogen and water vapour. In the presence of a hot platinum wire, ammonia can be oxidized catalytically by the exothermic reaction

$$4 NH_3(g) + 5 O_2(g) = 4 NO(g) + 6 H_2O(g);$$
$$\Delta H = -900 \text{ kJ at } 25°C.$$

The gas has an irritating smell even in low concentrations. It is evolved from decayed living matter, and may be detected with Nessler's reagent.

The ammonia molecule is an important **ligand** in complex ion formation. Many metal ions (mainly transition metals) form **ammine** complexes, generally with a change of colour; the well-known colour change of copper(II) ions from pale blue to dark blue on adding ammonia is due to ammine formation:

$$Cu^{2+}(aq) + 4 NH_3(aq) \rightleftharpoons Cu(NH_3)_4^{2+}$$
$$\text{pale blue} \qquad\qquad \text{tetra-ammine}$$
$$\text{copper(II) ion}$$
$$\text{dark blue}$$

The basic nature of ammonia causes it to form the ammonium ion, NH_4^+. The presence of this ion in aqueous solution has been mentioned above, but the ion forms readily in the presence of any proton donor (Brønsted acid):

$$H^+ + NH_3 = NH_4^+$$
$$\underbrace{\qquad\qquad\qquad}$$
$$\text{acid-base pair}$$
$$\text{(conjugate pair)}$$

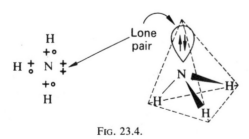

FIG. 23.4.

The ammonium ion is present in a series of salts, for instance in ammonium nitrate, $NH_4NO_3(c)$, and in ammonium sulphate, $(NH_4)_2SO_4(c)$. Ammonium chloride, $NH_4Cl(c)$, is formed as a white smoke when ammonia and hydrogen chloride gases are mixed:

$$NH_3(g) + HCl(g) \rightleftharpoons NH_4Cl(c);$$
$$\Delta H = -177 \text{ kJ at } 25°C.$$

On heating, the solid readily dissociates by the reverse reaction.

Another example of the basic nature of ammonia is its ability to donate its lone pair of electrons to any other electron-accepting compound. For instance, boron(III) fluoride, $BF_3(g)$, combines with ammonia to form an acid–base complex (Section 19.15).

Ammonia is the parent compound of an important series of organic substances called *amines**, derived from ammonia by the substitution of one or more hydrogen atoms by hydrocarbon groups.

The formula of ammonia has been established by completely decomposing it into its elements, and showing that the product is one-quarter nitrogen by volume and three-quarters hydrogen:

---

* The terms *amine* and *ammine* must be carefully distinguished.

2 vols. ammonia $\rightleftharpoons$ 3 vols. hydrogen + 1 vol. nitrogen

$\therefore$ by Avogadro's law,

2 moles ammonia $\rightleftharpoons$ 3 moles hydrogen + 1 mole nitrogen.

$\therefore 2\,NH_3 \rightleftharpoons 3\,H_2 + N_2.$

A large amount of manufactured ammonia is converted into nitric acid by catalytic oxidation, and thence to explosives and fertilizers. Fertilizer manufacture is an important application: it was first realized at the end of the nineteenth century that the supplies of nitrogen essential to life would ultimately give out (when the Chile deposits of sodium nitrate became exhausted), and this led chemists to tackle the problem of making nitrogen compounds from the nitrogen in the air, known as the **fixation** of nitrogen. Nowadays most nitrogen is "fixed" by conversion to ammonia, but the problem has always been made difficult by the high bond energy, and consequent low reactivity, of the element. There exist nitrogen-fixing bacteria capable of combining with elementary nitrogen at room temperature, but the precise mechanism by which these work remains a mystery. Much current work is devoted to the development of catalysts which might mimic the behaviour of these organisms,

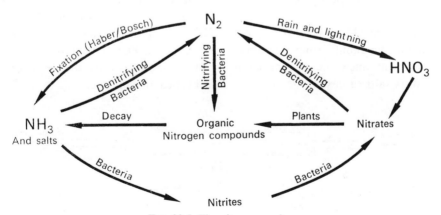

FIG. 23.5. The nitrogen cycle.

and future variants of the Haber process might ultimately work at near room temperature in almost 100% yield. In this context it is interesting to note that complexes containing the N_2 molecule as ligand have recently been prepared.

Figure 23.5 is a diagram of the **nitrogen cycle.** It illustrates how nitrogen is exchanged between the atmosphere, the Earths' crust and living matter.

23.6 Phosphine

Phosphine is a gas, but unlike ammonia it does not exhibit hydrogen bonding; consequently it has a lower boiling point and is almost insoluble in water. It is generally made by the hydrolysis of white phosphorus by sodium hydroxide solution—an example of a disproportionation reaction:

oxidized (O.N. increases, 0 to $+1$)

$$P_4 + 3OH^- + 3H_2O \longrightarrow 3H_2PO_2^- + PH_3$$

reduced (O.N. decreases, 0 to -3)

Phosphine is a much weaker base than ammonia, and the remaining Group V hydrides are weaker still, although the phosphonium ion, PH_4^+, and its salts, do exist.

23.7 Arsine and stibine

These gases are obtained when compounds of arsenic and antimony are subjected to vigorous reduction in aqueous solution. These reactions form the basis of the Marsh test for arsenic, and later modifications such as Gutzeit's test. In Gutzeit's test arsine or stibine, evolved from the reaction with granulated zinc in dilute sulphuric acid, is allowed to react with $Ag^+(aq)$, and black metallic silver is deposited.

The thermal stability of the Group V hydrides varies as follows:

$$NH_3 > PH_3 > AsH_3 > SbH_3 > BiH_3$$

In the older Marsh test, arsine was passed down a heated tube, and a mirror of metallic arsenic deposited on the cooler part of the tube. The lower stability of stibine means that metallic antimony is deposited even before the very hot portion of the tube is reached (Fig. 23.6).

23.8 Halides of Group V elements

All the Group V elements form halides XHa_3, where $X = $ Group V element and $Ha = $ halogen; in addition, some halides XHa_5 exist, such as PCl_5. Phosphorus forms a mixture of PCl_3 and PCl_5 when the element phosphorus is burned in chlorine.

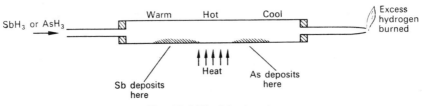

Fig. 23.6. The Marsh test.

Phosphorus trichloride is a colourless liquid, and is a typical non-metal chloride, fuming in moist air, and reacting with water to form phosphor*ous* acid. This reaction is the normal hydrolysis reaction in which chlorine atoms are displaced by hydroxyl groups (section 21.8).

$$PCl_3 + 3\,H_2O = H_3PO_3 + 3\,HCl; \quad \Delta H \text{ negative.}$$

Phosphorus pentachloride has a trigonal bipyramid shape in the vapour phase, (Figure 6.11) but in the solid this is replaced by what appears to be an ionic structure made up of PCl_4^+ (tetrahedral) and PCl_6^- (octahedral). It reacts with water (and with organic compounds containing hydroxyl groups) to give first phosphorus oxychloride:

$$PCl_3 + H_2O \longrightarrow POCl_3 + 2\,HCl;$$
$$\Delta H = -92 \text{ kJ at } 25°C.$$

Phosphorus oxychloride is a liquid consisting of tetrahedral molecules, this observation being in accord with the rules for molecular shape (section 6.2). Phosphorus oxychloride is relatively stable towards hydrolysis, but boiling water converts it to orthophosphoric acid.

In these hydrolysis reactions the oxidation number of phosphorus remains $+5$ throughout:

The formation of oxychlorides is not confined to phosphorus: Antimony and bismuth form oxychlorides where the oxidation number is $+3$.

$$SbCl_3(aq) + H_2O \rightleftharpoons SbOCl(c) + 2\,HCl(aq);$$

white precipitate (also Bi.)

The above reaction affords a good illustration of the principle of equilibrium and le Chatelier: addition of $H^+(aq)$ ions readily dissolves

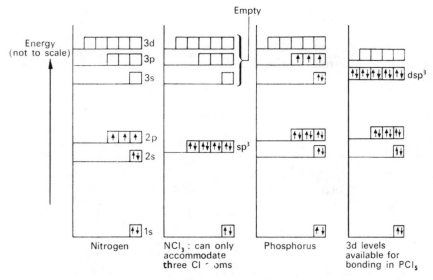

FIG. 23.7. To show that nitrogen cannot form a chloride NCl_5.

the antimony oxychloride precipitate, and dilution of this solution leads to reprecipitation.

Nitrogen is exceptional in that it forms only one chloride, NCl_3. There are insufficient energy levels to accommodate five halogen atoms and form NCl_5, as the energy group of principal quantum number 2 does not have d-levels.

23.9 Oxides and oxo-acids of Group V elements

The pattern presented by the oxides of Group V elements is more complicated than that of the chlorides, due to the appearance of oxidation states other than $+3$ and $+5$. Table 23.4 lists the main ones, together with their molecular formulae and physical states at room temperature.

Nitrogen(I) oxide, N_2O, and nitrogen(II) oxide, NO, do not display acidic or basic tendencies—they were classed as *neutral* oxides in section 22.11. With the remainder, a distinct trend from acidic to basic is observed:

N	P	As	Sb	Bi
Strongly acidic oxides	weakly → acidic oxides	amphoteric →oxides (Sb₂O₃ acidic)	→ weakly basic oxide	

The acidic tendency can be observed with nitrogen oxides, where nitrous and/or nitric acid are formed:

$N_2O_3 + H_2O = 2HNO_2$ (nitrous acid)
$N_2O_5 + H_2O = 2HNO_3$ (nitric acid)

$+4$ $+3$ $+5$

$N_2O_4 + H_2O = HNO_2 + HNO_3$
          (disproportionation)

Oxides of phosphorus form first the meta-acid (least water in the formula), and addition of more water leads to the ortho-acid (most water in the formula) (Fig. 23.8, 23.9).

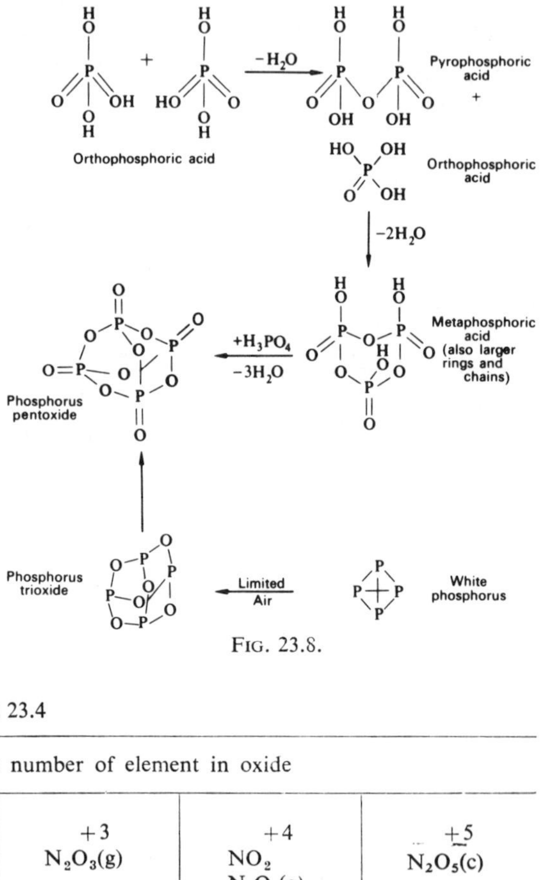

FIG. 23.8.

TABLE 23.4

Element	Oxidation number of element in oxide				
	$+1$	$+2$	$+3$	$+4$	$+5$
Nitrogen	$N_2O(g)$	NO(g)	$N_2O_3(g)$	NO_2 $N_2O_4(g)$	$N_2O_5(c)$
Phosphorus			$P_4O_6(c)$	$PO_2(c)^*$	$P_4O_{10}(c)$
Arsenic			$As_4O_6(c)$		$As_4O_{10}(c)$
Antimony			$Sb_2O_3(c)^*$	$SbO_2(c)^*$	$Sb_2O_5(c)^*$
Bismuth			$Bi_2O_3(c)^*$		

* indicates formula to be empirical: structure consists of larger molecules or giant lattice.

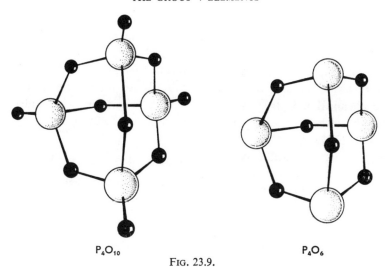

P₄O₁₀   P₄O₆

FIG. 23.9.

$\frac{1}{2}$ P$_4$O$_{10}$+H$_2$O = 2 HPO$_3$ (metaphosphoric acid)

HPO$_3$+H$_2$O = H$_3$PO$_4$ (orthophosphoric acid)

An oxide which can react either with an acid or a base is classed as amphoteric, and antimony(III) oxide comes into this category:

$$Sb_2O_3(c)+6\,H^+(aq) = 2\,Sb^{3+}(aq)+3\,H_2O.$$
$$Sb_2O_3(c)+6\,OH^-(aq) = 2\,SbO_3^{3-}(aq)+3\,H_2O.$$

Bismuth(III) oxide dissolves readily in dilute acids, but does not dissolve in alkali; it is therefore a true basic oxide although weak compared with, say, the oxide of a Group IA metal. Figure 23.10 is a plot of half-cell potentials of the oxides and oxo-acids in the +5 oxidation state. Note the following points:

(i) Phosphoric acid is not a strong oxidizing agent.

(ii) Bismuthic acid is an extremely strong oxidizing agent, and will even oxidize manganese(II) directly to manganese(VII).

Some of the oxo-acids of Group VB, and their salts, are of special importance, and they are considered in the sections which follow.

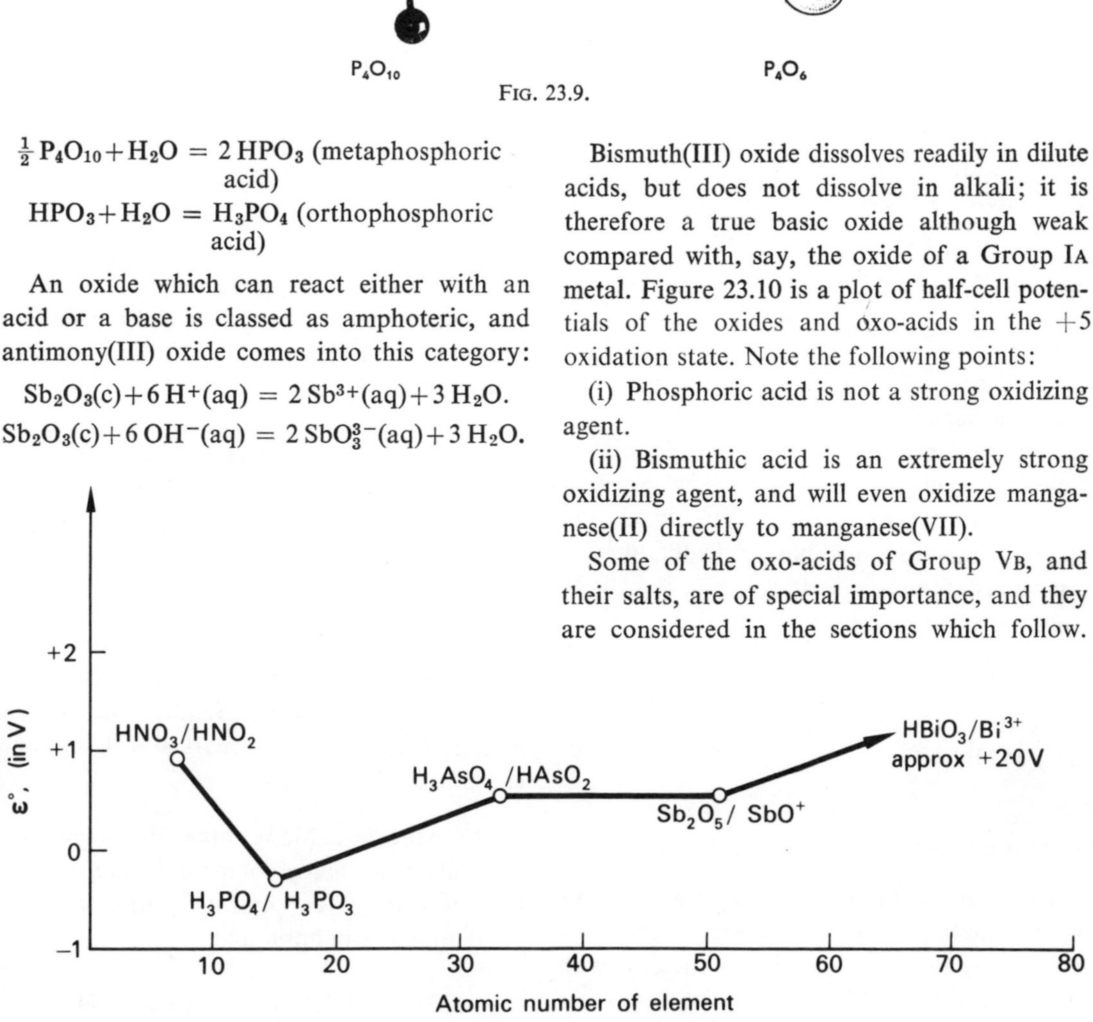

FIG. 23.10. Half-cell potentials for X(V) ⇌ X(III).

23.10 Nitric acid

This is the most important oxo-acid of nitrogen, and very large amounts are manufactured for the chemical industry. In dilute aqueous solution it ionizes forming $H^+(aq)$ and the planar ion, $NO_3^-(aq)$. Nitric acid is obtained by the following sequence of reactions:

ammonia $\xrightarrow[\substack{\text{catalytic} \\ \text{oxidation}}]{}$ nitrogen(II) $\xrightarrow[\text{mix with air}]{}$ oxide

nitrogen(IV) $\xrightarrow[\substack{\text{water and} \\ \text{oxygen}}]{}$ nitric oxide acid

Nitrogen(II) oxide is a colourless gas which reacts instantly with the oxygen in air to form brown $N_2O_4 \rightleftharpoons NO_2$. In the presence of excess oxygen and water in an absorption tower, nitric acid is produced.

$$3\,NO_2 + H_2O = 2\,HNO_3 + NO.$$

Nitric acid is a colourless liquid when pure, but is normally yellow due to dissolved oxides of nitrogen produced by its decomposition. So-called "concentrated" nitric acid used in the laboratory contains about 30% water, and the true anhydrous acid has markedly different properties (for instance it will not react with copper). The reactions of dilute nitric acid are again different from those of laboratory "conc" acid. Concentrated nitric acid is a very powerful oxidizing agent while dilute nitric acid is only mildly so.

REACTIONS OF DILUTE NITRIC ACID

Although most dilute acids react with metals to give hydrogen gas, due to the reduction of $H^+(aq)$ by the metal, this is not the case with dilute nitric acid which gives mainly oxides of nitrogen such as NO and NO_2. This shows that the metal has reduced the nitrate ion instead of hydrogen ion. A very dilute solution of acid will, however, give hydrogen with a reactive metal such as magnesium. Apart from this, dilute nitric acid behaves as a normal acid in its reactions with oxides, hydroxides and carbonates, forming nitrates.

REACTIONS OF CONCENTRATED NITRIC ACID

These may be considered under the following headings:

(i) *Acidic*. Concentrated nitric acid forms nitrates when added to a basic substance (oxide, carbonate, or hydroxide for instance).

(ii) *Oxidizing*. Concentrated nitric acid will oxidize very powerfully, generally giving nitrogen(II) oxide as its reduction product. It is usually easier to treat the oxidation reactions as simple oxygen transfer when balancing equations, even though they are also examples of electron transfer. Consider, for instance, the oxidation of hydrogen sulphide to sulphur:

$$
\begin{aligned}
2\,HNO_3 &= 2\,NO + H_2O + 3[O] \\
3\,H_2S + 3[O] &= 3\,S + 3\,H_2O \\
\hline
2\,HNO_3 + 3\,H_2S &\longrightarrow 3\,S + 2\,NO + 4\,H_2O
\end{aligned}
$$

Although concentrated nitric acid attacks all metals except the noble metals (gold, platinum, etc.) its reaction with many metals is slowed down because it renders the surface *passive*. This effect is very marked with iron and chromium.

(iii) *Nitrating*. Many organic compounds, particularly aromatic compounds, are nitrated by a mixture of concentrated nitric acid and concentrated sulphuric acid, e.g.

$$C_6H_6 + HNO_3 = C_6H_5NO_2 + H_2O$$
benzene   nitrobenzene

The reaction takes place because of the presence of nitronium ions, NO_2^+, produced by proton transfer between the two substances:

$$\overbrace{HNO_3 + \quad 2\,H_2SO_4}^{\text{protons gained}} = \underbrace{NO_2^+ + H_3O^+ + 2\,HSO_4^-}_{\text{protons lost}}$$

Nitration in the manufacture of dyes and explosives is the main industrial use of nitric acid.

23.11 Nitrates

Since nitric acid is a strong acid, most metals form nitrates. All of these are soluble in water; they are less stable to heat than other commonly occurring salts such as sulphates, chlorides and phosphates. The usual mode of decomposition is by loss of oxygen and oxides of nitrogen (mainly NO_2), leaving a residue of the metal oxide:

$$Pb(NO_3)_2(c) \longrightarrow PbO(c) + 2NO_2(g) + \tfrac{1}{2}O_2\ (g).$$

With the very electropositive metals however (metals below sodium in Group IA) the only gas given off is oxygen, for these metals form stable nitrites:

$$NaNO_3(c) \longrightarrow NaNO_2(c) + \tfrac{1}{2}\,O_2(g)$$

Where the oxide is itself unstable with respect to heating (where its free energy of formation, ΔG_f°, becomes positive, Chapter 13) the residue is the metal itself. This is only true of the metals near the bottom of the electrochemical series:

$$AgNO_3(c) \longrightarrow Ag(c) + NO_2(g) + \tfrac{1}{2}\,O_2(g).$$

Since all nitrates are soluble in water, it is not possible to devise a precipitation test for the nitrate ion: instead some other reaction

has to be used. Two tests are commonly employed:

(i) Devarda's alloy (Cu 50, Al 45, Zn 5%) will rapidly reduce nitrate ions to ammonia in the presence of aqueous alkali.

(ii) Take the suspected solution of $NO_3^-(aq)$ in a test tube, make acid with dilute sulphuric acid and add $Fe^{2+}(aq)$ ions, as iron(II) sulphate. Slowly add concentrated sulphuric acid so that it does not mix, but forms a separate layer beneath the aqueous layer. A brown ring forms at the interface between the two liquids if a nitrate is present. The colour is thought to be due to nitroso-complexes of iron(II) such as the following:

(octahedral)

Nitrates are oxidizing agents on account of their readiness to give up oxygen. Sodium nitrate occurs naturally in Chile, and other important nitrates include ammonium nitrate —a nitrogen-rich fertilizer—and potassium nitrate.

23.12 Nitrous acid

Nitrous acid, HNO_2, is thermally unstable, and exists in dilute solution only; it is a weak monobasic acid ($K_a = 4 \times 10^{-4}$ mol dm^{-3} at 25°C) and its salts are called nitrites. Nitrous acid reacts with all compounds containing the NH_2-group, giving nitrogen and an OH-group. The acid is made *in situ* from sodium nitrite and hydrochloric acid. Nitrous acid contains nitrogen in a lower oxidation state, and is not such a powerful oxidizing agent as nitric acid: it can in fact act both as an oxidant and as a reductant in different reactions.

23.13 Phosphoric acid

The commonest oxo-acid of phosphorus is orthophosphoric acid, H_3PO_4. It is a member of the iso-electronic series H_3PO_4, H_2SO_4 and $HClO_4$, and is the weakest member of this series (section 19.10). It is classed as a tribasic acid—that is, it has three replaceable hydrogen atoms—though the replacement of the third atom is not normally carried out in titration reactions. It was observed by Pauling that, where an acid is polybasic, its successive dissociation constants, in mol dm^{-3}, are in the rough ratio

$$k_1 : k_2 : k_3 = 1 : 10^{-5} : 10^{-10}.$$

Phosphoric acid provides a good illustration of Pauling's rule:

$H_3PO_4 \rightleftharpoons H^+ + H_2PO_4^-$; $k_1 = 0.75 \times 10^{-2}$; (not very strong)

$H_2PO_4^- \rightleftharpoons H^+ + HPO_4^{2-}$; $k_2 = 0.67 \times 10^{-7}$; (weaker than acetic acid)

$HPO_4^{2-} \rightleftharpoons H^+ + PO_4^{3-}$; $k_3 = 1.0 \times 10^{-12}$; (extremely weak acid)

The reason for this progressive weakening is partly electrostatic: it is less easy for $H_2PO_4^-$ to lose a proton than for H_3PO_4 since the former has an overall negative charge which can attract the proton.

The titration of phosphoric acid with sodium hydroxide is dealt with in section 19.13.

Pure orthophosphoric acid is a syrupy, relatively involatile, liquid, which decomposes below its boiling point. The syrupy nature is due to hydrogen bonding in which O—H...O bridges form (sulphuric acid exhibits the same effect). It does not show the oxidizing properties

associated with sulphuric and nitric acids: it can be used for instance to prepare hydrogen iodide by distillation with an iodide:

$$KI(c) + H_3PO_4(l) \longrightarrow KH_2PO_4(c) + HI(g).$$

Under comparable conditions using sulphuric acid, the hydrogen iodide would be oxidized to iodine. In writing this equation, it is important to write the most acid salt since the conditions of the reactions are strongly acidic and Na_2HPO_4 and Na_3PO_4 cannot form.

Careful heating of orthophosphoric acid gives two degrees of dehydration, prefixed *pyro* and *meta* respectively:

$2H_3PO_4 - H_2O \longrightarrow H_4P_2O_7$; (pyrophosphoric acid)

$2H_3PO_4 - 2H_2O \longrightarrow 2HPO_3$; (metaphosphoric acid)

$2H_3PO_4 - 3H_2O \longrightarrow \frac{1}{2}P_4O_{10}$ (complete dehydration to oxide).

Phosphoric acid is used for treating metal surfaces and as a food flavouring. A large amount is converted into phosphate fertilizers.

Calcium orthophosphate, $Ca_3(PO_4)_2$, is made by the action of lime on phosphoric acid, but unfortunately it is not a good fertilizer since it is insoluble in water and cannot be taken up by plants. In recent years it has been customary to convert it to the acid salt which is soluble in water and sold as *superphosphate*:

$$Ca_3(PO_4)_2 + 4 H_3PO_4 = 3 Ca(H_2PO_4)_2.$$

Sodium salts of phosphoric acid are manufactured for a variety of purposes which include emulsifying of processed cheese and manufacture of synthetic detergents. A polymer of sodium metaphosphate, $NaPO_3$, is sold as "Calgon", a water-softening agent. It forms soluble phosphate complexes with the Ca^{2+}(aq) and Mg^{2+}(aq) in hard water, thereby preventing insoluble stearate precipitates forming when soap is added.

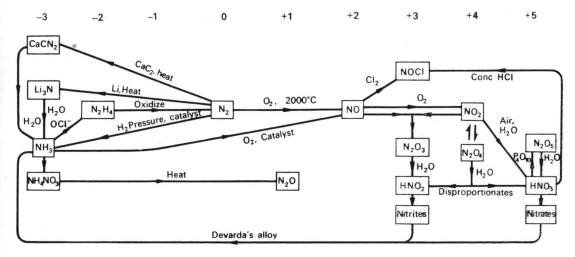

FIG. 23.11. Some reactions of nitrogen and its compounds.

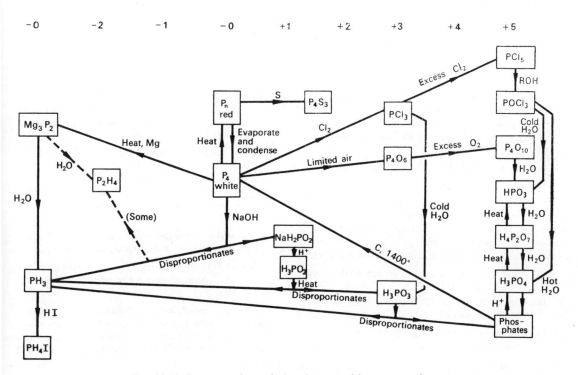

FIG. 23.12. Some reactions of phosphorus and its compounds.

23.14 Summary

The elements in Group V afford a particularly interesting example of Group trends. In particular:

(i) the range of observed chemical behaviour is wider than in Groups I and VII;

(ii) the differences between the first two members nitrogen and phosphorus is quite marked (cf. oxygen and sulphur, Chapter 22).

TRENDS DOWN THE GROUP

(1) Structure, bonding, and chemical properties show a complete change, from non-metallic $N_2(g)$ to metallic $Bi(c)$ which has a close-packed structure.

(2) The inert-pair effect is shown strongly by bismuth.

(3) The stability to heat, and the basic strength, of hydrides varies $NH_3 \gg PH_3$ $> AsH_3 > SbH_3 > BiH_3(?)$. (Cf. similar trends in hydrogen halides.)

(4) Oxides become more basic down the Group.

EXCEPTIONAL PROPERTIES OF NITROGEN

(a) The $N{\equiv}N$ bond is exceptionally strong (cf. section 22.14).

(b) Nitrogen cannot accept more than 8 electrons in its valence shell.

(c) It is the only member of the Group which is a gas at room temperature.

(d) It is the only member of the Group that forms an ion X^{3-}.

(e) It is the only member of the Group that forms hydrogen bonds.

(f) Its oxides, oxo-acids, and chlorides do not conform to the patterns of the remainder of the Group.

Study Questions

1. Describe what you would expect to see when the following compounds are heated. Give balanced equations in each case:

(a) Potassium nitrate.   (b) Mercury(II) nitrate.

(c) Copper(II) nitrate.   (d) Ammonium nitrite.

2. 25 cm³ of a gaseous hydride of nitrogen were passed over a heated iron catalyst and completely decomposed into 75 cm³ of a mixture of nitrogen and hydrogen. On passing this mixture over heated copper oxide, the volume dropped to 25 cm³. (All volumes at s.t.p.)

(a) How does the nitrogen-hydrogen mixture react with copper oxide?

(b) How many cm³ of hydrogen were produced in the decomposition?

(c) How many cm³ of nitrogen were produced in the decomposition?

(d) What is the empirical formula of the nitrogen hydride?

(e) What is the molecular formula of the nitrogen hydride?

†**3.** On heating, a white solid A sublimed as two gases, B and C. B had a powerful obnoxious smell, while C was an acid gas. At s.t.p., 448 cm³ of B weighed 0·68 g, and 112 cm³ of C weighed 0·64 g. The gases recombined when condensed on a cold surface, reforming A. Identify A, B and C.

4. (a) Write down the oxidation numbers of phosphorus in H_3PO_3, H_3PO_4 and PH_3.

†(b) H_3PO_3 disproportionates on heating, giving phosphine. Suggest an equation for this reaction.

5. If a mixture of nitric and hydrochloric acids (aqua regia) is heated, it is possible to distil off a gas, which can be liquefied using an ice-salt mixture. The gas has a molecular weight of 65·5 and contains 56% by weight of chlorine.

(a) Suggest a molecular formula for the gas.

(b) Suggest a structural formula for the gas.

(c) Suggest how the molecular weight of the gas could be measured.

(d) Suggest why an ice–salt mixture was used to liquefy the gas, rather than pure ice.

6. The half-cell potentials for As, $3H^+/AsH_3(g)$ and Ag^+/Ag are -0.60 and $+0.80$ V respectively.

(a) Could AsH_3 reduce $Ag^+(aq)$ completely to silver, or would there be an appreciable quantity of $Ag^+(aq)$ at equilibrium?

(b) Give a balanced equation for the reaction of AsH_3 with $Ag^+(aq)$.

(c) Do these data tell you anything about the rate of this reaction?

7. Give equations for the following reactions of nitrous acid:

(a) With $Fe^{2+}(aq)$ to give $NO(g)$.

(b) With acetamide, CH_3CONH_2.

(c) The spontaneous slow decomposition into nitric acid and nitric oxide.

8. Some bismuth nitrate was dissolved in concentrated hydrochloric acid. On diluting this solution, a white precipitate appeared, which redissolved on adding more concentrated hydrochloric acid. This cycle could be repeated. Write down the equation for this equilibrium.

9. (a) Of the elements in Group VB, which will:

    (i) have the lowest first ionization energy?

    (ii) form an ionic compound with calcium?

    (iii) form the most basic oxide?

    (iv) form the hydride, XH_3, with the lowest boiling point?

(b) With sulphuric acid, arsenic gives an oxide, while antimony and bismuth form sulphates; with nitric acid, arsenic and antimony form oxides, but bismuth forms a nitrate. How can you account for these differences?

10. Elements in the same Group show (a) similarities, (b) gradations in their physical and chemical properties. Discuss the truth of this statement when applied to the elements of Group VB.

11. "The properties of the first element of a Group often differ considerably from those of the remainder." Show the extent to which this is true for nitrogen.

†**12.** The standard heats of formation of $NH_4NO_3(c)$, $H_2O(g)$ and $N_2O(g)$ are -365, -242 and $+83$ kJ mol^{-1} respectively. The corresponding free energies of formation are -184, -228, and $+104$ kJ mol^{-1}.

(a) Write down the equations for the decomposition of ammonium nitrate into (i) nitrogen, oxygen and water (ii) nitrous oxide and water.

(b) What are the enthalpy and free energy changes associated with each equation?

(c) Is it possible to use these data to forecast which reaction will occur?

13. When 3·45 g of a white solid, X, were warmed gently with alkali, 0·51 g of ammonia were evolved. The resulting solution was acidified with nitric acid and the addition of excess ammonium molybdate solution caused a yellow precipitate to form in the cold.

(a) Suggest a formula for X.

(b) What would you expect the pH of an aqueous solution of X to be?

14. A white solid, Y, was dissolved in hydrochloric acid. When hydrogen sulphide was passed into the solution, a brown precipitate formed. When the original solution in hydrochloric acid was diluted, a white precipitate formed. The addition of concentrated sulphuric acid to solid Y caused acid brown fumes to be evolved.

(a) Identify Y.

(b) What would you expect the action of heat on Y to be?

15. The nitronium ion, NO_2^+, is formed when concentrated nitric and sulphuric acids are mixed.

(a) Which substance is the Brønsted base in this reaction?

(b) Predict the shape of NO_2^+.

16. (a) Meta-phosphoric acid, $(HPO_3)_n$, exists as both rings and chains. Draw diagrams of these structures.

(b) What is the structure of nitric acid, HNO_3?

(c) Can you suggest why nitric and meta-phosphoric acids have such different structures?

†**17.** For each of the following reactions, (a) give an equation, (b) explain the nature of the reaction, and (c) give the equation for the corresponding reaction where nitrogen replaces oxygen, and liquid ammonia replaces water as the solvent:

(i) Sodium hydroxide solution reacting with hydrochloric acid.

(ii) Anhydrous copper sulphate turning blue when water is added.

(iii) Sodium reacting with water and giving off hydrogen.

(iv) A solution of zinc ions, Zn^{2+}, giving a white precipitate with sodium hydroxide solution.

(v) The precipitate in (iv) redissolving on the addition of excess sodium hydroxide solution.

CHAPTER 24

Boron, carbon and silicon

DIAGONAL RELATIONSHIPS

At this point the book departs from the previous pattern of treating a complete Group of the periodic table in a single chapter. The non-metals boron, carbon and silicon form a closely related set and it is profitable to consider them together. Carbon is the building element of life—all organic molecules are based on skeletons of carbon atoms—and at first sight silicon and boron appear to behave differently. A closer investigation, however, reveals that silicon and boron have many properties in common with carbon. Carbon is often exceptional, for it is the only element apart from hydrogen where number of valence electrons is numerically equal to the number of valence orbitals. The similarities in the chemistries of boron and silicon are often referred to as a **diagonal relationship.**

24.1 Occurrence and extraction of boron, carbon and silicon

Carbon is the only element of the three to occur native. This is a remarkable fact for carbon is quite a strong reducing agent. In the presence of such large quantities of oxygen it may be wondered why all the carbon on Earth has not burned away; as it is, deposits of coal are quite extensive and graphite and diamond occur naturally. The reason for the

TABLE 24.1

Element	Principal quantum number of valence shell	Ground state electronic structure (abbreviated form)	Ground state electronic structure (long form)
B	2	2, 3	$1s^2$; $2s^2p^1$
C	2	2, 4	$1s^2$; $2s^2p^2$
Si	3	2, 8, 4	$1s^2$; $2s^2p^6$; $3s^2p^2$

existence of free carbon is that it is formed from living systems: Carbon is, as far as we know, the only element capable of building the vast complexity of molecules which constitute living matter. All carbon compounds are thermodynamically unstable in the presence of oxygen, but they are *kinetically* very stable due to the fact that their energy levels are completely filled. A carbon skeleton does not have electron vacancies or lone pairs.

The remainder of carbon in nature occurs mostly as carbon dioxide, evolved from living matter, or as carbonates, mainly $MgCO_3$ and $CaCO_3$, which constitute sedimentary rocks and are derived from living matter.

While carbon is the "organic" building element, silicon is its inorganic counterpart. The Earth's crust is made up very largely of complex silicates—giant lattices containing

–Si–O–Si–O–Si– linkages—and, after oxygen, silicon is the most abundant element in the Earths crust. Boron by comparison is relatively rare though the borates in which it occurs are very similar to silicates in structure and properties. Neither element is found free in nature.

Graphite. Impure forms of graphitic carbon which occur in nature, such as coal and anthracite are first converted to coke by heating in the absence of air. A finely powdered mixture with sand is then heated in an electric furnace, and crystals of graphite are formed, probably via the intermediate formation of silicon carbide, SiC.

Colloidal suspensions of graphite in oil and water are used as lubricants; the layer structure of graphite is partly responsible for its lubricant properties (Chapter 5). A mixture of graphite and clay is used in pencil "lead".

Diamond. Diamond crystals are formed from graphite at extreme pressures and fairly high temperatures (1200°C and 100 000 atm) and synthetic diamonds made this way can nowadays compete economically with natural diamonds for industrial purposes. Such diamonds are dark, sometimes being completely black, and are useful as abrasives, but modern technology has not yet succeeded in solving the problem of synthesizing large gem-stones in this way.

Silicon. Silicon is obtained by the reduction of silicon dioxide, though the free energy graph (Fig. 24.1) shows that only an element like magnesium is capable of achieving this. Carbon will also reduce silica at very high temperatures but the product contains a high proportion of silicon carbide. Very pure silicon ($>99\cdot99\%$) is required in the manufacture of transistors, and the technique of zone refining is employed. Zone refining is a procedure in which impurities are removed from a solid phase by melting and resolidifying. It is therefore essentially a recrystallization process. Figure 24.2 shows the

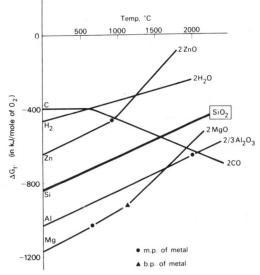

FIG. 24.1.

principle, though in practice more than one heater is used and the tube is moved through the stationary heaters.

Boron. Boron is a difficult element to obtain pure, and it is mainly produced by reduction of boron trioxide, B_2O_3, with an electropositive element like magnesium. The free element does not find many uses.

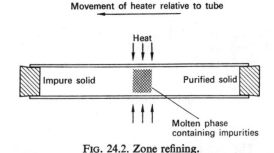

FIG. 24.2. Zone refining.

24.2 General survey of properties

The three elements are all macromolecular in structure, with high binding energies. Carbon and silicon have four electrons per atom available to form bonds while boron has three; these figures are reflected in the high heats of sublimation of the elements.

The structure of boron is complicated, the atoms being arranged in eicosahedra (20-faced). One would expect boron to be a metal, since this is usually the case when there are surplus orbitals with insufficient electrons to fill them. However, boron is a non-conductor and the electrons are not able to move freely through the lattice; a detailed explanation of why this should be so is outside the scope of this book.

Carbon is able to form a tetrahedral structure with a co-ordination number of four. It is also able to form a layer structure (graphite) with a coordination number of three. These structures are described in Chapter 5. The remaining elements of Group IVB do not form layer structures analogous to graphite, though all the elements apart from lead form diamond-like structures.

Table 24.2 is a comparison of some of the physical data for the three elements.

TABLE 24.2

	Boron	Carbon (graphite)	Silicon
Melting point	2300°C	3700°C	1420°C
Boiling point	2600°C	4800°C	2600°C
Density (g cm^{-3})	2·34	2·26	2·35
First ionization energy (kJ)	795	1085	785
Atomic radius (nm)	0·080	0·077	0·117

The elements all burn in air, though boron and silicon do not do so easily unless they are finely divided. Fluorine attacks all three elements, while the other halogens react with boron and silicon but not with carbon. This fact is fortunate, for many metals can be extracted by electrolysis of their fused chlorides using carbon anodes which are immune to attack by the evolved chlorine. Carbon will react directly with sulphur in an electric furnace:

$$C(\text{graphite}) + 2\,S(g) = CS_2(g);$$
$$\Delta H = +115\,\text{kJ at } 25°C.$$

Boron combines directly with nitrogen when heated in the gas, to form a nitride, BN(c), which is structurally similar to graphite and isoelectronic with it; (the ions B^- and N^+ are isoelectronic with the carbon atom). Recently a diamond-like form called *borazon*, an even harder substance than diamond, has been prepared.

Both boron and silicon form a number of hydrides, though these are nothing like as stable kinetically as those of carbon, and are not formed by direct combination of the elements. They are discussed in section 24.5.

All three elements combine directly with most metals to form compounds of various structures and stoichiometry (section 24.4).

24.3 Bond formation and reactivity

Figure 24.3 compares the energy level occupancy of boron, carbon and silicon, and shows that carbon is the only element which, when it forms electron pair bonds, exactly fills all its valence orbitals. This is the factor which accounts for the kinetic stability of carbon compounds.

The point is well illustrated by comparing the three chlorides BCl$_3$(l), CCl$_4$(l) and SiCl$_4$(l). The first and last named fume strongly in moist air and are highly reactive towards attack from negative ions such as OH$^-$ and NH$_2^-$, and from molecules containing atoms with lone pairs of electrons such as water and ammonia. Carbon tetrachloride in contrast is highly inert and even prolonged boiling with concentrated alkali has almost no effect upon

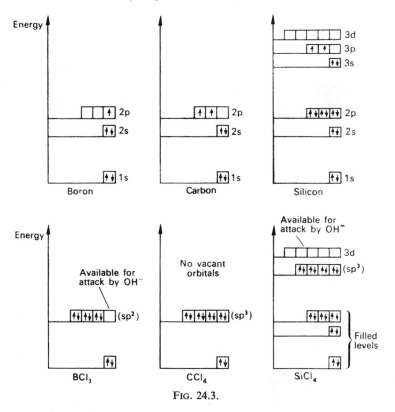

FIG. 24.3.

it. In terms of free energy all three chlorides would be expected to hydrolyse, as the following standard free energy changes at 25°C show:

$$BCl_3 + \tfrac{3}{2}\,H_2O \longrightarrow \tfrac{1}{2}\,B_2O_3 + 3\,HCl;$$
$$\Delta G° = -142 \text{ kJ}$$

$$SiCl_4 + 2\,H_2O \longrightarrow SiO_2 + 4\,HCl;$$
$$\Delta G° = -142 \text{ kJ}$$

$$CCl_4 + 2\,H_2O \longrightarrow CO_2 + 4\,HCl;$$
$$\Delta G° = -275 \text{ kJ}.$$

Carbon tetrachloride does not react, despite the favourable free energy change. It is not a *thermodynamic* factor which stabilizes carbon tetrachloride, but a *kinetic* factor: the reason that carbon tetrachloride does not react is that the required energy of activation cannot be attained. Figure 24.4 illustrates how the energies of activation differ for the attack of the three molecules by OH⁻ ion.

A similar explanation applies to the hydrides of carbon and silicon: these are superficially very similar, and both elements form **homologous series:**

$$CH_4, \quad C_2H_6, \quad C_3H_8, \quad C_4H_{10},$$
$$\ldots, C_nH_{2n+2}\text{---the alkanes}$$
$$SiH_4, \quad Si_2H_6, \quad Si_3H_8, \quad Si_4H_{10},$$
$$\ldots, Si_nH_{2n+2}\text{---the silanes.}$$

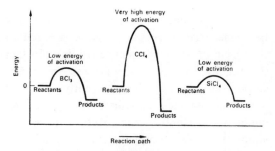

FIG. 24.4. Ease of hydrolysis of chlorides depends on energy of activation.

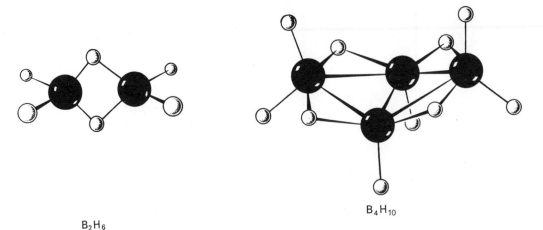

B_2H_6

B_4H_{10}

FIG. 24.5. Structures of two boron hydrides.

Whereas hydrocarbons are stable and obtainable in almost unlimited chain lengths, silicon hydrides inflame spontaneously in air and have not been isolated above Si_8H_{18}.

A different set of conditions applies to silicates however. Here the roles are reversed, and carbonates are *less* stable to heat than silicates. Calcium carbonate and calcium silicate have similar heats of formation. Calcium carbonate decomposes because it forms a gas, with consequent increase in disorder (entropy, Chapter 12):

$$CaCO_3(c) = CaO(c) + CO_2(g);$$
increase in entropy

$$CaSiO_3(c) = CaO(c) + SiO_2(c);$$
solid phases throughout, ∴ negligible entropy change.

Boron hydrides are reactive compounds for the same kind of reason as BCl_3 is reactive: they have unoccupied valence orbitals. In fact the bonding in boron hydrides was not understood until comparatively recently, when it was found that electron pairs in boron hydrides could bind several atoms together instead of simply two. The complex structures of boron hydrides, and their peculiar formulae, were thus explained (Fig. 24.5).

24.4 Ion formation by boron, carbon and silicon

By analogy with aluminium we might expect boron to form an ion B^{3+}, but in fact this ion would be too small to exist, and too much energy of ionization would be needed to form it. For the same reasons, C^{4+} and Si^{4+} do not exist (Fig. 24.6). There is no evidence for positive ion formation by these elements except in a discharge tube, and this is one reason why they are classed as non-metals.

All three elements form oxo-anions. In the case of carbon the carbonate ion, CO_3^{2-}, is particularly simple, being a planar equilateral

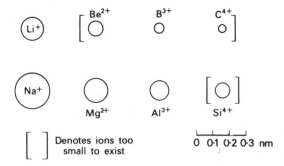

FIG. 24.6.

triangle in shape (section 20.12). Boron forms the analogous orthoborate ion, BO_3^{3-}, and also a series of ions representing intermediate degrees of hydration of the parent acid such as the ion $B_4O_7^{2-}$ which occurs in borax, $Na_2B_4O_7 \cdot 10 H_2O(c)$, and metaborate ions such as $B_3O_6^{3-}$. This tendency to form poly-ions (ions with more than one atom of the element in question) is shared with silicon, where the property is extremely marked. Simple silicate ions SiO_3^{2-}, analogous to carbonate, do not exist but instead silicon tends to form extended polymeric structures.

Carbon forms the rather unexpected ion C_2^{2-}, which occurs in compounds such as calcium carbide. On adding water to calcium carbide, acetylene, $C_2H_2(g)$, is evolved and this observation (together with X-ray data on solid calcium carbide) leads to the conclusion that the carbide ion contains a triple bond:

$$[C\equiv C]^{2-} \qquad H-C\equiv C-H$$
carbide ion (acetylide)   acetylene

Boron and silicon do not form analogous ions: the reluctance of second-row elements to enter into multiple-bond formation has already been noted, and in the case of boron there are insufficient valence electrons available to form a triple bond.

24.5 Hydride formation by boron, carbon and silicon

The most striking fact in dealing with the hydrides of these three elements is the vast range of "stable" hydrocarbons. Very many occur naturally on Earth: North Sea gas consists largely of methane, CH_4, and crude oil deposits consist largely of paraffin hydrocarbons of general formula C_nH_{2n+2}, where $n =$ up to about 40. Hydrocarbons are also present in wood and coal. They owe their stability to the fact that

reactions such as combustion require a high energy of activation. Silanes, Si_nH_{2n+2}, inflame spontaneously in air but paraffins need to be "triggered off" with a spark.

Structurally the following features may be noted:

(i) The structures of the boron hydrides (boranes) are different from those of carbon and silicon hydrides due to the different bonding (section 24.3).

(ii) Hydrocarbons frequently exhibit structures where there are multiple bonds, $C=C$ and $C\equiv C$. For instance:

ethylene

benzene*

Silicon can only form single-bonded structures.

A mixture of silanes can be prepared by the action of water on magnesium silicide, $Mg_2Si(c)$. For instance:

$$Mg_2Si(c)+4H_2O(l) = SiH_4(g)+2Mg(OH)_2(aq).$$

Similarly a mixture of boranes can be made by the action of water on magnesium boride. This method is of general use (section 18.6), and can also be applied to the preparation of certain hydrocarbons though it rarely is in practice.

If a pure sample of a hydride is required, the usual procedure is to reduce the corresponding chloride with the powerful reducing agent, lithium aluminohydride, $LiAlH_4$. This reagent is frequently employed for reducing organic

---

* Benzene really forms a structure with six equal bonds, due to delocalization of the electrons; the above structure is now regarded as an oversimplification, though it is frequently used for convenience.

compounds, and its use here can be summarized in the form of the following equation:

$$SiCl_4 + LiAlH_4 \longrightarrow SiH_4 + LiCl + AlCl_3.$$

Physical properties. The boiling points and melting points of these hydrides increase with molecular weight within a given series. Figure 6.4 shows a plot of boiling point against number of carbon atoms for the paraffin hydrocarbons, and similar trends are observed for boron and silicon hydrides. Hydrogen bonds do not form among these compounds: hence methane is considerably more volatile than NH_3, H_2O and HF, and is considerably less soluble in water (Fig. 18.4). The absence of hydrogen bonds is explained by the fact that these elements are not very electronegative, and also that they do not possess lone pairs in their compounds.

24.6 Halides of boron, carbon, and silicon

Carbon tetrachloride. The exceptional unreactivity of carbon tetrachloride, $CCl_4(l)$, has already been commented upon (section 24.3). This compound has found uses as a non-polar solvent, suitable for dissolving oils and fats; proprietary dry-cleaning fluids were until recently based upon it. Its solvent properties are similar to that of petrol (a hydrocarbon mixture) but it has the added advantage of being non-inflammable. It is now falling into disuse on account of its toxicity; in the presence of a naked flame it reacts with oxygen to give carbonyl chloride, *phosgene*, $COCl_2(g)$, which is highly poisonous. Nowadays, other chlorine-substituted hydrocarbons such as $CH_3.CCl_3$ are taking the place of carbon tetrachloride, and substances such as these also find uses in fire extinguishers, particularly where there is a risk of electrical fires.

Polyvinyl chloride. The carbon-chlorine bond is quite strong (bond energy = 330 kJ mol⁻¹) and is in most cases fairly unreactive. The plastic polyvinyl chloride (PVC) has a structure analogous to polyethylene, but with chlorine atoms in place of some hydrogen atoms:

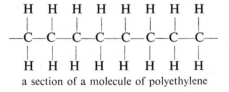

a section of a molecule of polyethylene

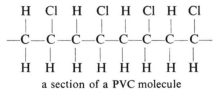

a section of a PVC molecule

PTFE. The abbreviation PTFE stands for polytetrafluoroethylene, $(C_2F_4)_x$. It is quite extraordinarily unreactive, due to the high energy of the C–F bond (438 kJ mol⁻¹), and also possesses a very low coefficient of friction with most other substances. It is a difficult material to work and is expensive, but finds uses where frictionless bearings and non-stick surfaces (e.g. cooking pans) are required. Strong heat causes it to depolymerize into smaller structural units of the same empirical formula.

Freons. The general term 'Freon' refers to gaseous compounds such as $CClF_3$ and CCl_2F_2 which are used as refrigerant gases. On a large scale ammonia is used, but for domestic units "Freon" is less toxic though more expensive.

Silicon halides. Silicon tetrachloride fumes in moist air, and is hydrolysed by water, depositing hydrated silica in a hydrated form:

$$SiCl_4(l) + 4\,H_2O \longrightarrow [Si(OH)_4] + 4\,HCl(g)$$

$$\downarrow\, -H_2O$$

$$SiO_2.xH_2O(c).$$

The same hydrolysis reaction is used in the manufacture of *silicones*. If $(CH_3)_2SiCl_2(l)$ is hydrolysed, the product is a mixture of ring and chain polymers:

$$(CH_3)_2SiCl_2 + 2H_2O \rightarrow [(CH_3)_2Si(OH)_2] + 2HCl$$

$$\downarrow -H_2O$$

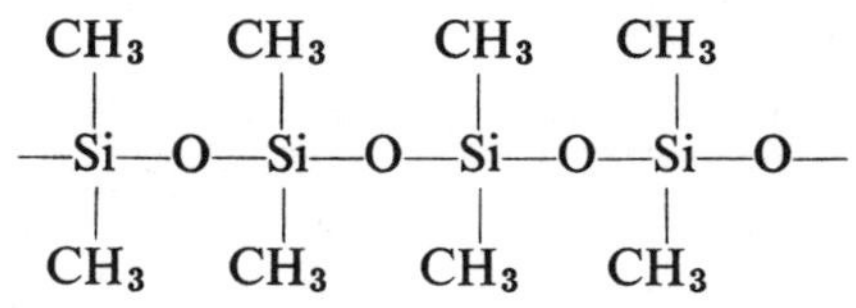

The physical properties of a silicone can be varied by introducing other compounds: $(CH_3)_3SiCl$ can be included to introduce chain ends, and CH_3SiCl_3 leads to branched chains:

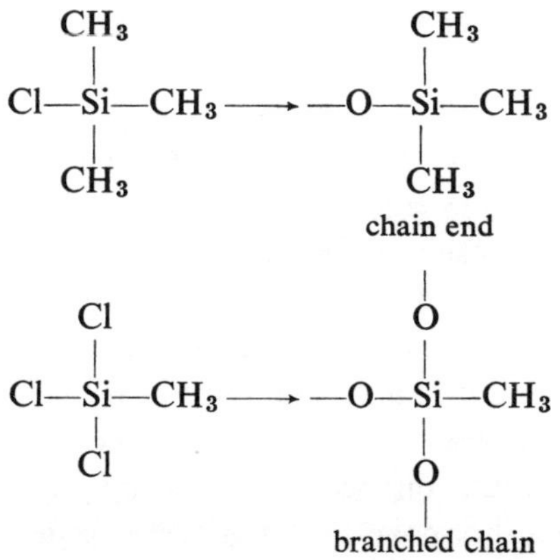

chain end

branched chain

Silicone polymers can thus be "tailor made" with a wide choice of properties: some are oils, and others with a higher degree of cross linking form useful plastics. Silicone products are strongly water-repellent.

24.7 Oxides and oxo-acids of boron, carbon and silicon

The principal oxides of the three elements which will be considered are $B_2O_3(c)$, $CO(g)$, $CO_2(g)$ and $SiO_2(c)$.

TABLE 24.3

OXIDES OF BORON AND SILICON

Formula	Trivial name	Systematic name	Melting point
$CO(g)$	carbon monoxide	carbon(II) oxide	$-205°C$
$CO_2(g)$	carbon dioxide	carbon(IV) oxide	$-78°C$ (subl.)
$SiO_2(c)$	silicon dioxide, silica	silicon(IV) oxide	$1728°C$
$B_2O_3(c)$	boron trioxide or boron sesquioxide	boron(III) oxide	$577°C$

All except carbon monoxide are very weakly acidic, dissolving in alkali to form salts:

$$CO_2(g) + 2\,OH^-(aq) \longrightarrow CO_3^{2-}(aq) + H_2O;$$
$$\text{(rapid)}$$

$$B_2O_3(c) + 6\,OH^-(aq) \longrightarrow 2\,BO_3^{3-} + 3\,H_2O;$$
$$\text{(slow)}$$

$$SiO_2(c) + 2\,OH^-(aq) \longrightarrow SiO_3^{2-}(aq) + H_2O;$$
$$\text{(slow)}$$

Boron(III) oxide and silicon dioxide are solids with macromolecular structures which only dissolve in alkali when finely ground and heated; carbon dioxide is a gas which is absorbed very rapidly by alkali. Carbon dioxide also forms well-defined acid salts, known as bicarbonates or hydrogen carbonates:

$$CO_2(g) + OH^-(aq) \longrightarrow HCO_3^-(aq)$$

Carbon monoxide is normally classed as a neutral oxide (section 22.11) though it does react with alkali to form the *formate* ion at high temperature and pressure:

Carbon dioxide is present in the air (about 0·03%) due to its evolution from animals and plants, and by processes such as burning and fermentation. Its concentration remains remarkably constant, and it is continually removed from the air by photosynthesis. The gas is obtained by:

(i) the action of $H^+(aq)$ on a carbonate or bicarbonate;
(ii) action of heat on carbonates and bicarbonates;
(iii) oxidation of carbon, carbon monoxide, or carbon compounds;
(iv) fermentation.

The gas is liquefied by pressure, and the liquid is stored in cylinders. Rapid expansion of the gas leads to pronounced cooling, and solid carbon dioxide (sold as "dry-ice") is thus obtained. The gas is used in refrigeration, "fizzy" drinks, fire extinguishers, fruit preservation some metallurgical processes, and the manufacture of some carbonates (e.g. the white pigment $PbCO_3(c)$).

When carbon dioxide is dissolved in water, some carbonic acid, $H_2CO_3(aq)$, is formed, but the anhydrous acid cannot be obtained.

Carbonates. The *s*-block elements (except beryllium) form stoichiometric carbonates; the thermal stability of these was discussed in section 20.12.

B-metals and transition metals form non-stoichiometric *basic* carbonates when $CO_3^{2-}(aq)$ is added to an aqueous solution of the metal ion. For instance copper carbonate forms $xCu(OH)_2.y\,CuCO_3$. Some metals do not form carbonates at all, e.g. aluminium and beryllium.

Apart from the Group IA carbonates, all carbonates are insoluble in water, and decompose on heating giving the oxide and $CO_2(g)$.

Bicarbonates. See section 20.12.

Carbon monoxide is formed whenever carbon or its compounds react with oxygen at a high temperature, or where the supply of oxygen is insufficient. It also forms when a metal oxide reacts with carbon at high temperature. Essentially its formation depends upon the following endothermic reaction which becomes favoured above about 700°C:

$$C(graphite) + CO_2(g) \rightleftharpoons 2\,CO(g);$$
$$\Delta H = +172 \text{ kJ}$$

Carbon monoxide can be made in the laboratory in this way but a purer product is obtained by dehydrating formic acid with concentrated sulphuric acid:

$$H.COOH - H_2O \longrightarrow CO.$$

Carbon monoxide is very poisonous due to its ability to combine with haemoglobin in the blood to form a stable addition compound; this prevents the normal exchange between haemoglobin and oxygen from taking place. *Town gas* used to contain about 15% of carbon monoxide in the days when it was made exclusively from coal, but the modern product is considerably safer. An industrial fuel called *producer gas*, consisting of carbon monoxide diluted with nitrogen, is made by passing air over red-hot coke. A better fuel is *water gas*, a mixture of carbon monoxide and hydrogen made by passing steam over heated coke. The latter reaction is endothermic, but by mixing air and steam the overall process can be made approximately thermoneutral.

$$2\,C(graphite) + O_2(g) = 2\,CO(g);$$
$$\Delta H = -222 \text{ kJ}$$

$$C(graphite) + H_2O(g) = CO(g) + H_2(g);$$
$$\Delta H = +130 \text{ kJ}$$

Water gas is also used as a source of hydrogen, carbon monoxide, and methanol.

Carbon monoxide forms volatile compounds, which are known as carbonyls, with some transition metals (section 26.13).

Silica differs markedly in structure from carbon dioxide, due to the reluctance of silicon to form double bonds (Fig. 24.7). This difference in structure is also reflected in differences between silicates and carbonates. Sand and quartz are naturally occurring forms of silica, and a

$$O = C = O$$
Linear molecule
CO_2

Three-dimensional macromolecule
$(SiO_2)_x$

FIG. 24.7.

variety of other crystalline forms—polymorphs or allotropes—also occur. If silica is melted (m.p. 1728°C) and recooled it becomes **vitreous**, or glass-like; it loses its regular crystalline lattice and becomes amorphous. This glass eventually devitrifies and becomes crystalline once more.

Vitreous silica is used in special laboratory apparatus; it has a low coefficient of thermal expansion and consequently does not shatter like ordinary glass when subject to violent thermal shock. Acids do not attack it, but alkalis slowly react forming silicates. Aqueous hydrogen fluoride dissolves it, forming the ion SiF_6^{2-}:

$$SiO_2(s) + 6 HF(aq)$$
$$\longrightarrow 2 H^+(aq) + SiF_6^{2-}(aq) + 2 H_2O.$$

Silicic acid, $SiO_2 . xH_2O$, is the name given to the colloidal gel produced when $H^+(aq)$ is added to a solution of silicate ion. A continuous range of substances of varying degrees of hydration can be obtained by heating this. One such hydrate, a dry powder called silica gel, is widely used as a drying agent.

Sodium silicate, Na_2SiO_3 (approximate formula), is formed as a glass when sodium carbonate is fused with silica:

$$Na_2CO_3 + SiO_2 \longrightarrow Na_2SiO_3 + CO_2.$$

Prolonged heating with water under pressure converts it to a treacly aqueous solution known as "water glass", formerly used as an egg preservative; its main uses now are as an additive for detergents, and as a cheap adhesive.

It is probable that the extended lattice formed by silicon and oxygen atoms in silica persists in water glass, thereby giving it a high viscosity.

Silicates in the Earth's crust. The empirical formulae of naturally occurring silicates are complex, and their structures fall into three distinct types:

(i) **Fibrous structures,** in which chain molecules exist, e.g. the *amphiboles*.
(ii) **Layer structures,** in which the solid readily splits into flat sheets, e.g. *mica*.
(iii) **Solid structures,** in which the macromolecules are three-dimensional, e.g. *felspars* (components of granite).

In these structures there may be varying proportions of water; specially processed *zeolites* treated to remove the water have been found to act as **molecular sieves.** For instance a sieve is manufactured which will allow straight chain hydrocarbons to pass through it, while blocking branched chains. Other zeolites find uses as ion exchangers (section 16.1) though they have been largely superseded by synthetic polystyrene resins.

Glass. Glass has been known for many centuries, and ordinary cheap varieties are made by fusing together sand, limestone and an alkali such as sodium hydroxide or sodium carbonate. The resultant non-stoichiometric compound is glass, in effect a mixture of silicates. Soda-glass, made by using sodium carbonate, is described as soft glass as it has a low softening point on heating; by substituting potassium carbonate a harder glass is obtained. Glasses never have sharp melting points, but always soften over a range of temperature due to their amorphous nature.

Coloured glasses are made by adding small traces of transition metal oxides which form coloured ions; the green colour of cheap glass is due to traces of iron impurity. Glass of high refractive index is made by using lead oxide in place of calcium oxide. Modern laboratory glassware uses oxides of boron and aluminium in place of calcium oxide, and this is known as *borosilicate glass* ("Pyrex" is a well-known example).

The *borax bead test* used in analysis consists of the formation of a coloured glass when borax is fused with a trace of a transition metal salt.

24.8 Borides, carbides and silicides

Metals form binary compounds with boron, carbon and silicon, the two main classes being **ionic** (*saline* or salt-like), and **interstitial** (where the non-metal atoms occur within a metal lattice).

Ionic. (a) The carbides of the alkali and alkaline-earth metals contain discrete C_2^{2-} units. These give acetylene, C_2H_2, on hydrolysis.

(b) Be_2C and Al_4C_3 contain discrete carbon atoms or ions, and give methane, CH_4, on hydrolysis.

The difference in property can be understood by considering the sizes of the ions which might be expected. It has previously been remarked for carbonates (section 20.12) and polyhalides (section 21.5), that large anions form more stable lattices with large cations. Here the larger ons of Groups I and II favour the larger C_2^{2-} ion.

Interstitial borides, carbides and silicides (and also nitrides and phosphides) are formed by many transition metals. These are metal-like (they conduct electricity) and are usually non-stoichiometric. The lattice is based upon the metallic element. A reverse state of affairs occurs with certain compounds based on the graphite structure, in which metal atoms are "sandwiched" interstitially between the layers, e.g. KC_8, KC_{24} and KC_{36}. Borides such as CaB_6 occur also.

24.9 Cyanides

Cyanides are salts of the acid hydrogen cyanide, HCN (b.p. 26°C). This gas dissolves in water giving a very weak acid:

$$HCN(g) + aq \rightleftharpoons \underbrace{H^+(aq) + CN^-(aq)}_{\text{hydrocyanic acid}};$$

$$K_a = 7 \cdot 2 \times 10^{-10} \text{ mol dm}^{-3} \text{ at } 25°C.$$

Hydrogen cyanide is in some respects similar to the hydrogen halides, and indeed the cyanide ion has properties similar to the halide ions in general. The gas *cyanogen*, $(CN)_2$, bears a formal resemblance to the halogens and is often referred to as a *pseudo-halogen*.

Hydrogen cyanide and cyanides are extremely poisonous substances which require careful handling. The gas smells of almonds. The antidote is a solution of iron(II) ions which reacts with CN^-(aq) to form the harmless complex ion $Fe(CN)_6^{4-}$ (see below).

The HCN molecule is linear, and it is isoelectronic with acetylene, possessing a triple bond:

H—C≡N   or $H\overset{\times}{\underset{\bullet}{}}C\overset{\times}{\underset{\times}{}}N$

The cyanides of the alkali metals are probably ionic in the solid state, but the structure of most others suggests that they are covalent. The majority are insoluble in water, but dissolve in excess $CN^-(aq)$ forming complex ions.

Sodium cyanide is manufactured by the following reactions:

$$Na + NH_3 \longrightarrow NaNH_2(c) + \tfrac{1}{2} H_2$$
sodium amide
(sodamide)

$$NaNH_2 + C \longrightarrow NaCN + H_2.$$

Since hydrogen cyanide is very weak, the conjugate base $CN^-(aq)$ is fairly strong, and can react with water to produce an alkaline solution:

proton accepted

$$CN^-(aq) + H_2O \rightleftharpoons HCN(aq) + OH^-(aq)$$

proton lost

Cyano complexes. The $CN^-(aq)$ ion is a powerful ligand, that is to say, it can join to metal ions forming stable complexes. In many cases the ion is so stable that the properties of the free ions cannot be detected in solution; this is true of the $Fe(CN)_6^{4-}$ ion mentioned above:

$$Fe^{2+}(aq) + 6 CN^-(aq) \rightleftharpoons Fe(CN)_6^{4-}(aq);$$

iron(II)   hexacyanoferrate(II)
ion   (ferrocyanide) ion

Transition metals and B-metals form stable complexes with CN^-; *s*-block metals do not, and neither does aluminium.

24.10 Thiocyanates

Another pseudohalide ion, closely related to cyanide, is the thiocyanate ion, CNS^-. This is also a ligand, and one of its special uses is in the test for the presence of iron(III). When $Fe^{3+}(aq)$ and $CNS^-(aq)$ are mixed, a variety of species can be obtained depending on the relative ratios of the simple ions present, but all have a deep blood red colour. This is used to estimate iron(III) colorimetrically: the intensity of absorption of light can be compared with a known standard in a colorimeter. Alternatively, photoelectric measurements can be made.

24.11 Summary

Boron and silicon afford one of the clearest examples of a diagonal relationship. Hence:

(a) The elements are similar in physical properties, existing in both amorphous and crystalline forms.

(b) Their oxides are similar—weakly acidic, macromolecular, giving rise to similar "polymeric" oxo-ions and readily forming glasses.

(c) Both elements form a series of hydrides which are kinetically unstable in the presence of oxygen (i.e. they ignite spontaneously).

(d) Both form readily hydrolysable molecular chlorides.

OTHER EXAMPLES OF DIAGONAL RELATIONSHIPS

Lithium and magnesium react similarly with air, water and non-metals. Lithium resembles Group II rather than Group I in having a thermally unstable carbonate, in forming a hydrated chloride when the aqueous solution is evaporated, and in forming an ion $Li^+(aq)$ which hydrolyses in aqueous solution.

Beryllium and aluminium form similar hydroxides which are amphoteric; the metals themselves dissolve in alkali giving hydrogen. Similar complexes are formed, e.g. $BeCl_4^{2-}$ and $AlCl_4^-$. (But note BeF_4^{2-} and AlF_6^{3-}, due to expansion of aluminium valence shell.) Beryllium and aluminium chlorides have analogous structures.

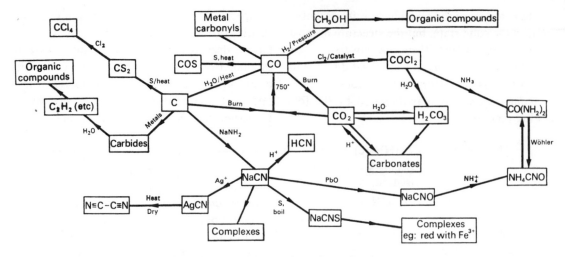

FIG. 24.8. A summary of reactions of carbon.

Study Questions

1. List the ways in which the chemistry of aluminium is (a) similar to, and (b) different from the chemistry of boron. How can you account for the differences?

2. (a) Aluminium forms an ion AlF_6^{3-}. Why is BF_6^{3-} not formed?

  †(b) Suggest why $BCl_3(g)$ is monomeric, while the chloride of aluminium in the gas phase consists of Al_2Cl_6 dimers.

3. (a) Compare and contrast
  (i) CO_2 and SiO_2.
  (ii) $CaSiO_3$ and $CaCO_3$.
  (iii) $SiCl_4$ and CCl_4.
  (iv) $(CH_3)_2CO$ and $(CH_3)_2SiO$.

  (b) What reasons can you suggest for the differences that you mention?

4. (a) List as many similarities as you can in the chemistries of boron and silicon.

  (b) Explain, in terms of fundamental principles, why the chemistries of boron and silicon are similar.

  (c) Is there any evidence for a diagonal relationship between carbon and phosphorus?

5. (a) List the typical properties of a halogen.

  (b) Cyanogen has been described as a 'pseudo-halogen'. How many of the halogen properties does cyanogen possess?

6. When 10·01 g of a white solid, X, were heated, 2·2 g of an acid gas, A, that turned lime water milky, were given off, together with 0·9 g of a gas, B, which condensed to a colourless liquid. The solid that remained, Y, weighing 6·91 g, dissolved in water to give an alkaline solution, which with excess barium chloride solution gave 9·85 g of a white precipitate, Z. Z effervesced with acid, giving off carbon dioxide.

  (a) Identify Z, A and B.
  (b) How many moles of Z, A and B were produced?
  (c) Deduce the nature of Y.
  (d) Deduce the nature of X and write down an equation for its thermal decomposition.

7. A litre of M sodium carbonate was evaporated and 286 g of white crystals were obtained. These were left in a dry laboratory and were later found to weigh 124 g.

  (a) What are the formulae of the two solids?
  (b) What is the name given to the process by which the crystals lost weight?

8. When a solution of sodium bicarbonate is added to a solution of zinc ions, zinc carbonate, $ZnCO_3$, is precipitated.

  (a) What else is formed in this reaction?
  (b) Give a balanced equation for the reaction.
  (c) Zinc carbonate is readily decomposed by heat. Suggest what the products of the thermal decomposition might be.

9. In qualitative inorganic analysis, substances are sometimes heated on a charcoal block in a blowpipe flame.

(a) Which metal oxides would be reduced to the metal in this way?

(b) How would you expect a nitrate or chlorate to react in this test?

10. Give as many reasons as you can why carbon is often preferred to silicon or boron as a reducing agent.

11. When HF(g) reacts with glass in the presence of water, fluorosilicic acid, "H_2SiF_6", is formed.

(a) Suggest an equation for this reaction.

(b) What shape would you expect the SiF_6^{2-} ion to be?

(c) Can you suggest why $SiCl_6^{2-}$ has yet to be made?

(d) When barium hexafluorosilicate is heated, SiF_4 is evolved. Suggest an equation for this reaction.

†12. (a) The four heavy atoms in $(SiH_3)_3N$ are co-planar. Why is this surprising?

(b) Would you expect $(SiH_3)_3N$ to be an electron donor?

CHAPTER 25

The B-metals

25.1 Classification and electronic structure

The B-metals are normally taken to be those metals which follow a transition series in a given period. Thus aluminium is not normally classified as a B-metal and it has properties which are generally more akin to those of s-block metals. Zinc, cadmium and mercury have properties similar to those of the metals on their right in the periodic table and will be considered as B-metals in this book.[†]

TABLE 25.1

IB	IIB	IIIB	IVB	VB	VIB	VIIB	0
		Al*					(Ar)
Cu§	Zn‡	Ga	[Ge]	[As]			(Kr)
Ag§	Cd‡	In	Sn	Sb	[Te]		(Xe)
Au§	Hg‡	Tl	Pb	Bi	Po	[At]	(Rn)

* Aluminium does not follow a transition series.
§ These elements are more like transition metals than B-metals.
[ ] These elements are borderline between metal and non-metal.
‡ Some books treat these elements as transition metals.

† For a note on the nomenclature adopted in this book, see also section 20.1.

The ground state electronic structures of B-metal atoms may be derived by taking the noble gas structure at the end of the period, and removing an appropriate number of electrons. Table 25.2 shows how this is done. For instance, the configuration of an atom of tin may be written down by taking the configuration for xenon, and writing p^2 instead of p^6;

$$1s^2; \ 2s^2p^6; \ 3s^2p^6d^{10}; \ 4s^2p^6d^{10}; \ 5s^2p^2$$

or, in the abbreviated form,

$$2, 8, 18, 18, \underline{4}.$$

It is normally the valence electrons which are of greatest interest to the chemist, and the following table summarizes how these vary in this region of the periodic table, often known as the **p-block.**

TABLE 25.3

Group	No. of electrons in valence shell	Configuration of valence shell
IIIB	3	s^2p^1
IVB	4	s^2p^2
VB	5	s^2p^3
VIB	6	s^2p^4
VIIB	7	s^2p^5
0	8	s^2p^6

TABLE 25.2

II p^0	III p^1	IV p^2	V p^3	VI p^4	VII p^5	VIII p^6	Abbreviated electronic configuration of noble gas	Electronic configuration of noble gas written out in full
Zn	Ga	Ge	As	Se	Br	Kr	2, 8, 18, <u>8</u>	$1s^2$; $2s^2p^6$; $3s^2p^6d^{10}$; $4s^2p^6$
Cd	In	Sn	Sb	Te	I	Xe	2, 8, 18, 18, <u>8</u>	$1s^2$; $2s^2p^6$; $3s^2p^6d^{10}$; $4s^2p^6d^{10}$; $5s^2p^6$
Hg	Tl	Pb	Bi	Po	At	Rn	2, 8, 18, 32, 18, <u>8</u>	$1s^2$; $2s^2p^6$; $3s^2p^6d^{10}$; $4s^2p^6d^{10}f^{14}$; $5s^2p^6d^{10}$; $6s^2p^6$

25.2 Occurrence and extraction of B-metals

Most of the B-metals occur as oxides or sulphides, particularly the latter. It is a characteristic of B-metals that their ions are "soft" (easily polarized or distorted), and they therefore occur with soft anions such as sulphide. Some mercury occurs native. The sulphide ores are converted into the oxide by roasting, and the oxide is then reduced with carbon. The essentials of these extraction processes are described in Chapter 13.

This chapter will be mainly concerned with Group IIB (zinc, cadmium and mercury) and Group IVB (germanium, tin and lead). The Group VB elements have been considered in Chapter 23, and the Group IIIB elements (gallium, indium and thallium) are relatively rare and will not be considered in any detail. Germanium, and many of the elements in this

TABLE 25.4

Element	ε° value		Name of ore	Formula of ore	Method of extraction
Zn	Zn^{2+}/Zn,	-0.763 V	Zinc blende	ZnS	Roast; C reduction
Cd	Cd^{2+}/Cd,	-0.403 V	E.g. greenockite	CdS	C reduction; extracted from Zn ores
Sn	Sn^{2+}/Sn,	-0.136 V	Cassiterite	SnO_2	C reduction
Pb	Pb^{2+}/Pb,	-0.126 V	Galena	PbS	Roast; C reduction
Sb	Sb_2O_3/Sb,	$+0.152$ V	Stibnite	Sb_2S_3	Roast; C reduction
Bi	Bi^{3+}/Bi,	$+0.32$ V	Bi_2O_3, Bi_2S_3 and $(BiO)_2CO_3$		By-product of Sb, etc., extraction
Hg	$Hg_2^{2+}/2Hg$,	$+0.789$ V	Cinnabar and native	HgS	Roast in air; Hg vapour condenses

borderline region between metal and non-metal, have assumed technological importance on account of their semi-conductor properties. Germanium itself is widely used in the manufacture of transistors, and compounds between elements of Groups IIIB and VB such as gallium arsenide, GaAs, are also important. The requirements of the electronics industry have forced chemists to manufacture these elements and compounds to exceptional purity standards. Germanium for transistor manufacture needs to be at least 99·99% pure, and the technique known as zone refining is used.

Table 25.4 gives a summary of the methods of extraction used for the main B-metals, the metals being arranged in order of the electrochemical series.

25.3 Properties and uses of the B-metals

The B-metals are generally low melting (compared with the transition metals) and fairly soft. Zinc, cadmium and mercury are the most volatile metals in this part of the periodic table.

Chemically, the B-metals are generally fairly "weak", that is, they have less tendency than the s-block metals to combine with non-metals such as oxygen and chlorine, and this is reflected in their $\varepsilon°$ values (Table 25.4). One of the most reactive B-metals is zinc, and even this is less reactive with non-metals than any s-block metal.

Gallium and germanium are used in the electronics industry. Zinc is used for protecting iron in "galvanized" articles; it is "sacrificially" corroded when an electrolyte bridges the two metals (Chapter 11). Brass is a copper–zinc alloy and zinc is widely used as a building material and in the preparation of pigments. Cadmium is a soft low-melting metal which is used in some anti-friction bearings, and in low melting alloys such as Wood's metal (m.p. between 60° and 70°C). A cadmium–mercury amalgam is used in the Weston standard cell. Mercury is the only metal which is liquid at room temperature, and is widely used in thermometers, manometers, and other scientific apparatus.

Metallic tin is used mainly as tin-plate for protecting steel containers from corrosion; food cans are coated in this way. So-called "tin foil" is more likely these days to be made from aluminium, but some tin is used in the manufacture of alloys such as pewter (lead–tin), bronze (copper–tin), bearing metals, type metal and solder (lead–tin).

Most of the lead produced at the present time is used for the manufacture of accumulators, especially in motor vehicles. Other uses are the manufacture of anti-knock agents such as lead tetramethyl, $Pb(CH_3)_4$, electric cable sheaths and lead pipes. Lead blocks are used in radioactive screening: the high nuclear charge of lead makes it an efficient absorber of radiation.

25.4 Ion formation and the inert-pair effect

The ions of B-metals, although comparable in size with those of s-block metals (e.g. Zn^{2+} has a similar radius to Mg^{2+}, Fig. 25.1), are nevertheless very different. The main differences between s-block ions and B-metal ions may be summarized as follows:

(a) B-metals enter into chemical combination less readily than s-block metals, because B-metals have higher ionization energies. Figure 25.2 shows this effect.

(b) B-metal ions are more readily polarized (softer, more easily distorted) than s-block metal ions. Thus when compounds do form, they are more likely to be covalent:
   (i) they enter into complex formation readily;

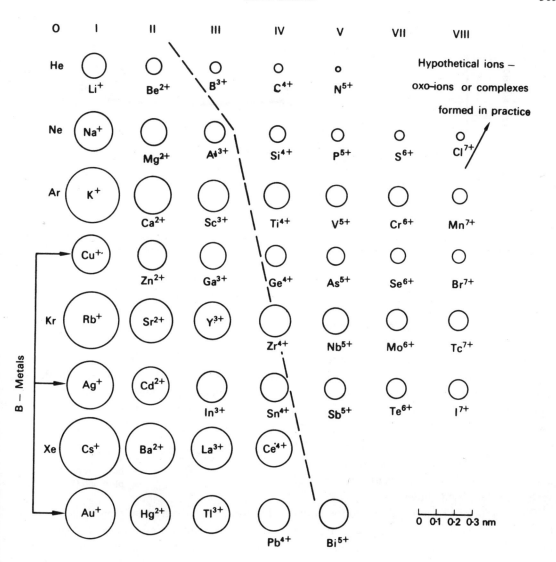

FIG. 25.1.

(ii) the structures of compounds do not fit the ionic model, i.e. they are frequently layered or distorted;

(iii) physical properties of compounds, such as solubility in water and volatility, are often not consistent with the existence of true metal ions.

(c) B-metals show variable valency, exhibiting two states:

(i) oxidation number = Group number.

(ii) oxidation number = Group number minus two.

This is the **inert-pair** effect (see below). Loss of electrons from a B-metal atom can take place in two ways:

(i) All the outer s and p electrons may be lost: in this case a compound exists in which the oxidation number equals the Group number, for instance,

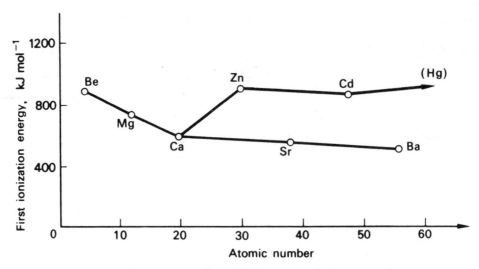

FIG. 25.2. B-metals have higher ionization energies than A-metals.

tin(IV) and antimony(V). Such compounds are not ionic: electron-pair bonds are formed, either in neutral molecules or complex ions.

(ii) The outermost s electrons are frequently not lost, and the oxidation number of the element is then equal to *Group number minus two*, for instance, thallium(I), lead(II) and bismuth (III). The pair of *s*-electrons is said to have become inert. The inert-pair effect is most strongly marked for the heavier elements of a Group such as those named above.

The ions formed by B-metals are quite unlike those formed by *s*-block metals: they are very much more polarizable. The "hard-soft concept" has been used to illustrate this: we say that *s*-block metal ions are hard, but B-metal ions are soft. The consequence of this is that B-metals enter into complex formation and electron-pair bond formation much more readily than *s*-block metals of similar atomic radius. Ions of very high charge are not often formed in any case, and one would not expect to observe simple ions such as Bi^{5+}, partly because their size would be too small and partly because their energy of ionization would be too great; in

aqueous solution such ions would certainly hydrolyse.

Ligands of negative charge, which can neutralize the charge present on the simple ion, form particularly stable complexes. Hence although the simple ion Pb^{4+} does not occur, complex ions such as $PbCl_6^{2-}$ are stable:

$$Pb^{4+} + 6\,Cl^- \rightleftharpoons PbCl_6^{2-}$$
$$\text{lead(IV)} \qquad\qquad \text{hexachloro-}$$
$$\text{plumbate(IV)}$$

Figure 25.1 is a diagram summarizing the atomic and ionic radii of s-block and p-block elements. The radii shown to the right of the dotted line are hypothetical only: the ions are too small and highly charged to exist free, and oxo-ions or complexes occur instead.

25.5 Oxides and hydroxides of B-metals

The B-metals react directly with oxygen, though in general they do not occur very high up the electrochemical series and reaction is usually slow. With elements such as mercury reaction only occurs on heating the metal in oxygen.

Oxides in this region of the periodic table often have structures of an irregular kind suggesting that covalent bonds exist between atoms. Symmetrical structures determined by such factors as the radius ratio of the ions are encountered less often. B-metal oxides are insoluble in water, though a very weak equilibrium is set up in which metal ions and hydroxide ions are formed in aqueous solution.

Many of the B-metal oxides and hydroxides are **amphoteric,** that is to say, they react either as acids or as bases. When an amphoteric oxide dissolves in an acid it gives a solution containing

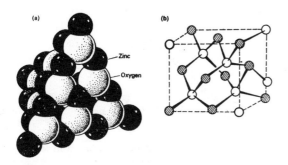

Fig. 25.3.

cations $M^{n+}(aq)$; when it dissolves in an alkali it gives a solution containing hydroxo-complexes of the type $M(OH)_n^{y-}$.

Zinc. Zinc oxide, ZnO, is a white solid with a tetrahedral structure showing 4 : 4 co-ordination (Fig. 25.3). It is formed when zinc is heated in oxygen, and is insoluble in water. It dissolves in acids and alkalis:

$$ZnO(c) + 2 H^+(aq) \longrightarrow Zn^{2+}(aq) + H_2O$$
$$ZnO(c) + 2OH^-(aq) + H_2O \longrightarrow Zn(OH)_4^{2-}(aq)$$
$$\text{tetrahydroxozincate}$$

Zinc hydroxide is precipitated when alkali is added to a solution of $Zn^{2+}(aq)$, but excess reagent redissolves the precipitate:

$$Zn^{2+}(aq) + 2OH^-(aq) \rightleftharpoons Zn(OH)_2(c);$$
$$K_s = 2 \times 10^{-14} \text{ mol}^3 \text{ dm}^{-9};$$
$$Zn(OH)_2(c) + 2OH^-(aq) \longrightarrow Zn(OH)_4^{2-}(aq)$$

Zinc oxide and hydroxide are therefore amphoteric. Zinc metal is itself amphoteric in nature, reacting with alkali to give hydrogen:

$$Zn(c) + 2 OH^-(aq) + 2 H_2O$$
$$\longrightarrow Zn(OH)_4^{2-} + H_2(g)$$

Zinc forms only one oxidation state, with oxidation number $= +2$. It is not necessary to write zinc(II) in the names of compounds because there can be no ambiguity.

Cadmium. Cadmium forms a brown oxide, CdO(c), and a white hydroxide $Cd(OH)_2(c)$ under conditions similar to those for zinc. These compounds are not amphoteric: they dissolve in acids but not in alkalis.

Mercury. Mercury forms two series of compounds containing respectively the $+1$ and $+2$ oxidation states of the metal. On heating mercury in oxygen, mercury(II) oxide, HgO, is obtained as a red solid which readily dissociates into its elements on strong heating (ΔG_f positive, Chapter 13). There is also a yellow form of mercury(II) oxide, which differs from the red form only in the size of its constituent particles. Addition of alkali to a solution of $Hg^{2+}(aq)$ gives mercury(II) oxide: the hydroxide does not exist though the oxide is hydrated when first precipitated. Mercury(I) oxide, Hg_2O, does not exist though salts of it are readily obtained.

Tin. Ordinary tin does not oxidize easily, though molten tin will combine with oxygen to form tin(IV) oxide, SnO_2. This same compound is present in *cassiterite*. The same oxidation state of tin is formed when steam is passed over the heated metal, though water itself has a negligible effect on tin:

$$Sn(c) + 2 H_2O(g) \longrightarrow SnO_2(c) + 2 H_2(g)$$

Tin(IV) oxide is more acidic than basic in character; in this respect it is more like the oxide of a non-metal. High oxidation states of a metal behave in a "non-metallic" fashion, and other examples will be encountered, for instance manganese(VII) and chromium(VI). Concentrated sulphuric acid slowly dissolves tin(IV) oxide to give the sulphate:

$$SnO_2 + 2\,H_2SO_4 \longrightarrow Sn(SO_4)_2 + 2\,H_2O$$

Alkalis dissolve it forming a stannate ion:

$$SnO_2(c) + 2\,OH^-(aq) + 2H_2O \longrightarrow Sn(OH)_6^{2-}$$
hexahydroxo-
stannate(IV)

Hydroxides of tin do not appear to exist: the addition of alkali to a solution of $Sn(SO_4)_2$ precipitates SnO_2.

Tin shows the inert-pair effect: it forms a series of compounds with an oxidation state $+2$, which is "Group valency minus two". Tin(II) oxide is precipitated in hydrated form by adding alkali to Sn^{2+}(aq) ions; on heating this precipitate in absence of air a brown powder is obtained, approximating to SnO in composition. On exposure to air this oxide quickly reacts to form SnO_2. It is amphoteric, dissolving in acids to give tin(II) ions, Sn^{2+}(aq), and in alkali to give stannate(II) hydroxo-complexes, often known as stann*ites*. Sn^{2+}(aq) is a good reducing agent ($\varepsilon°$ for $Sn\,|\,Sn^{2+}$ $= +0\cdot14$ V).

Lead. Like tin, lead forms two series of compounds, in which the oxidation states are $+2$ and $+4$ respectively.* Lead shows the inert pair effect more strongly than tin, and consequently lead(II) compounds are the more common while lead(IV) compounds are powerful oxidizing agents ($\epsilon°$ for $Pb^{2+}|Pb^{4+} = +1\cdot5$ V). Lead(II) oxide, *litharge*, PbO(c), is formed

---

* The older names are tin(II) = stann*ous*, tin(IV) = stann*ic*; lead(II) = plumb*ous*, and lead(IV) = plumb*ic*.

as a yellow powder when lead is strongly heated in air, or when a higher oxide of lead is heated strongly. It is also formed when lead(II) nitrate or lead(II) carbonate is heated in air. It is amphoteric and forms Pb^{2+}(aq) when dissolved in acids, and $Pb(OH)_4^{2-}$ when heated with concentrated alkali. True lead hydroxide is not formed by adding OH^-(aq) to Pb^{2+}(aq): instead a hydrated oxide approximating to $2PbO.H_2O$ is precipitated.

The structure of lead(II) oxide is strange, and quite clearly not derived from the ions Pb^{2+} and O^{2-}. The environment of the lead atoms is shown in Fig. 25.4: it is thought that the inert pair of electrons might occupy the apex of the pyramid.

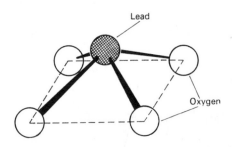

FIG. 25.4. The environment of a lead atom in lead(II) oxide (cf. ZnO, Fig. 25.3).

If lead(II) oxide is heated in air for some time at about 450°C, a red powder of formula Pb_3O_4, *red lead*, is formed. Its constitution suggests that it is a "salt" compounded of the acidic oxide PbO_2 and the basic oxide PbO, i.e. $2PbO.PbO_2$. Its systematic name is therefore lead(II,II,IV) oxide. If red lead is warmed with dilute nitric acid dark brown lead(IV) oxide, PbO_2(c), is formed: it is as if the "basic" part of the Pb_3O_4 had been dissolved out:

$$Pb_3O_4 + 4\,HNO_3(aq)$$
$$\longrightarrow 2\,Pb(NO_3)_2(aq) + PbO_2(c) + 2\,H_2O$$
cf. $PbO(c) + 2\,HNO_3(aq)$
$$\longrightarrow Pb(NO_3)_2(aq) + H_2O$$

Lead(IV) oxide is an acidic oxide: it will not dissolve in acids except by some other type of reaction, e.g. oxidation:

$$PbO_2(c) + 4\,HCl(aq)$$
$$\longrightarrow PbCl_2(aq) + Cl_2(g) + 2\,H_2O$$

Lead(IV) oxide is used in accumulators, where energy is derived from the process

$$Pb(IV) + Pb(0) \rightleftharpoons 2\,Pb(II)$$

Antimony and bismuth. The oxides of these metals are discussed in Chapter 23, and they should be compared with the other B-metal oxides discussed in this section.

25.6 Halides of the B-metals

Although they are not so reactive as the *s*-block metals, the B-metals will enter into direct combination with all the halogens and most of the other non-metals. A few of the more important halides are considered in detail below.

Zinc halides. Zinc reacts directly with all the halogens to form $ZnX_2(c)$. These compounds are all white crystalline solids soluble in water; in solution the ions $Zn^{2+}(aq)$ and $2\,X^-(aq)$ are produced. The hydrated zinc ion is hydrolysed by water and the solution therefore has an acidic pH (section 19.13). Another method of preparing the halide is by the action of $HX(aq)$ on zinc metal. For instance dilute hydrochloric acid produces a solution of zinc chloride:

$$Zn(c) + 2\,HCl(aq) \longrightarrow ZnCl_2(aq) + H_2(g)$$

It is difficult to obtain the anhydrous salt from this solution by evaporation, because of hydrolysis, but dry hydrogen chloride gas produces the same reaction giving the anhydrous salt.

Zinc fluoride is much less soluble in water than the other zinc halides. This is probably a lattice energy effect (cf. lithium fluoride, Chapter 20).

Mercury halides. If chlorine is passed over heated mercury the higher oxidation state, mercury(II) chloride, is produced:

$$Hg(l) + Cl_2(g) \longrightarrow HgCl_2(c);$$
$$\Delta H = -230 \text{ kJ at } 25°C$$

It is relatively easy to oxidize mercury to the $+2$ state, and all the halogens are capable of doing this. The $+1$ state is produced by heating a mixture of metallic mercury and the mercury(II) compound. For instance:

$$Hg(l) + HgCl_2(c) \longrightarrow Hg_2Cl_2(c)$$
$$\text{mercury(I)}$$
$$\text{chloride}$$

Although the *empirical* formula of mercury(I) chloride is HgCl, the molecular formula is Hg_2Cl_2. In the solid state and in the vapour phase linear molecules exist:

$$Cl-Hg-Cl \quad \text{and} \quad Cl-Hg-Hg-Cl$$

There is little evidence for the presence of ions in the solid, and the shape of these molecules is that predicted from the repulsion of electron pairs (mercury has a completely filled $5d$ shell and no lone pairs).

The Hg–Hg bond present in mercury(I) chloride persists in aqueous solutions of the mercury(I) ion. Measurements of the e.m.f.s of concentration cells and of ionic conductances point conclusively to the existence of the ion Hg_2^{2+}. Mercury(I) chloride is used in the *calomel electrode*, used as a reference electrode for pH measurement.

Like silver(I) halides, mercury(I) halides are insoluble in water. If dilute hydrochloric acid is added to an aqueous solution of a metal salt, silver(I), mercury(I) and lead(II) are the only common ions which will form precipitates—a fact which is often made use of in analysis. It is interesting that this insolubility occurs despite the fact that the ionic radii of Ag^+ and Pb^{2+} are comparable with K^+ and

Ba^{2+}, and indicates that these B-metal halides are not really ionic.

Mercury readily forms complex ions where the ligand is a halide ion. Addition of $I^-(aq)$ to $Hg^{2+}(aq)$ results first in the precipitation of red HgI_2, but excess iodide redissolves the precipitate to form a colourless solution containing a complex ion:

$$HgI_2(c) + 2I^-(aq) \longrightarrow HgI_4^{2-}(aq)$$
$$\text{red} \qquad\qquad\qquad\qquad \text{colourless}$$

Tin halides. Tin forms halides with all the halogens, and the oxidation number may be either $+2$ or $+4$: direct action of the halogen on the metal produces the $+4$ state, except in the case of iodine, where tin(II) iodide, SnI_2, is formed because iodine is not a sufficiently powerful oxidizing agent. Except for tin(IV) fluoride, the $+4$ halides are molecular in structure; the $+2$ halides have higher melting points but their structures are layer-like, suggesting that they are macromolecular rather than ionic.

Compounds of tin(IV) are very strongly hydrolysed by water forming hydrated tin(IV) oxide. $SnCl_4$ is a fuming liquid which reacts vigorously with water. The $+2$ halides are not so strongly hydrolysed, though their solutions are acidic.

Tin forms colourless complex ions with the halogens, such as hexachlorostannate(IV) $SnCl_6^{2-}$. The formation of such ions is typical of B-metals.

Lead halides. Lead shows the inert-pair effect more strongly than tin, and hence only fluorine is capable of oxidizing the metal directly up to the $+4$ state. Chlorine reacts with heated lead, to form lead(II) chloride, $PbCl_2$, and bromine and iodine react similarly though rather more slowly. Lead(IV) chloride is a yellow, unstable oil which fumes in air and reacts with water giving $PbO_2(c)$ (cf. tin(IV) chloride).

The usual way of making the $+2$ halides is by precipitation: all are practically insoluble in water, the most soluble being the chloride which dissolves quite readily in hot water.

$$Pb^{2+}(aq) + 2\,Cl^-(aq) \longrightarrow PbCl_2(c).$$

e.g. lead(II)
nitrate or acetate

Stable halide complexes are formed, and these are soluble in water; concentrated hydrochloric acid will dissolve lead(II) chloride with the formation of tetrachloroplumbate(II), $PbCl_4^{2-}$ (aq). In a similar fashion, potassium iodide solution will precipitate lead(II) ions as yellow PbI_2, which redissolve on adding excess reagent with the formation of $PbI_4^{2-}(aq)$.

25.7 Sulphides of the B-metals

Sulphur combines directly with most metals; among the B-metals, the more reactive ones such as zinc combine violently when finely divided. B-metal sulphides are insoluble in water, and their structures are macromolecular rather than ionic. Most of them are best prepared by precipitation reactions, rather than by direct combination of the elements.

Zinc sulphide, ZnS. This compound occurs naturally in two crystalline forms, known as *zinc blende* and *wurtzite*. Both lattices have a co-ordination number of $4:4$ (tetrahedral) though they differ in symmetry. A white precipitate of zinc sulphide is obtained on adding sulphide ions (or hydrogen sulphide) to a solution of $Zn^{2+}(aq)$, in neutral or slightly alkaline solution. Zinc sulphide is almost completely insoluble in water:

$$ZnS(c) \rightleftharpoons Zn^{2+}(aq) + S^{2-}(aq);$$
$$K_s \simeq 10^{-24} \text{ mol}^2 \text{ dm}^{-6} \text{ at } 25°C.$$
$$S^{2-}(aq) + 2H^+(aq) \rightleftharpoons H_2S(aq);$$
$$pK_a = 20, \text{ at } 25°C.$$

If $[H_2S] = 0.1 \text{ mol dm}^{-3}$ for a saturated solution, and $[H^+] = 1$, then $[S^{2-}]$ becomes 10^{-21} mol dm^{-3}. Hence in theory a concentration of $Zn^{2+} > 10^{-24}/10^{-21}$ mol dm^{-3} ought to cause precipitation if hydrogen sulphide is passed into an acidified solution of $Zn^{2+}(aq)$ ions. In practice such a precipitate takes about a month to form. This is another example of a kinetic factor overriding a thermodynamic factor.

Mercury(II) sulphide, HgS. Like zinc sulphide this compound can exist in two crystalline forms; the form commonly found in nature is a red ore called *cinnabar* which has a macromolecular structure in which each mercury atom has two near sulphur neighbours.

If hydrogen sulphide or sulphide ion is added to a solution of $Hg^{2+}(aq)$, a series of colour changes through orange and brown usually occurs, resulting finally in the formation of a black precipitate of HgS with an amorphous structure. This is one of the least soluble sulphides, with a solubility product quoted as 10^{-54}; it will not dissolve in $H^+(aq)$ and is not attacked even by concentrated nitric acid.

Tin sulphides. Direct action of sulphur on heated, finely divided tin results in the exothermic formation of tin(II) sulphide, SnS. Sulphur is not a powerful enough oxidizing agent to oxidize tin to the $+4$ state. Tin(IV) sulphide does however exist. Both sulphides can be prepared by precipitation:

$$Sn^{2+}(aq) + S^{2-}(aq) \longrightarrow SnS(c); \quad \text{(black)}$$

$$Sn(OH)_6^{2-}(aq) + 2\,S^{2-}(aq)$$
$$\longrightarrow SnS_2(c) + 6\,OH^-(aq); \text{ (yellow)}$$

Tin readily forms complex ions with S^{2-} in the $+4$ oxidation state. A solution of ammonium sulphide, which can be regarded as containing both sulphur and sulphide ions,

will dissolve both the sulphide precipitates as thiostannate complexes:

$$SnS + S + S^{2-} \longrightarrow SnS_3^{2-}$$
$$SnS_2 + S^{2-} \longrightarrow SnS_3^{2-}$$

Arsenic and antimony sulphides behave in a similar fashion.

Lead(II) sulphide. Lead(II) sulphide forms as a black precipitate when $Pb^{2+}(aq)$ and $S^{2-}(aq)$ ions are mixed, or when the elements are heated together. It occurs naturally as *galena*. The solubility product is less than that of zinc sulphide by a factor of about 1000 ($K_s = 3 \times 10^{-27}$ mol^2 dm^{-6} at 25°C) and it does not dissolve in dilute acids. Hot dilute nitric acid, being an oxidizing agent, does dissolve lead(II) sulphide:

2 N reduced from $+5$ to $+4$

$$PbS(c) + 4\,HNO_3(aq) \longrightarrow$$
$$Pb(NO_3)_2(aq) + S + 2\,NO_2(aq) + 2\,H_2O$$

S oxidized from -2 to 0

Group VB sulphides. The trends observed for Group IVB sulphides (tin and lead) are repeated in a parallel fashion in Group VB (antimony and bismuth). In particular:

(i) Antimony, like tin, forms sulphides which can dissolve in ammonium sulphide as thiocomplexes, such as SbS_3^-. These two metals can be identified in analysis by this property. Bismuth, like lead, fails to form thio-complexes.

(ii) The inert-pair effect shows itself in the same sort of way. Bismuth fails to form a sulphide Bi_2S_5 but forms instead the "Group valency minus two" oxidation state in Bi_2S_3. The same effect is observed with lead and thallium.

25.8 Complex ion formation by the B-metals

Whereas it was a characteristic of *s*-block metals that their ions are generally simple, simple ions of the B-metals are rarely encountered. The ions Ba^{2+}(aq) and Zn^{2+}(aq) for instance are not very similar: in the case of the barium ion the state symbol (aq) denotes an indefinite number of water molecules loosely attached around a simple barium ion as a hydration sheath, while in the case of the zinc ion there is evidence for the presence of four water molecules fairly firmly held in tetrahedral positions. The zinc ion is really a complex ion $Zn(H_2O)_4^{2+}$, and the same effect occurs with other B-metal ions (it is also true of aqueous ions of transition metals, Chapter 26).

Stepwise replacement of ligands. Other ligands apart from the water molecule can attach themselves to B-metal ions by displacing the water molecules present. For instance ammines are formed if ammonia is added to an aqueous solution of zinc ions; in the first instance only one ligand is replaced:

$$Zn(H_2O)_4^{2+} + NH_3(aq) \longrightarrow \left[Zn_{(NH_3)}^{(H_2O)_3}\right]^{2+} + H_2O$$

Successive replacements of H_2O by NH_3 lead finally to $Zn(NH_3)_4^{2+}$—in fact, this is the ion normally present in an aqueous solution containing zinc ions in the presence of excess ammonia.

Negatively charged ligands behave similarly: addition of Cl^-(aq) ions to a solution of Zn^{2+}(aq) leads mainly to $ZnCl_3^-$(aq) and $ZnCl_4^{2-}$(aq).

Ligands which will add to B-metals. In addition to H_2O and Cl^- mentioned above, the B-metals will form complexes with other halide ions, such as F^-, Br^- and I^-. Ions such as cyanide, CN^-, and thiocyanate, CNS^- (referred to previously as pseudohalides, section

24.9) will also form stable complexes with some, though not all, B-metals. Zinc for instance forms a stable tetracyanozincate ion, $Zn(CN)_4^{2-}$, but the corresponding complexes of tin and lead are less stable.

Ligands with more than one point of attachment to the central metal atom are known as **polydentate** and polydentate ligands form complexes which often have greater stability than simple ligands. Ethylenediamine, $H_2N.CH_2.CH_2.NH_2$ is an example: both nitrogen atoms can donate an electron pair to the same metal atom by forming a ring structure (known as a **chelate**):

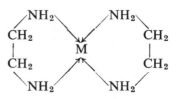

Such a complex is more stable than a simple ammine complex, because even if one point of attachment comes adrift the other end of the ligand remains attached. Tridentate ligands are even more stable than bidentate, and so on. One of the most powerful chelating agents of all is EDTA (ethylene-diaminetetra-acetic acid), the anion of which has *six* points of attachment. Such a ligand can even form reasonably stable complexes with "non-complexing" ions such as Ca^{2+} and Li^+, and if added to a B-metal it forms a complex species of great stability. The simple ion is literally "wrapped up" in the ligand, and removed from the solution just as effectively as if it had been precipitated out. If EDTA is added to a solution of zinc ion, it is impossible for instance to precipitate the sulphide of the metal with hydrogen sulphide.

Antimony and tin form soluble thio-complexes SbS_3^- and SnS_3^{2-} when a solution of ammonium sulphide (an alkaline solution containing S^{2-}) is added to a sulphide of antimony or tin. The solubility of these sulphides in ammonium sulphide is used as an analytical test for antimony and tin.

Properties of B-*metal complexes.* Complex ions of B-metals have the following properties:

(i) They are colourless, except when there is colour associated with the ligand.

(ii) They are generally more easily decomposed than complex ions of transition metals, though far more stable than complex ions of s-block metals (where these exist).

(iii) They are usually soluble in water, and do not readily form precipitates.

This last property requires some explanation, for a precipitate is in effect a complex of zero charge, which happens to be insoluble in water. In the stepwise loss of protons which occurs on adding alkali to $Zn(H_2O)_4^{2+}$, the following species are obtained with increasing pH:

$$[Zn_{(OH)}^{(H_2O)_3}]^+ \longrightarrow Zn(OH)_2 . xH_2O(c)$$
$$\text{soluble} \qquad\qquad\qquad \text{insoluble}$$
$$\longrightarrow [Zn_{(OH)_3}^{(H_2O)}]^- \longrightarrow Zn(OH)_4^{2-}$$
$$\text{soluble} \qquad\qquad \text{soluble}$$

The solubility of precipitates and the stability of complexes are therefore closely connected.

Stability constants. The tendency of a complex ion to decompose in aqueous solution is expressed quantitatively by its **stability constant** at a given temperature, which is the equilibrium constant for the equilibrium:

$$M + nL \rightleftharpoons M_nL$$

(M = metal; L = ligand; charges omitted)

The reciprocal of K, which is the equilibrium constant for the reaction from right to left, is sometimes termed the instability constant.

B-metal complex ions in general have lower stability constants than those of transition metals, though there are exceptions. (See also section 26.11).

25.9 Some common salts containing B-metals

Simple binary salts (halides and sulphides) have been dealt with and they will not be considered further. Other salts may be classified into

(a) *Those in which the metal is the cation:* e.g: lead nitrate, zinc sulphate etc.

(b) *Those in which the metal is in the anion:* e.g: sodium metazincate, Na_2ZnO_2.

Many B-metal oxides (ZnO, SnO, SnO_2, PbO_2, Sb_2O_3) will dissolve in alkalis such as sodium hydroxide, forming hydroxo-complexes. On evaporation of the solution the solid which is obtained is the *meta* salt formed by loss of water. B-metal salts, when soluble, form colourless solutions provided there is no other ion present which might confer colour. Many solid binary salts of B-metals are coloured due to charge transfer absorption. Charge transfer absorption is the absorption of electromagnetic energy due to the transfer of an electron from one atom to another. Where this energy of absorption corresponds to a wavelength in the visible region, the substance will be highly coloured. Thus halides, oxides and sulphides are frequently brightly coloured or black.

When a B-metal salt is soluble in water, the solid obtained on crystallization is often hydrated. This is generally due to the hydrated cation of the metal being itself included in the crystal lattice. Zinc sulphate is soluble in water and crystallizes out as $ZnSO_4.7H_2O$. Such behaviour is also observed with s-block metals and magnesium, for example, forms $MgSO_4.7H_2O$.

All nitrates of metals are soluble in water. B-metal nitrates decompose on heating to give the oxide, oxygen, and nitrogen(IV) oxide.

Addition of $CO_3^{2-}(aq)$ to a solution of a

B-metal ion generally results in the precipitation of a non-stoichiometric basic carbonate such as $xPbCO_3.yPb(OH)_2$. In some cases the normal carbonate or something approximating to it can be obtained by using *bi*carbonate ions, $HCO_3(aq)$, as the precipitating agent. All B-metal carbonates, whether normal or basic, are white in colour and insoluble in water.

Most of the B-metal sulphates are soluble in water and crystallize out in hydrated form but lead(II) sulphate forms as an unhydrated white precipitate when the relevant ions are mixed. This reaction is sometimes used as a test for lead ions (among the common metals only barium forms a sulphate whose insolubility is comparable), and for the gravimetric estimation of lead.

25.10 Summary

The B-metals have characteristic properties which distinguish them from s-block metals and transition metals.

(1) They are generally fairly soft, dense, with relatively low melting points.

(2) Unlike s-block metals, they have little tendency to form truly ionic lattices; this is often interpreted by saying that the ions which they form are more readily polarized.

(3) Like transition metals, they readily form complex ions, but these differ from transition metal complexes in that

(a) they are colourless (unless the ligand causes the colour);

(b) they do not contain unpaired electrons, and are thus diamagnetic.

(4) Their compounds show two oxidation numbers, "Group number" and "Group number minus two", especially at the bottom of the periodic table (inert-pair effect). This is in contrast to transition metals, where adjacent oxidation states are more commonly one unit apart, and s-block metals where variable oxidation state is not shown.

For a table summarizing the properties of metals see section 26.18.

Study Questions

1. In what ways does zinc resemble

(a) A Group IIA metal?
(b) A transition metal?
(c) A typical B-metal such as lead or tin?

2. Show how the properties of lead and tin differ from the corresponding properties of silicon. In what ways do the three elements resemble each other?

3. How do (a) the oxides, (b) the sulphides of antimony and bismuth compare with those of tin and lead?

4. Thallium forms two oxides.

(a) What would you expect their formulae to be?
(b) Would you expect them to be coloured?
(c) Which oxide would be the more basic?

5. Addition of NaOH(aq) to $CdCl_2(aq)$ gives a white precipitate insoluble in excess alkali.

(a) Suggest a formula for the precipitate.

(b) How does the behaviour of cadmium differ from that of zinc in this case?
(c) How would you expect the precipitate to react with HCl(aq)?

6. (a) Why does lead(II) sulphide dissolve in hot concentrated hydrochloric acid?
(b) Why does arsenic(III) sulphide dissolve in ammonium sulphide?
(c) Why does lead(II) bromide dissolve in potassium bromide solution?

7. (a) Suggest why lead appears to react only very slightly with hydrochloric acid.
(b) Find out why lead tetraethyl is used as an antiknock compound in petrol.
(c) Find out why lead is often used on roofs.
(d) Find out why lead is used in car batteries.

8. (a) Why are mercury(I) salts diamagnetic?

(b) HgF_2 boils at 650°C and $HgCl_2$ boils at 303°C. The former is insoluble, but the latter soluble in organic solvents. Comment on the likely nature of these compounds.

9. (a) Use Appendix 1 to determine whether

   (i) $Hg^{2+}(aq)$ will react with $Sn^{2+}(aq)$.
   (ii) $Hg^{2+}(aq)$ will react with $Pb^{2+}(aq)$.

   (b) What would you expect to *observe* if solutions of mercury(II) chloride and tin(II) chloride were mixed?

   (c) What would you expect to *observe* if solutions of mercury(II) chloride and lead(II) nitrate were mixed?

†10. What evidence is there for the inert-pair effect in Group IIB?

11. When a brown solid, A, was heated, oxygen was evolved and a red solid, B, and eventually a yellow solid, C, were formed. C dissolved in dilute nitric acid to give a solution of D, which with hydrochloric acid gave a white precipitate of E. Dry E reacted with chlorine to give the thermally unstable liquid, F. Molten E could be electrolysed to give the metal G at the cathode. When the solution of D was treated with H_2S, a black precipitate, H, formed.

   (a) Identify the lettered compounds.
   (b) Give equations for the reactions.
   (c) How can H be converted into G?
   (d) How can B be converted directly into D?

12. When a white solid J was heated, carbon dioxide was evolved and the solid became yellow. On cooling, the solid K became white. K dissolved in hydrochloric acid to give a solution of L. When L was treated with sodium hydroxide solution, a white precipitate M formed, but this then dissolved in excess base to give N.

   (a) Identify the lettered compounds.
   (b) Give equations for the reactions.

13. Aluminium is a Group III element. In what ways does it resemble (a) an *s*-block metal (b) a B-metal?

CHAPTER 26

The transition metals

THE transition metals occupy the central region of the periodic table known as the d-block. There are three transition series, each marking the filling up of d-orbitals in the atoms, but this chapter will be mainly concerned with the **first transition series,** from scandium to copper. Zinc is treated in some books as a transition metal, being the next member of the series after copper, but as was shown in Chapter 25, it is more like the B-metals to its right in the periodic table.

26.1 Electronic structures of the first transition series

It was stated in Chapter 3 that, except for hydrogen, the sub-levels s, p, d, etc., within a quantum shell differ in energy. Energy level diagrams such as those in Fig. 3.7. illustrate this. As the atomic number increases, and orbitals interact with each other to a greater extent, this splitting becomes even more marked, and before the first transition series is reached, the principal quantum levels actually begin to overlap in energy. It is for this reason (tacitly assumed in earlier chapters) that the $3d$ energy level does not begin to fill with electrons immediately after the $3p$ is full. The element which follows argon in the periodic table is potassium ($1s^2$; $2s^2p^6$; $3s^2p^6$; $\underline{4s^1}$) and the $3d$ level remains unfilled because it is higher in energy than the $4s$.

Once the $4s$ energy level is filled, the $3d$ can begin to fill; here the pattern of behaviour differs from that in the previous period. The transition metal electronic structures are given in Table 26.1.

The filling up is not entirely regular, due to the fact that the $3d$ and $4s$ levels are very close in energy along this series. Although the $3d$ level normally fills preferentially, chromium and copper are exceptional in that a $4s$ electron is "lost" to the $3d$. It appears that half-filled and filled d-shells are favoured.

26.2 Physical properties and uses of the transition metals

The metals themselves differ from the s-block elements and the B-metals, because of the high binding energy which results from incompletely filled d-levels. The chief characteristics are:

(i) high melting and boiling points, and high heats of vaporization;
(ii) considerable mechanical strength, especially in alloys.

Figure 26.1 is a plot of melting and boiling points for the metals of the first long period. Boiling point of an element is a more reliable

TABLE 26.1

Element		Electronic structure in ground state (short form)	Electronic structure (full form)	
s-block	Ar	2, 8, 8	$1s^2$; $2s^2p^6$; $3s^2p^6$	
	K	2, 8, 8, 1	[Ar core]; $4s^1$	
	Ca	2, 8, 8, 2	[Ar core]; $4s^2$	
d-block	Sc	2, 8, 9, 2	[Ar core] $3d^1$; $4s^2$	
	Ti	2, 8, 10, 2	[Ar core] $3d^2$; $4s^2$	
	V	2, 8, 11, 2	[Ar core] $3d^3$; $4s^2$	
	Cr	2, 8, 13, 1	[Ar core] $3d^5$; $4s^1$	half-filled d-shell
	Mn	2, 8, 13, 2	[Ar core] $3d^5$; $4s^2$	
	Fe	2, 8, 14, 2	[Ar core] $3d^6$; $4s^2$	
	Co	2, 8, 15, 2	[Ar core] $3d^7$; $4s^2$	
	Ni	2, 8, 16, 2	[Ar core] $3d^8$; $4s^2$	
	Cu	2, 8, 18, 1	[Ar core] $3d^{10}$; $4s^1$	filled d-shell
	Zn	2, 8, 18, 2	[Ar core] $3d^{10}$; $4s^2$	

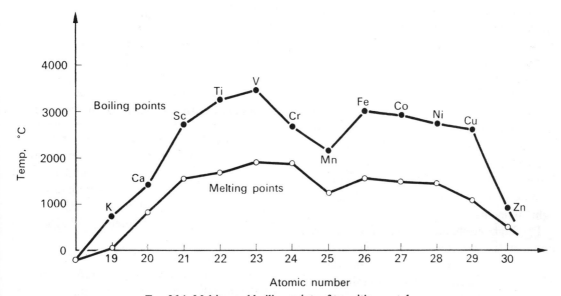

FIG. 26.1. Melting and boiling points of transition metals.

guide to binding energy than melting point, since the latter is rather dependent upon crystal structure, yet both properties are seen to follow the same trend. Figure 26.2 shows how binding energy, as measured by the heat of vaporization, varies along the same series. It is very noticeable that zinc has a low melting point, boiling point and binding energy; all this suggests that relatively weak forces must exist between the atoms in the metal. In Chapter 5 a correlation between binding energy and number of valence electrons available for bond formation was noted: if this conclusion is valid then we must further conclude that the d-electrons in zinc play little part in metallic bond formation. A completely filled d-shell, $3d^{10}$, is stable and

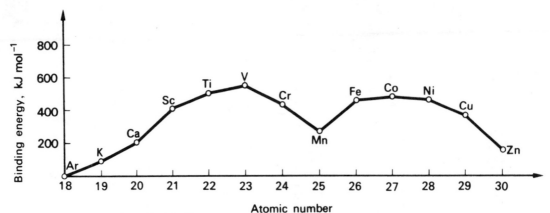

FIG. 26.2. Binding energy, as measured by heat of vaporization.

to some extent "inert"—it can be treated as part of the inner core of electrons.

The dip in binding energy (also m.p. and b.p.) at manganese coincides with the configuration $3d^5$; half-filled d-shells have some sort of stability associated with them, but this is not so great as that for filled shells.

The boiling points of the first transition series are lower than those of corresponding metals in the second and third transition series. It is evident that stronger interatomic forces operate in the later series; tungsten, for instance, is the least volatile metal known, and because of its involatility it is used in electric light filaments.

The high mechanical strength of the transition metals make some of them valuable structurally: iron is the most abundant and the most widely used transition metal; titanium has in recent years become a structural metal in its own right—it is lighter than steel but superior to aluminium in strength, corrosion resistance and heat resistance. Unfortunately despite its widespread occurrence titanium is very costly to extract. Some of the transition metals find specialized uses—nickel and chromium because of their resistance to chemical attack and copper because its high electrical conductivity—and the remainder, such as manganese and cobalt, are widely used in special steels. Mechanical properties cannot be explained simply in terms of crystal structure and binding energy, since they depend more upon such factors as grain structure and lattice defects.

26.3 Occurrence and extraction of transition metals

The transition metals are less electropositive than the s-block metals as shown by their higher first ionization energies (Fig. 26.3), and occur mainly as oxides or sulphides. The "noble" metals, gold, platinum, osmium, iridium, etc., are extremely unreactive and usually occur native.

Apart from titanium, for which special methods of extraction are required, the oxides of the metals can be reduced with carbon at high temperatures. Since carbon is a cheap reducing agent, this is the method normally employed, though for special purposes aluminium may be used as reducing agent (Thermit process). The sequence of operations for extracting the metal are therefore:

(i) Mechanical concentration of the ore (flotation, magnetic separation, etc., where appropriate).

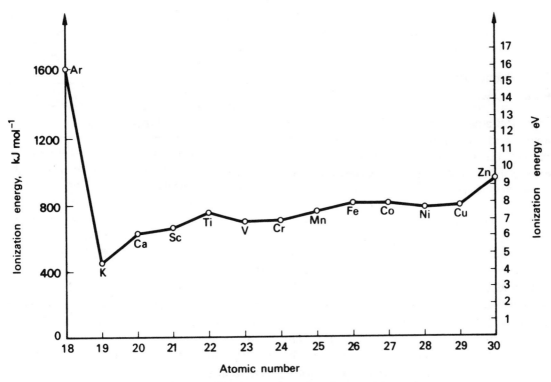

FIG. 26.3. First ionization energies along first transition series.

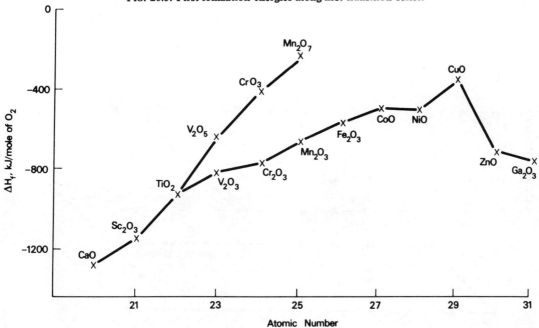

FIG. 26.4. Heats of formation of transition metal oxides.

(ii) Roasting of sulphide ores to convert to the oxide, as sulphides are not readily reduced by carbon directly (see Chapter 13).

(iii) Reduction of the ore with carbon in the form of coke.

(iv) Refining of the metal. This may involve lowering of the carbon content by blowing air through the molten metal ($\rightarrow$ CO). In many cases refining is done by electrolysis (copper, chromium, silver).

For many purposes a pure metal is not required. In the case of chromium, for instance, the pure metal is obtained by electrolytic deposition (chromium plate) but the alloy *ferrochrome*, composed of chromium and iron, is obtained by reducing the ore *chromite*, $FeCr_2O_4$, and is quite suitable for making stainless steel.

Figure 26.4 is a plot of the heat of formation of the common oxides of the transition metals against atomic number of the metal. The ex-

traction of some transition metals (iron, chromium, nickel and copper) from their ores is dealt with in Chapter 13.

26.4 Ion formation in the first transition series

One of the main features which distinguishes a transition metal from other metals is the property of **variable oxidation state.** The B-metals are capable of forming compounds in two oxidation states due to the inert-pair effect, but the transition metals can show *several* states, frequently differing from each other by only one unit. Figure 26.5 is an oxidation number diagram for the first transition series: the oxidation states shown are either simple hydrated ions (lower oxidation states) or oxo-ions (higher oxidation states).

The main features shown in Fig. 26.5 are as follows:

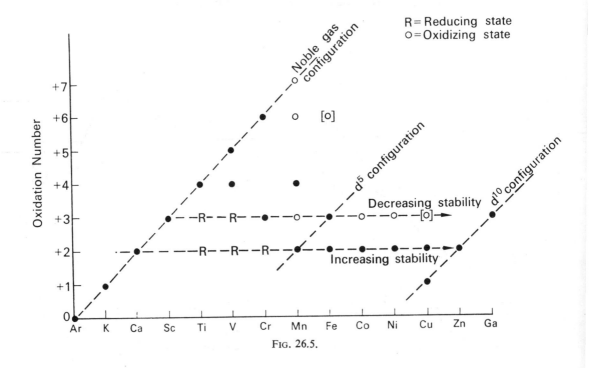

FIG. 26.5.

(i) Practically all the transition metals possess oxidation states of $+2$ and $+3$. Both nickel(III) and copper(III) compounds are rather rare, while scandium(II) compounds do not exist.

(ii) There is a sequence of ions which is *formally* related to the noble gas argon. Since very small, highly charged ions exert such a powerful electrostatic field however, V^{5+}, Cr^{6+} and Mn^{7+} are never observed free, and in aqueous solutions oxo-ions are formed.

Simple ions of the transition metals do not exist in aqueous solution: the ions which appear to be simple are in fact complexed with water, generally accepting six H_2O ligands at octahedral positions. Other ligands such as Cl^-, NH_3, CN^-, CNS^-, and $H_2N.CH_2.CH_2.NH_2$ readily form complexes with the transition metals (see later sections).

Complexes of the transition metals have properties which markedly distinguish them from those of B-metals:

(a) *They are generally coloured;* exceptions are those with completely filled d-orbitals, e.g. those of copper(I), and silver(I), or with empty d-orbitals, e.g. scandium(III).

(b) *They are frequently paramagnetic,* showing that there must be unpaired electrons present. Ions of s-block metals and B-metals do not show paramagnetic behaviour.

(c) *They often act as homogeneous catalysts in solution.* The metals and their solid compounds are also good heterogeneous catalysts for many gas phase reactions.

The transition metals occupy similar positions in the electrochemical series to the B-metals—in fact there is considerable overlap—indicating that the tendency of the metals to form aqueous ions is generally less strong than for the s-block metals.

Table 26.2 gives the colours of the $+2$ and $+3$ hydrated ions.

TABLE 26.2

	$+2$ ion	$+3$ ion
Ti	—	violet
V	violet	green
Cr	blue	dark green
Mn	pale pink	red
Fe	pale green	yellow-brown
Co	pink	—
Ni	green	—
Cu	blue	—
Zn	colourless	—

26.5 Redox potentials

It was noted in section 26.2 that the d^5 configuration (half-filled $3d$ shell) was stable relative to other configurations: the redox potentials for the formation of the $+2$ and $+3$ ions provide further evidence for this effect (Fig. 26.6 and 26.7).

When a transition metal forms an ion, it is the s electrons which are lost first, not the d. The electronic configurations of the $+2$ and $+3$ ions are shown in Table 26.3.

The redox data indicate that Mn^{2+}, Fe^{3+}, Cu^+ and Zn^{2+} compounds are more stable relative to their adjacent oxidation states than

TABLE 26.3

Metal	Electronic configuration (outer shell only; core = argon)	
	$+2$ state	$+3$ state
Sc	—	[argon]
Ti	$3d^2$	$3d^1$
V	$3d^3$	$3d^2$
Cr	$3d^4$	$3d^3$
Mn	$\mathbf{3d^5}$	$3d^4$
Fe	$3d^6$	$\mathbf{3d^5}$
Co	$3d^7$	$3d^6$
Ni	$3d^8$	$3d^7$
Cu	$3d^9$ [+1 state: $3d^{10}$]	—
Zn	$\mathbf{3d^{10}}$	—

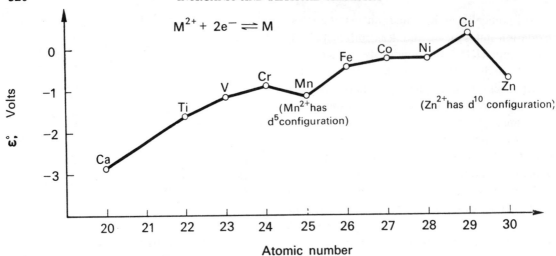

FIG. 26.6. Half-cell potentials (compare this plot with Fig. 26.1 and 26.2).

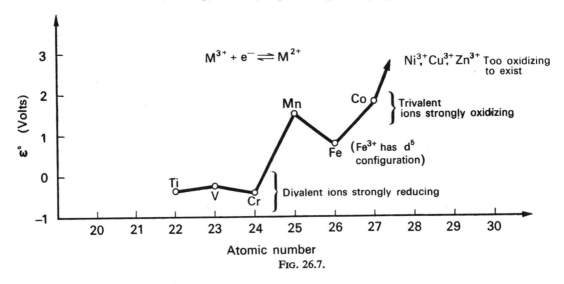

FIG. 26.7.

their position in the transition series would have suggested. Fig. 26.6 should be compared with Fig. 26.1 and 26.2.

26.6 Oxides and hydroxides of the transition metals

All the transition metals form oxides. The affinity of transition metals for oxygen is in roughly the same order as the electrochemical series, and hence the metals like silver and gold have very low affinities for oxygen and do not combine easily with the gas.

Addition of OH^-(aq) to the aqueous solution of a transition metal salt results in precipitation of the hydroxide in most cases. Amphoteric character is less common than among B-metals, though one or two hydroxides, for instance chromium(III), redissolve in excess alkali to form an anionic hydroxo-complex. Table 26.4 summarizes the action of alkali on aqueous

TABLE 26.4

Cation (aqueous)	Precipitate on adding dilute alkali	Colour of precipitate	Comments
Cr^{3+}	$Cr(OH)_3.xH_2O$	Green	Highly insoluble in water but dissolves in conc. alkali (amphoteric)
Mn^{2+}	$Mn(OH)_2.xH_2O$	Buff	Not amphoteric
Fe^{2+}	$Fe(OH)_2.xH_2O$	Grey-green	Readily oxidized to $Fe(OH)_3$
Fe^{3+}	$Fe(OH)_3.xH_2O$	Brown	Not amphoteric
Co^{2+}	$Co(OH)_2.xH_2O$	Pink	Dissolves in conc alkali
Ni^{2+}	$Ni(OH)_2.xH_2O$	Green	Not amphoteric
Cu^{2+}	$Cu(OH)_2.xH_2O$	Pale blue	Rapidly loses water forming black hydrated CuO
Ag^+	$Ag_2O.xH_2O$	Greyish-white	True hydroxide does not exist

solutions of some common transition metal cations.

The trivalent hydroxides are more insoluble in water than the divalent ones. Iron(III) and chromium(III) hydroxides are precipitated by a buffer solution of pH = 10, such as a mixture of ammonium chloride and ammonium hydroxide, whereas the divalent hydroxides do not precipitate under these conditions. This fact is made use of in analysis, for separating and identifying these two metals. The reason for the higher insolubility is probably the higher lattice energy of the hydroxides containing triply-charged ions.

The transition metal hydroxides form fairly gelatinous precipitates containing loosely bonded water in non-stoichiometric amounts, and if they are heated to remove this water decomposition to the oxide always occurs. Copper(II) hydroxide decomposes at room temperature on standing, and silver(I) hydroxide is not formed.

Transition metals form oxides which illustrate well the property of variable valency. Figure 26.8 gives the formulae of some of the common ones. In general these oxides are:

(i) basic for low oxidation number, acidic for high oxidation number;

(ii) insoluble in water;

(iii) highly coloured, or black;

(iv) macromolecular in structure, rather than truly ionic (section 22.12).

Basic or acidic character. Oxides lying below the line in Fig. 26.8 will dissolve in acids; the rate of dissolution is often slow unless the solid is finely divided. For instance, prolonged boiling with concentrated hydrochloric acid is necessary to dissolve chromium(III) oxide, Cr_2O_3, although this is a *rate* effect and the $Cr^{3+}(aq)$ ions are perfectly stable once formed.

Oxides lying on the line are amphoteric in character. Manganese(IV) oxide MnO_2, reacts slowly with fused alkali forming the manganate

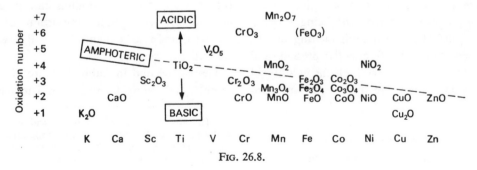

FIG. 26.8.

(IV) ion, sometimes known as manganite:

$$MnO_2 + 2\,OH^- \longrightarrow MnO_3^{2-} + H_2O$$

It reacts even more readily in the presence of oxygen, or an oxygen-donating substance such as potassium nitrate, for then it is oxidized up to the +6 state, forming manganate(VI):

$$MnO_2 + 2\,OH^- + [O] \longrightarrow MnO_4^{2-} + H_2O.$$

Manganese(IV) oxide dissolves in ice-cold concentrated hydrochloric acid forming the chloride $MnCl_4$, but this is unstable and the more usual reaction which occurs at room temperature is the redox process:

$$MnO_2 + 4\,HCl(aq) \longrightarrow Mn^{2+}(aq) + 2\,Cl^-(aq)$$
$$+ Cl_2(g) + 2H_2O.$$

The properties of manganese(IV) oxide are predominantly acidic and it will not react with acids unless it can oxidize them.

Chromium(VI) oxide, CrO_3, is a typical oxide of a transition metal in a high oxidation state. Unlike the lower oxides, it is soluble in water, in which it dissolves forming chromic acid:

$$H_2O + 2\,CrO_3 \longrightarrow H_2Cr_2O_7.$$

The solution which results contains mainly the dichromate ion, $Cr_2O_7^{2-}$.

Structure of transition metal oxides. Many of the oxides have roughly cubic structures, but they are macromolecular rather than ionic. There is no evidence for ions being present, and the solids have very high melting points. Layer structures are frequently present and this accounts for the very high insolubility of these oxides; they are often non-stoichiometric.

Reactions of transition metal oxides. The chemical inertness of transition metal oxides

may have an important bearing on the properties of the metals themselves, though the subject is far from being fully understood. Certain metals, notably iron and chromium, become "passive" in contact with nitric acid at certain concentrations, that is, they refuse to dissolve in it. The formation of a protective oxide coat may play a part in the mechanism of passivity.

All the transition metal oxides can be reduced by carbon though certain ones such as titanium(IV) oxide require an exceedingly high temperature (Chapter 13). Aluminium is also a reducing agent, which undergoes a Thermit reaction with transition metal oxides. Hydrogen reduces the oxides of metals low in the electrochemical series, but fails with metals higher up. For instance copper(II) oxide is reduced while chromium(III) oxide is not.

Oxides of mixed oxidation state occur among the transition metals, and these may be regarded as "salts" of two oxides (one basic and one acidic) of the same metal. For instance:

$$Fe_3O_4 = FeO.Fe_2O_3$$
$$\text{iron(II,III,III) oxide}$$

$$Mn_3O_4 = 2\,MnO.MnO_2$$
$$\text{manganese(II,II,IV) oxide}$$

26.7 Oxo-ions of transition metals

The higher oxidation states of transition metals exist only as anions, which means that the names of these compounds end in -ate (or occasionally -ite).* Some important oxo-salts are listed in Table 26.5 and these will be considered in this section.

The salts listed in Table 26.5 are all water-soluble, and are oxidizing agents. The oxidizing

---

* The unsystematic ending -ite can lead to ambiguity in the names of transition metal compounds, and the use of I.U.P.A.C. nomenclature (-ate followed by the oxidation number) is often preferable.

TABLE 26.5

Name and formula of anion	Name and formula of a typical salt	Oxidation state of transition metal
Vanadate, VO_3^-	Ammonium vanadate NH_4VO_3	+5
Chromate, CrO_4^{2-}	Potassium chromate K_2CrO_4	+6
Dichromate, $Cr_2O_7^{2-}$	Sodium dichromate $Na_2Cr_2O_7$	+6
Manganate, MnO_4^{2-}	Potassium manganate K_2MnO_4	+6
Permanganate, MnO_4^-	Potassium permanganate $KMnO_4$	+7

reactions of the anions are important in volumetric analysis.

Vanadates. Vanadates are salts of the acidic oxide V_2O_5, which is a yellow solid formed during the extraction of the metal. Dilute alkali dissolves V_2O_5, forming ortho-, meta- and pyro-vanadates (corresponding in formula to the phosphates, section 23.9), of which the meta-vanadates are the most stable. Action of NH_4^+(aq) on a solution of a metavanadate, VO_3^-(aq), precipitates ammonium meta-vanadate, NH_4VO_3. Solutions of vanadates are used for volumetric analysis, and vanadium pentoxide itself is used as a catalyst in the manufacture of sulphur trioxide and sulphuric acid.

Chromates and dichromates. Chromium(VI) forms a series of oxo-anions depending on the pH of the solution; in alkali, the (mono)chromate ion is formed when a Cr^{3+} salt is heated with sodium peroxide:

$$2\,Cr^{3+} + 4\,OH^- + 3\,O_2^{2-} \longrightarrow 2\,CrO_4^{2-} + 2\,H_2O$$

green   from peroxide   yellow

On acidification of an aqueous solution of a chromate, the ion 'condenses' to form di-

chromate ion, and further to form trichromate and tetrachromate if the pH is low enough:

$$2CrO_4^{2-} + 2H^+ \rightleftharpoons Cr_2O_7^{2-} + H_2O;\;(\rightarrow Cr_3O_{10}^{2-}\,\text{etc.})$$

yellow chromate   orange dichromate   orange trichromate

The structure of these poly-ions is similar to those formed by non-metals, such as silicate, phosphate and pyrosulphate, in that the element is alternately linked with oxygen atoms:

$$\begin{bmatrix} & O & & O & \\ & \uparrow & & \uparrow & \\ O\!\leftarrow\!Cr\!&\!-\!O\!-\!&\!Cr\!\rightarrow\!O \\ & \downarrow & & \downarrow & \\ & O & & O & \end{bmatrix}^{2-}$$

$$\begin{bmatrix} & O & & O & & O & \\ & \uparrow & & \uparrow & & \uparrow & \\ O\!\leftarrow\!Cr\!-\!O\!-\!Cr\!-\!O\!-\!Cr\!\rightarrow\!O \\ & \downarrow & & \downarrow & & \downarrow & \\ & O & & O & & O & \end{bmatrix}^{2-}$$

Potassium chromate is a yellow crystalline solid which forms a useful analytical reagent. It forms precipitates with a number of metal ions, such as lead(II) chromate, $PbCrO_4$, which is the pigment *chrome yellow*, and silver(I) chromate, Ag_2CrO_4, which is a brick red solid formed at the end point when potassium chromate is used as indicator for silver nitrate titrations.

Sodium dichromate is an orange solid highly soluble in water and deliquescent. Its deliquescence makes it unsuitable as a standard for volumetric analysis and for this purpose the less soluble potassium dichromate is used. Potassium dichromate does not contain water of crystallization and when used as an oxidizing agent one mole of it will accept six moles of electrons from a reducing agent:

$$Cr_2O_7^{2-} + 14\,H^+ + 6e^- \rightleftharpoons 2\,Cr^{3+} + 7\,H_2O;$$
$$\epsilon^\circ = +1\cdot33\text{ V}$$

Although there is a colour change from orange to green at the endpoint this is not sufficiently

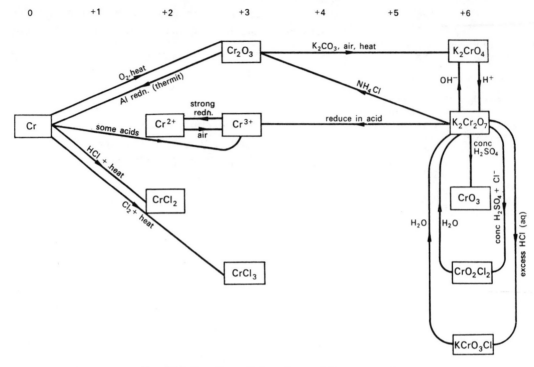

FIG. 26.9. Reactions of chromium and its compounds.

distinct, and an indicator such as barium diphenylamine sulphonate is generally added.

Conversion of chromate to dichromate is *not* a redox process since the metal is in the same oxidation state in both ions.

Manganates. The manganate(VI) ion is dark green, and a dark green mass of K_2MnO_4 is formed when manganese(IV) oxide is fused with potassium nitrate (section 26.5). This ion is only stable in alkali, for in acid it disproportionates into manganate(VII), known as *permanganate*, and manganese(IV):

$$3\ MnO_4^{2-}(aq) + 4\ H^+(aq) \longrightarrow$$
dark green

$$MnO_2(c) + 2\ MnO_4^-(aq) + 2\ H_2O$$
black   purple

This is the usual way of preparing potassium permanganate: the salt crystallizes out on evaporation. An alternative way, which gives complete conversion instead of just two-thirds, is to oxidize the manganate(VI) ions with chlorine:

$$2\ MnO_4^{2-}(aq) + Cl_2(g)$$
$$\longrightarrow 2\ MnO_4^-(aq) + 2\ Cl^-(aq)$$

Figure 26.10 shows some of the reactions of the various oxidation states of manganese, showing how they are linked together. The +7 state of manganese is normally only accessible via the +4 and +6, though very powerful oxidizing agents are capable of oxidizing directly from manganese(II) to the permanganate ion. A test for manganese is to add a solution of sodium bismuthate, "$NaBiO_3$", in nitric acid to the suspected solution; a pink colour of permanganate ion shows the presence of manganese. It is the inert-pair effect which makes sodium bismuthate such a powerful oxidizing agent:

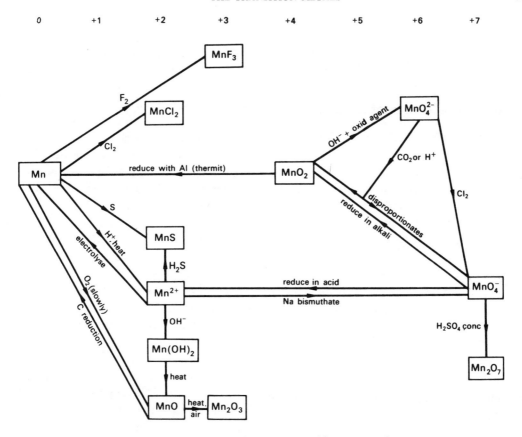

FIG. 26.10. Reactions of manganese and its compounds.

bismuth has a strong tendency to be reduced from the +5 state in bismuthate, to +3.

Potassium permanganate. Stable permanganates of the alkali metals exist, but the one used for volumetric analysis is potassium permanganate, $KMnO_4$. It is less soluble than the sodium salt, and is easier to obtain and keep in the pure state. The extremely dark colour of permanganate solutions is deceptive, for the potassium salt is not very soluble in water (a saturated solution is about 0·4 M at 20°C).

One mole of the permanganate ion will take *five* moles of electrons from a reducing agent:

$$MnO_4^-(aq) + 8\,H^+ + 5e^- \rightleftharpoons Mn^{2+} + 4\,H_2O;$$
$$\varepsilon^\circ = +1·51\ V$$

The permanganate ion is one of the most powerful oxidizing agents available for volumetric analysis, and its extremely high half-cell potential suggests that it ought to oxidize water to oxygen (ε° of O_2/H_2O couple = +1·229 V); in fact this reaction does occur slowly in the presence of light. Potassium permanganate solutions do not alter in concentration over a period of weeks if kept in dark bottles.

Among the oxidations for which potassium permanganate solutions may be used are (a) iron(II) to iron(III) (b) oxalate ion to carbon dioxide, (c) hydrogen peroxide to oxygen. All these reactions are used in volumetric analysis.

Along a transition series the highest oxidation

state becomes progressively more oxidizing. Hence the order of increasing oxidizing power is $Sc(III) < Ti(IV) < V(V) < Cr(VI) < Mn(VII)$.

26.8 Transition metal halides

The transition metals readily form compounds with the halogens, the order of reactivity being roughly fluorine > chlorine > bromine > iodine. Fluorine reacts directly with all metals, even the noble metals, and forms a compound with the metal in a high oxidation state. Chlorine attacks all the elements of the first transition series; bromine and iodine also form compounds, though the reactions with iodine generally require the assistance of heat. Fluorine is more powerfully oxidizing than chlorine and sometimes produces a higher oxidation state, for instance MnF_3 and CoF_3, but $MnCl_2$ and $CoCl_2$.

Chlorides. Nearly all the anhydrous chlorides are soluble in water. Among later transition elements, copper(I) chloride, CuCl, silver chloride, AgCl, and mercury(I) chloride, Hg_2Cl_2, are notable for being insoluble in water.

Hydrated chlorides are formed if aqueous solutions of transition metal chlorides are crystallized, but these are hydrolysed on evaporation. Iron(II) chloride is a typical example, for on evaporation it yields the basic chloride, Fe(OH)Cl:

$$FeCl_2(aq) + H_2O \rightleftharpoons Fe(OH)Cl(c) + HCl(g).$$

This property is not exclusive to the chlorides of transition metals: most metal ions are in fact hydrolysed in aqueous solution and hence give basic salts on evaporation, and indeed the salts of the alkali metals and alkaline earth metals are to some extent exceptional in *not* being hydrolysed.

Chlorine and iron filings react to give iron(III) chloride, a red solid which is volatile

owing to its molecular structure. Molecules of formula Fe_2Cl_6 exist in the anhydrous solid, exactly analogous to Al_2Cl_6, but on adding water a solution containing $Fe^{3+}(aq)$ and $Cl^-(aq)$ ions is formed. The Fe^{3+} ions are hydrolysed.

The chlorides of the transition metals are coloured in the solid state. Ions are produced when they are dissolved in water; there is often a change of colour, for the aqueous solution takes its colour from the aqueous ion of the metal, while the solid chloride generally has a colour characteristic of a chloro-complex of the metal. Thus copper(II) chloride, $CuCl_2$, is green-blue in the solid state but dissolves to form a blue solution.

Chlorides of very low boiling point are molecular, e.g. $TiCl_4$ and Fe_2Cl_6, while the remainder have generally macromolecular structures. Ions may be produced on melting, as indicated by the fact that most of the fused chlorides can be electrolysed, though the structures of the solids suggest that the bonding there is more covalent in nature.

Figure 26.11 is a graph of heat of formation of the divalent chlorides, MCl_2, against atomic number. This graph shows a similar pattern to Fig. 26.2 because the heats of atomization of the metals are important in determining heat of formation of compounds.

Other halides. The properties of fluorides are such as to suggest that their bonding is generally predominantly ionic in nature.

Bromides and iodides of transition metals are often less soluble in water than the chlorides. Note the trend:

AgF	$\gg$	AgCl	>	AgBr	>	AgI
soluble in water and in dilute $NH_3(aq)$		almost insoluble in water: soluble in dilute $NH_3(aq)$		soluble in conc $NH_3(aq)$		insoluble in water; insoluble even in conc $NH_3(aq)$

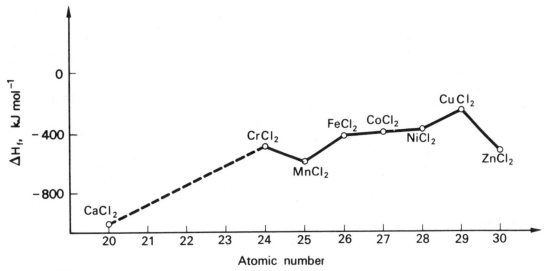

Fig. 26.11. Heat of formation of divalent chlorides (compare this plot with Figs. 26.1, 26.2 and 26.6).

TABLE 26.6

Aqueous ion	Halide complexes	Colour
Ti(IV)	$[TiCl_5.H_2O]^-$ and $TiCl_6^{2-}$	Colourless
Fe(III)	$[Fe(H_2O)_5Cl]^{2+}$,	
	$[Fe(H_2O)_4Cl_2]^+$	Yellow
Co(II)	$CoCl_4^{2-}$	Deep blue
Cu(II)	$CuCl_4^{2-}$	Yellow
Zn(II)	$[ZnCl_3.H_2O]^-$ and	
	$ZnCl_4^{2-}$	Colourless

Complex ions. The transition metals readily form halide complexes, and these usually have different colours from the aquo-complexes, for instance $CuCl_4^{2-}$ (yellow), produced by adding concentrated HCl(aq) to a solution of Cu^{2+}(aq).

26.9 Sulphides of the transition metals

It is a commonly observed phenomenon that elements which form polarizable ions tend to occur together in compounds in the Earth's crust. Transition metal ions are polarizable and so are sulphide ions, so it is not surprising that many of the ores of transition metals contain sulphide. Common ores are *iron pyrites*, FeS_2(c), and *copper pyrites*, $CuFeS_2$. Manganese, nickel, silver and mercury also occur as sulphides. These sulphides are highly coloured or black, macromolecular, and insoluble in water, the colour being due to charge transfer absorption.

All the metals of the first transition series react directly with sulphur to form a sulphide of fairly low oxidation state. Sulphur is not such a powerful oxidizing agent as oxygen or fluorine, so iron reacts to form iron(II) sulphide, FeS, whereas oxygen gives iron(III) oxide, Fe_2O_3.

The sulphide ores of the metals are often used for extraction. It is not practicable to reduce the sulphide ore directly with carbon (the theoretical reasons for this are outlined in Chapter 13) and so the sulphide ore is first converted to the oxide by **roasting** in air. This reaction forms the metal oxide and sulphur dioxide.

26.10 Ammine complexes

Complex ions in which the ligand is ammonia are known as ammines. Ammines are formed by all the transition metals, but many of them are not particularly stable in aqueous solution. Particularly stable ones are formed by the later members of the first transition series, and these are readily distinguished from the aquo-complexes by their different colours.

Cobalt ammines. Although $Co^{3+}(aq)$ is a very powerful oxidizing agent (Fig. 26.7), the addition of ammonia alters the redox potential of the cobalt(II)/cobalt(III) couple completely and stabilizes the $+3$ state.

$$Co^{3+}(aq) + e^- \rightleftharpoons Co^{2+}(aq);$$
$$\epsilon^\circ = +1{\cdot}84 \text{ V}$$

$$Co(NH_3)_6^{3+} + e^- \rightleftharpoons Co(NH_3)_6^{2+};$$
$$\epsilon^\circ = +0{\cdot}10 \text{ V}$$

If ammonia is added to a solution of $Co^{2+}(aq)$, a blue precipitate of a basic cobalt(II) salt is first formed. On adding excess ammonia this precipitate dissolves. In the absence of air, a brownish-yellow solution of a cobalt(II) ammine is formed, but in the presence of air this is rapidly oxidized to a red solution of the cobalt(III) ammine.

This reaction affords a good illustration of the stabilization of an oxidation state by a ligand; cobalt(III) complexes are quite common even though the simple aquo-complex is unstable due to its ready reduction. If cobalt(II) chloride is reacted with ammonia and air a series of mixed complexes is formed, depending on the conditions:

$Co(NH_3)_6^{3+} 3\,Cl^-$—all the Cl^- precipitated by silver nitrate.

$[Co(NH_3)_5Cl]^{2+} 2\,Cl^-$—two-thirds of the chloride precipitated by silver nitrate.

$[Co(NH_3)_4Cl_2]^+ Cl^-$—one third of the chloride precipitated by silver nitrate.

This series of compounds illustrates the stepwise replacement of one ligand by another.

Copper ammines. On adding ammonia to a solution of copper(II) ions, a pale blue precipitate is first formed, consisting of a basic salt of copper(II). Excess ammonia dissolves this to form a very deep blue solution containing tetra-ammine copper(II) ion, $Cu(NH_3)_4^{2+}$. Like other complexes, this one is in fact formed stepwise by the addition of ligands in succession, though in equilibrium the complex with four ligands is the one which predominates.

The ion $Cu(NH_3)_4^{2+}$, sometimes referred to as cuprammonium for short, is tetragonal (square planar) in shape. Although this seems to be a violation of the rules for molecular and ionic shapes (section 6.2) the reason is that there is a non-bonding electron left in the d-shell which repels the ligands away from tetrahedral positions.

26.11 Formulae and stability constants of complex ions

In order to find the formula of a complex ion, the equilibrium between the complex and its component species must be investigated quantitatively. Various methods are available, for instance:

   (i) if the complex is highly coloured, a colorimetric method may be used;

   (ii) if the ligand is soluble in an organic solvent, its concentration in the aqueous system may be determined by a partition method;

   (iii) a solubility method may sometimes be used.

Method (iii) may be illustrated by the determination of the formula of the silver(I) ammine, which can be taken as $Ag(NH_3)_x^+$. This ammine is quite stable and soluble, and is formed when a precipitate of silver chloride is dissolved in ammonia. Silver chloride is very sparingly soluble in pure water ($K_s = [Ag^+] [Cl^-]$ = about 10^{-10} mol^2 dm^{-6}) but dissolves readily in aqueous ammonia. The stability constant of the complex ion may be written as follows:

$$Ag^+(aq) + xNH_3(aq) \rightleftharpoons Ag(NH_3)_x^+;$$
$$K = \frac{[complex]}{[Ag^+][NH_3]^x}$$

The complex must be fairly stable, since addition of ammonia lowers the concentration {Ag^+} sufficiently for the solubility product of silver chloride not to be exceeded.

The procedure is to measure the solubility of silver chloride in a series of ammonia solutions of varying concentrations; the more concentrated the ammonia, the more silver chloride is required to saturate the solution. The formula is then calculated as follows:

Since $K_s = [Ag^+][Cl^-]$,
$$K = \frac{[complex][Cl^-]}{K_s.[NH_3]^x}$$

Since we know that the solubility of silver chloride in pure water is very low, there cannot be many Ag^+ ions present, therefore almost all the silver dissolved must be present as complex. For every mole of silver dissolved, one mole of chloride dissolves, therefore [complex] = [Cl$^-$]

Therefore, $K = \dfrac{[Cl^-]^2}{K_s.[NH_3]^x}$

Therefore, $[Cl^-]^2 \propto [NH_3]^x$

[Cl$^-$] may be determined readily by extracting a portion of the solution and acidifying it to reprecipitate the silver chloride which can then be weighed. [NH$_3$] can be assumed to be the same as the concentration of the ammonia solution originally used, proved a large excess is employed. A graph of $log_{10}[Cl^-]$ against $log_{10}[NH_3]$ gives the value of x, which is found by experiment to equal 2. Substitution of actual values of [Cl$^-$] and [NH$_3$] leads to the numerical value of K, which turns out to be about 10^7 at 25°C.

26.12 Thermodynamic and kinetic stability of complexes

Complexes of cobalt(III) are stable in aqueous solution, usually because there are no suitable vacant electron orbitals available to accept an attacking species: they are said to be *kinetically* stable. Kinetic and thermodynamic stability can be distinguished as follows:

Kinetically stable: rate of decomposition slow, due to high energy of activation of reactions,

Thermodynamically stable: free energy of reaction under the given conditions (e.g. in presence of air, water, etc.) is positive.

There is an analogy between cobalt(III) complexes in the presence of water, and carbon compounds in the presence of air: both are often thermodynamically unstable but kinetically stable.

26.13 Bonding in transition metal complexes

The precise nature of the bonds between ligands and the central metal ion is not fully understood. Some ligands such as ammonia, water and chloride ion appear to be attached by what are essentially electrostatic forces, rather like the forces that hold together an ionic lattice. On the other hand there are ligands such as carbon monoxide and the cyanide ion where transfer of electrons appears to occur between the ligand and the metal. In such cases a pair of electrons is often donated by the ligand to form a weak bond, and this bond can be strengthened by the metal donating some of its d-electrons back to the ligand. This back-donation can only occur with multiple-bonded ligands, such as $C\equiv N^-$, $C=O$, nitrogen(II) oxide and olefins.

Back-donation from the metal to the ligand occurs most readily from metals with a large number of d-electrons, namely those at the end of a given transition series. Examples of such compounds are ferrocene, $Fe(C_5H_5)_2$, dibenzenechromium(0), $Cr(C_6H_6)_2$, tetracarbonylnickel(0), $Ni(CO)_4$, and Zeise's salt, $K[PtCl_3(C_2H_4)]H_2O$, first made in 1829.

Ligands that are not unsaturated, but which have a large number of electrons available for donation, tend to form complexes with those metals with few d-electrons, namely those at the beginning of a given transition series. Thus titanium, vanadium and chromium all form strong complexes with the peroxide ion, $[O-O]^{2-}$, which can be used in the qualitative analysis of these elements.

Apart from the importance of nickel carbonyl in the production of nickel, the property of the later transition metals forming complexes with unsaturated hydrocarbons makes them important catalysts. Nickel, palladium and platinum are widely used as hydrogenation catalysts in organic chemistry.

26.14 Cyano-complexes

All the transition metals form complex ions where the ligand is cyanide, CN^-, but the most stable ones are formed by the elements at the end of the $3d$-series, since the cyanide ion contains a multiple bond $[C\equiv N]^-$, (section 26.13).

Iron forms two hexa-cyano complexes, with oxidation states of $+2$ and $+3$. The redox potential of the iron(II)/iron(III) couple is considerably altered by the addition of CN^- ligands:

$$Fe^{3+}(aq) + e^- \rightleftharpoons Fe^{2+}(aq);$$
$$\epsilon^\circ = +0.77 \text{ V}$$

$$Fe(CN)_6^{3-} + e^- \rightleftharpoons Fe(CN)_6^{4-};$$
$$\epsilon^\circ = +0.36 \text{ V}$$

If a solution of iron(II) sulphate is boiled with potassium cyanide, $Fe(CN)_6^{4-}$, hexacyanoferrate(II), is the product. This is readily oxidized to the iron(III) complex by passing chlorine into the solution.

Most cyano-complexes are sufficiently stable for the reactions of the free ions to be masked entirely. For instance, solutions of $Fe(CN)_6^{3-}$ and $Fe(CN)_6^{4-}$ do not give precipitates with alkali or with hydrogen sulphide, and addition of acid does not liberate $HCN(g)$.

The cyano-complexes of iron are octahedral in shape, the metal–carbon–nitrogen nuclei being collinear. The metal–carbon bonds have some multiple-bond character.

If iron(II) ions are added to $Fe(CN)_6^{3-}$, or alternatively iron(III) ions are added to $Fe(CN)_6^{4-}$ (that is, the oxidation states are mixed) a dark blue precipitate, *prussian blue* is formed. These reactions form the basis of sensitive tests for the two oxidation states of iron. The precipitate is non-stoichiometric in composition, and its constitution depends upon its method of preparation.

Most transition metals, and some non-transition metals, form precipitates with $Fe(CN)_6^{4-}$. A brown precipitate of copper(II) ferrocyanide, $Cu_2Fe(CN)_6(c)$, has been used as a semi-permeable membrane in osmosis, (section 15.10).

26.15 Thiocyanate complexes

The ligand CNS^-, thiocyanate, is similar to cyanide in its ability to form complex ions. If a solution of potassium or ammonium thiocyanate is added to $Fe^{3+}(aq)$, a very deep blood red colour is obtained due to complexes formed. Iron(II) does not give this reaction, so the reagent may be used to test for iron(III).

The complexes form by stepwise replacement of H_2O ligands by CNS^-. If a trace of $Fe^{3+}(aq)$ is added to a large excess of $CNS^-(aq)$, the main species formed will be $Fe(CNS)_6^{3-}$, while if an excess of $Fe^{3+}(aq)$ is used with a trace of $CNS^-(aq)$, the main species will be $Fe(CNS)^{2+}(aq)$. At an intermediate stage the zero-charged species $Fe(CNS)_3$ will be formed, and this, being molecular in nature can be extracted into an organic solvent such as ether.

The formation of thiocyanate complexes of iron(III) is used as the basis of a method of quantitative analysis based upon **colorimetry,** (section 24.10). A dilute solution of the iron(III) salt is taken, and excess potassium or ammonium thiocyanate is then added to form the complex. A standard solution of iron(III) is similarly treated. The intensity of colour in the two solutions is then compared, either by using a photoelectric meter to measure the transmitted light, or by comparing the relative depths of the two solutions required to give the same degree of absorption of light.

26.16 Polydentate ligands

Transition metals form many stable complexes with ligands which have more than one point of attachement. Ethylenediamine, $H_2N.CH_2.CH_2.NH_2$, has already been mentioned as an example of a bidentate ligand. Another example is the oxalate ion, $C_2O_4^{2-}$, which forms chelate complexes with elements such as chromium. The salt potassium tri-oxalatochromate(III), $K_3[Cr(C_2O_4)_3]$, is an example.

Chelates (complexes in which the ligand forms a ring with the central metal atom) are generally more stable than simple complexes with monodentate ligands. The reason is that for the ligand to become detached it is necessary for all the bonds holding it to the metal atom to be broken simultaneously. If only one bond breaks, the complex as a whole does not decompose. Among the most stable of all complexes are those formed by the ethylenediamine-tetra-acetate ion, EDTA, which has *six* points of attachment with the central atom.

26.17 Inner transition metals

The third and fourth transition series both start with a set of f-block elements, in which the $4f$ and $5f$ shells are filling with electrons. The $4f$-series, lanthanum to ytterbium, is known as the **rare-earth series** or **lanthanide series.** The term "rare-earth" is better avoided, for many of these elements are far from rare; the term arose because for many years pure compounds of these elements were difficult to obtain. Since their outer electron configurations are the same throughout the series, their chemical properties are extremely similar and they are very difficult to separate. Ion exchange and countercurrent distribution are nowadays used when pure samples of the elements are required. Solutions of cerium(IV) can be used as an alternative to potassium permanganate for volumetric oxidations.

The $5f$-series is known as the **actinide series,** and it includes the artificially made transuranic elements (elements 93 onwards). All the actinides are radioactive and the series was completed with the synthesis of lawrencium, element 103, in 1966.

26.18 Summary of the properties of metals

TABLE 26.7

	s-block	B-metals	Transition metals
Physical properties	Soft, low m.p.	Harder and less easily melted than s-block	Considerably harder, with higher melting points, than B-metals
Oxidation states	Not variable—Group number only	Inert-pair effect	Variable, frequently differing by one unit
Properties of ions	Simple ions, noble-gas structures, not easily polarized; complexes not readily formed (except Be.)	Simple ions have completed d-shell; easily polarized to form complex ions	Simple ions not formed; complexes readily formed
Complex ions	Not stable. Simple ions loosely hydrated, colourless	Formed in preference to simple ions; colourless	Readily formed, usually coloured
Bonding in compounds	Usually ionic	Usually covalent or complex ions	Usually covalent or complex ions
Reactivity with water	Often vigorous	Not with cold water except slowly	Not with cold water except slowly
Reactivity with non-metals	Vigorous	Less vigorous than s-block	About the same as B-metals
Reactivity with hydrogen	Saline hydrides formed	None as a rule	Some form interstitial hydrides

Study Questions

1. Give the outer electronic configuration of the following ions Ti^{4+}, Cr^{2+}, Mn^{2+}, Ni^{2+}, Ag^+.

2. (a) In older versions of the periodic table, chromium and sulphur were often placed in the same Group. In what ways are these elements and their compounds (i) similar, (ii) different?

(b) Can you discover any other reasonable analogies between transition metals and s- or p-block elements?

3. Of the transition elements (Ti–Cu) which would you expect

(a) To have the highest second ionization energy?

(b) To form a colourless singly charged cation?

(c) To form a carbonyl of formula $M(CO)_5$?

(d) To form the smallest M^{2+} ion?

(e) To form an ion $M_2O_7^{4-}$?

(f) To form a volatile bromide of formula MBr_4?

(g) To form a compound MO_3F?

4. (a) Which oxide of manganese is typically basic?

(b) Which oxide of manganese is molecular and acidic?

(c) Which ions are isoelectronic with the permanganate ion?

(d) Are any ions isoelectronic with the manganate ion?

5. Give balanced equations for the reactions of potassium permanganate in acid solution with (a) iron(II) giving iron(III); (b) oxalate ion giving carbon dioxide; (c) hydrogen peroxide giving oxygen; (d) iron(II) oxalate.

6. (a) How many isomers are there of the following?
(i) $Co(NH_3)_6^{3+}$; (ii) $[Co(NH_3)_5Cl]^{2+}$; (iii) $[Co(NH_3)_4Cl_2]^+$; (iv) $Co(NH_3)_3Cl_3$, (Assume octahedral geometry·about the cobalt atom.)

(b) There are two isomers of $Pt(NH_3)_2Cl_2$. What does this tell you about the geometry of this compound?

†**7.** "The oxidation state +2 becomes more stable, while the oxidation state +3 becomes less stable with increasing atomic number across the first transition series".

(a) Which metals provide exceptions to this statement?

(b) Suggest reasons for the exceptions.

8. M $KMnO_4$(aq) was titrated against a given volume of reducing agent, first in acid, second in neutral, third in alkaline solution. In acid, 10 cm³ of $KMnO_4$ were required, in neutral solution 16·7 cm³ and in alkaline solution, 50 cm³.

(a) What is the oxidation number of the reduction product in each case?

(b) Suggest a balanced equation for each half-reaction.

(c) How many cm³ of M $K_2Cr_2O_7$ would have been required to react with the same volume of reducing agent in acid solution?

9. A dark brown solid, A, was fused with KOH and KNO_3 to give a melt from which a green solution of B was extracted. With carbon dioxide, B gave A and a purple solution of C. With reducing agents in acid, C gave solutions of D. With ammonium sulphide, D gave a buff precipitate of E. With sodium hydroxide solution, D gave a precipitate of F, which oxidized rapidly to G.

Identify the lettered compounds and suggest equations for the reactions where possible.

10. A green solution, H, of a transition metal chloride was heated with sodium hydroxide and hydrogen peroxide solutions to give a yellow solution of J. On acidification, J gave K, an orange solution. When crystals of K were treated with sodium chloride and concentrated sulphuric acid, a red volatile liquid, L, was

obtained. On adding ammonia to a strongly acid solution of K, orange crystals of M formed. M decomposed violently on heating to give steam, nitrogen and a green solid, N.

Identify the lettered compounds and give equations for the reactions where possible.

11. On heating, a pale green solid, P, gave carbon dioxide and a black solid, Q. With hydrochloric acid, P gave carbon dioxide and a green solution of R, which, with ammonia, eventually gave a blue solution of S. Electrolysis of R gave the grey metal, T, at the cathode.

(a) Identify the lettered compounds.

(b) How would you confirm your answers to (a)?

12. A transition metal A was heated in a stream of hydrogen chloride to give hydrogen and B, which dissolved in water to give a green solution. With potassium cyanide solution, B gave a brown precipitate of C, which dissolved in excess potassium cyanide solution to give a yellow solution of D. B reacted with chlorine to give E, which dissolved in water to give a reddish solution. When solutions of D and E were mixed, an intense blue precipitate of F formed. With potassium thiocyanate solution, E gave a deep red colour due to G, but B was unaffected. A solution of E can be converted to a solution of B by using tin(II) chloride.

Identify the lettered compounds and where possible give equations for the reactions.

†**13.** A white solid A was dissolved in water and then reduced by zinc in hydrochloric acid to give, consecutively, a blue solution of B, a green solution of C and a violet solution of D. When A was heated, ammonia, water and a yellowish solid, E, formed. E is widely used as a catalyst. On heating with carbon and chlorine E gave a volatile liquid, F, of molecular weight 173. Using bromine instead of chlorine, G, a similar volatile liquid with a molecular weight of 307 was formed.

Suggest the nature of the lettered compounds.

14. When a blue solution, A, was treated with potassium iodide solution, iodine and a white compound, B, were formed. On dropping an iron nail into A, a red deposit of C formed on the nail. Addition of ammonia to A gave initially a pale blue precipitate of D, but, with excess ammonia, a deep blue solution of E. When solution A was evaporated to dryness and the resulting solid heated, oxygen, brown acidic fumes and a black solid F were obtained. F could be reduced in a stream of hydrogen to give C.

(a) Identify the lettered compounds.

(b) Give equations for the reactions that took place.

†**15.** Some blue crystals of A dissolved in water to give a pink solution of B. With ammonium sulphide solution, B gave a black precipitate of C. C did not form when the solution of B was acid. When 2·6 g of A were dissolved in dilute nitric acid and then treated

with silver nitrate solution, 5·73 g of a white precipitate were obtained. This precipitate dissolved in dilute ammonia.

Identify the lettered compounds.

†16. (a) What factors are important in determining the heat of formation of the chlorides, MCl_2? (Draw an energy level cycle.)

(b) Can the abnormal heat of formation of $MnCl_2$ be accounted for by the abnormal heat of sublimation of Mn? (Figs. 26.2 and 26.11.)

(c) Why does manganese so often behave in an exceptional manner?

(d) Use a data book to list other abnormal properties of manganese and its compounds.

17. Explain why

(a) Copper is used in electrical wires.

(b) Tungsten is used in electric light bulbs.

(c) Nickel is used as a hydrogenation catalyst for unsaturated hydrocarbons.

(d) Gold is used in wedding rings.

A-level Examination Questions

THE Chapters which may be helpful in answering the questions use given in square brackets after each question.

1. What is meant by the terms *atomic number, atomic weight* and *isotopes?*

Why do the chemical properties of elements correlate better with atomic numbers than with atomic weights?

Explain briefly, in terms of electron structures, the occurrence of *Periods, Groups* and *Transition Series* in the Periodic Table.

(O. & C., 1967) [Chs. 1, 3]

2. Explain four of the following observations:

(a) The atomic weight of fluorine (18·9984) is much nearer to a whole number than that of chlorine (35·453).

(b) An aqueous solution of copper sulphate, itself blue, turns blue litmus red.

(c) The freezing point of a molar solution of potassium iodide is hardly affected by the addition of 50 g of iodine to a litre of the solution.

(d) A blood cell swells and eventually bursts when placed in pure water, but appears to be unaffected when put into a solution containing 0·85 g of sodium chloride per 100 cm³ of water.

(e) The surfaces of some finely divided metals have profound effects on the speed with which some gases react.

(Oxf., 1967) [Chs. 1, 15, 17, 19]

3. Draw diagrams to show the bonding in aluminium chloride and sodium aluminium fluoride. Explain how the physical properties of these compounds are related to their structures.

Neutron irradiation of aluminium gives an isotope X, each aluminium atom, $^{27}_{13}Al$, absorbing one neutron and then emitting an α-particle in the process. Isotope X then decays by β emission to another element Y. Deduce the mass numbers and atomic numbers of X and Y.

If, when some of the aluminium had changed to Y, the product were dissolved in dilute hydrochloric acid, explain how you would show the presence in the solution of ions from both aluminium and Y.

(Camb., 1967) [Chs. 2, 6, 20]

4. Explain what you understand by ionic, covalent and metallic bonds. Illustrate your answer by considering the bonding in: sodium chloride, ammonium chloride, methane, diamond, any metal and any complex ion of your own choice.

How do we account for the shape of the sodium chloride crystal and of the methane molecule?

(Oxf., 1967) [Chs. 5, 6]

5. Discuss the general arrangement of elements in the modern (extended) form of the periodic table provided, explaining the underlying principles of the classification, and describing the broad trends observed in the characteristics of the elements.

A metallic element M forms two stable chlorides with formulae MCl_2 and MCl_4 respectively. Where would you expect the element to be placed in the classification? Give reasons.

(Southern, 1967) [Chs. 3, 25]

6. Define (a) specific conductance, (b) either molar or equivalent conductance of a solution.

Describe how you would measure in the laboratory the electrical conductance of a solution of a strong electrolyte.

The measured resistance of a conductance cell containing 0·555 g of calcium chloride per litre at 25°C was 1050 Ω. In a calibration experiment at the same temperature, the measured resistance of the same cell containing 0·02 M potassium chloride solution was 457 Ω. If the specific conductance of 0·02 M potassium chloride solution is $0·00277\ \Omega^{-1}\ cm^{-1}$, calculate

(i) the cell constant;

(ii) the specific conductance of the calcium chloride solution;

(iii) the molar conductance of calcium chloride at this concentration.

($CaCl_2 = 111.$)

(Southern, 1967) [Ch. 6]

7. State Faraday's laws of electrolysis. Discuss the electrolysis with platinum electrodes of each of the following: (a) a solution of potassium acetate; (b) molten sodium hydride; (c) an aqueous solution containing the sulphates of copper, magnesium and aluminium.

A 100 cm³ portion of a potassium chloride solution was electrolysed using a silver anode. A constant current of 16·1 mA for 5 min quantitatively precipitated the chloride as silver chloride. Calculate the molarity of the potassium chloride solution.

(1 faraday = 96,487 coulombs.)

(Welsh, 1967) [Chs. 6, 18]

8. (a) Describe how the molecular weight of glucose may be determined by finding the depression of the freezing point of water in which it is dissolved.

(b) When determined by this method, the value of the apparent molecular weight of sodium nitrate in a solution containing 85 g dm⁻³ is 62·7. What may be deduced from this?

(c) What definition of atomic weight was the basis for Cannizzaro's method of finding atomic weights? Show how this method was used to determine the atomic weight of carbon.

(AEB, 1965) [Chs. 7, 15]

9. (a) What analogies exist between the gas laws and the laws concerning the osmotic pressure of dilute solutions?

(b) Outline the principles underlying the process of distillation in steam. Show how the ratio of the weights of the components of the distillate obtained by this process may be determined by vapour pressure and molecular weight considerations.

(c) A certain volume of an arsenious sulphide sol is coagulated by 5 cm³ of M/8 sodium sulphate solution or by 5 cm³ of M/800 aluminium sulphate solution, but not by 5 cm³ of M/80 sodium sulphate solution.

Comment on these observations.

(AEB, 1965) [Chs. 8, 14, 15]

10. Describe one method for preparing a sample of ozonized oxygen.

The equilibrium between oxygen and ozone is represented by

$$3 O_2 \rightleftharpoons 2 O_3 \text{ and } \Delta H = +285 \text{ kJ.}$$

Explain why only a negligible quantity of ozone is formed when an electric spark is passed through oxygen.

The times for the *same* volumes of pure oxygen and pure ozone to effuse through a small hole under the same conditions, were 44 and 54 s respectively. What information about the composition of ozone do these data provide?

(AEB, 1967) [Chs. 8, 10, 22]

11. Describe how you would investigate the distribution of benzoic acid between two immiscible solvents such as chloroform and water.

At 18°C the distribution ratio of butyric acid between diethyl ether and water is 3·50 to 1·00 and is independent of concentration. Calculate the weight of butyric acid extracted by shaking up 100 cm³ of water containing 10·0 g of butyric acid with 100 cm³ of ether. What would be the weight of butyric acid removed in each of two further extractions using 100 cm³ of ether each time? Compare the total weight of butyric acid extracted in these three successive extractions with that in a single extraction using 300 cm³ of ether.

(Use logarithm tables for the calculations and give results to two places of decimals.)

(O. & C., 1967) [Ch. 10]

12. Explain the following observations.

(i) Hydrogen fluoride is both less volatile and a weaker acid than hydrogen chloride.

(ii) Silver chloride dissolves in ammonia solution; silver iodide does not.

(iii) Chlorine and methane react rapidly when exposed to strong sunlight.

(iv) Pure hydrogen bromide is not readily obtained by the action of concentrated sulphuric acid on sodium bromide.

(v) Sulphuric acid and hydrochloric acid are equally strong, yet hydrogen chloride is evolved by the action of concentrated sulphuric acid on sodium chloride.

(Camb., 1967) [Chs. 9, 12, 18, 21]

13. (a) The cooling system of a motor-car engine contains 8000 cm³ of water to which 1 kg of ethylene glycol ($C_2H_6O_2$) has been added. Calculate the temperature at which the radiator solution would begin to freeze. What weight of sodium acetate ($C_2H_3O_2Na$) would have to be used in lieu of ethylene glycol in order to give the same depression of the freezing point? Mention any assumption(s) which you have made in arriving at your answer(s) and state the law on which you base your calculations.

[H = 1, C = 12, O = 16, Na = 23. The molecular freezing point depression constant for water = 1·86 K mol⁻¹ dm⁻³.]

(b) 141·4 cm³ of an inert gas diffused through a porous plug in the same time as it took 50 cm³ of oxygen to diffuse through the same plug under identical conditions. Calculate the atomic weight of the inert gas and state the law involved in your calculation.

Explain why diffusion in solution is very much slower than in gases. Indicate *very briefly* how you would demonstrate diffusion in solution by means of a simple experiment.

(Welsh, 1964) [Chs. 8, 15]

14. Arrange the following in electrochemical order commencing with the metal that has the highest negative electrode potential: copper, iron, silver, sodium, tin, zinc.

How are the methods of (a) producing the metals copper and sodium, (b) protecting iron by zinc and by tin, related to the position of the metals in the electrochemical series?

A porous pot containing a copper rod and molar copper sulphate solution is placed in a glass vessel containing molar zinc sulphate solution and a zinc rod. *Explain* what happens when the rods are connected with a wire.

(AEB, 1965) [Chs. 11, 13]

15. Give brief explanatory accounts, illustrated with examples, of the following terms:
   (i) acids;
   (ii) neutralization;
   (iii) indicator;
   (iv) buffer solutions;
   (v) solubility product.

(AEB, 1966) [Chs. 10, 19]

16. Write a concise account of the main features of the methods of separation listed below. Give an example and discuss the advantages of the method in each case.
   (a) Ion exchange.
   (b) Steam-distillation.
   (c) Fractional distillation.

(O. & C., 1967) [Chs. 14, 16]

17. Write short accounts, with examples, of four of the following: (a) allotropy, (b) chain reactions, (c) constant-boiling mixture, (d) hydrolysis of salts, (e) energies of activation.

(O. & C., 1967) [Chs. 14, 17, 19]

18. Describe, with adequate experimental detail, how you would verify that the molecular freezing point depression constant of benzene is 5·227 K per 1000 cm³ of solvent.

The freezing point of benzene is 5·478°C and that of 0·10 M $AlBr_3$ solution in benzene is 5·218°C. Deduce the molecular state of aluminium bromide in benzene and write the corresponding structural formula.

(Welsh, 1967) [Chs. 15, 20]

19. The isomerization of cyclopropane(gas) to give propylene(gas) is a first order reaction. From the experimental results given below calculate the velocity constant of the reaction at 433°C.

The disintegration rates (measured on a Geiger counter) at 25°C of a sample of radioactive sodium-24 at various times are given in the table below. Utilizing all the experimental observations, determine the half-life of sodium-24 and from this calculate the disintegration constant (λ).

Comment on the effects of an increase in temperature on the rates of these two processes.

Time in hours	0	2	5	10	20	30
ISOMERIZATION OF CYCLOPROPANE at 433°C						
Percentage of cyclopropane remaining	100	91	79	63	40	25
RADIO-ACTIVE DECAY OF ^{24}Na at 25°C						
Rate of disintegration (counts s^{-1})	670	610	530	421	267	168

$$(\log_e 2 = 0\cdot693)$$

(O. & C., 1967) [Chs. 2, 17]

20. Compare the chemistry of a metal in an A subgroup of the Periodic Table with that of a metal in the corresponding B sub-group, with particular reference to calcium (Group IIA) and zinc (Group IIB). Give reasons for the similarities and differences in chemical behaviour. (Limit your answer to a discussion of the properties of the elements and any three compounds of each element with a common anion.)

(O. & C., 1967) [Chs. 20, 25, 26]

21. None of the reactions listed below takes place. For any five examples, by explaining or defining the physical principles involved, explain why.

(a) $NaNO_3 + H_2O \longrightarrow NaOH + HNO_3$

(b) $Cl_2 + 2 HF \longrightarrow F_2 + 2 HCl$

(c) $K_4Fe(CN)_6 + 2 KOH \longrightarrow Fe(OH)_2 + 6 KCN$

(d) $10\ MgSO_4 + 2\ KMnO_4 + 8\ H_2SO_4$
$\longrightarrow 5\ Mg_2(SO_4)_3 + K_2SO_4 + 2\ MnSO_4 + 8\ H_2O$

(e) $CaCl_2 + H_2S \longrightarrow CaS + 2 HCl$

(f) $2\ Ag + H_2SO_4 \longrightarrow Ag_2SO_4 + H_2$

(g) In electrolysis:
$$SO_4^{2-} - 2e^- \longrightarrow SO_4$$
followed by
$$2\ SO_4 + 2\ H_2O \longrightarrow 2\ H_2SO_4 + O_2$$

(Southern, 1967) [Chs. 18, 19, 20, 21, 22, 26]

22. (a) An ammonium salt X dissolves in water to yield a solution of orange colour which changes to yellow when the solution is made alkaline. When hydrogen sulphide is bubbled into the acidified orange solution, its colour changes to green and a pale cream colloidal precipitate appears.

(b) Excess of sodium hydroxide solution was added to 0·1801 g of an ammonium salt Y and the ammonia released was distilled into 25 cm³ of 0·2 M hydrochloric acid. The excess acid required 27·50 cm³ of 0·1 M sodium hydroxide solution for its neutralization. Excess of zinc dust was then added to the alkaline residue in the distillation flask and a further quantity of ammonia was released, equal to the amount obtained previously.

Identify X and Y from the data provided, giving reasons for your conclusions. Give equations for the reactions which are described and for the reactions which would occur if X and Y were separately heated in the dry state. Specify a suitable indicator for the titration.

$$[H = 1{\cdot}01; N = 14{\cdot}01; O = 16{\cdot}00.]$$

(Welsh, 1966) [Chs. 19, 23, 26]

23. Iodine is obtained from the mother liquors remaining after sodium nitrate has been extracted from Chile saltpetre. Describe the process. How may a sample of iodic acid be prepared from iodine? Explain what happens when iodine reacts with potassium hydroxide solution and describe *two* chemical tests for identifying the iodine ion.

(AEB, 1966) [Ch. 21]

24. Metals may react with (a) acids, (b) alkalis or (c) steam, liberating hydrogen in each case. Give one example of each type of reaction.

Account for the types of bonds present in (i) sodium hydride, (ii) water, (iii) the hydrated hydrogen cation.

Explain the meaning of reduction. A solution of iron(III) chloride acidified with hydrochloric acid is unaffected when hydrogen is bubbled through it, but when zinc dissolves in this solution hydrogen is evolved and the iron(III) ions are reduced. Explain this behaviour in terms of electron transfer.

(AEB, 1967) [Chs. 18, 19, 20]

Appendix I *Half-cell potential data**

THESE values apply to solutions of pH = 0 (acid solutions), at 25°C. Although the data apply to standard conditions, they may be used as a rough guide as to whether reactions will proceed under normal experimental conditions, though of course they give no guide as to the *rate* of reaction.

It is important to bear in mind that the formation of a complex ion may have a considerable effect on the half-cell potential, e.g.

$$Fe^{3+}(aq)+e^- \rightleftharpoons Fe^{2+}(aq); \quad E° = +0.771 \text{ V}$$
$$Fe(CN)_6^{3-}+e^- \rightleftharpoons Fe(CN)_6^{4-}; \quad E° = +0.36 \text{ V}$$

The formation of stable, highly insoluble compounds is analogous to complex formation. For example, the redox data predict that $Cu^+(aq)$ is unstable. Nevertheless, *insoluble* compounds of copper(I) and complexes such as $Cu(CN)_4^{2-}$ and $CuCl_4^{2-}$, can be prepared.

$E°$ (in V)	oxidized species $+ ne^- \rightleftharpoons$ reduced species
−3·05	$Li^+ + e^- \rightleftharpoons Li$
−2·925	$K^+ + e^- \rightleftharpoons K$
−2·925	$Rb^+ + e^- \rightleftharpoons Rb$
−2·923	$Cs^+ + e^- \rightleftharpoons Cs$
−2·90	$Ba^{2+} + 2e^- \rightleftharpoons Ba$
−2·89	$Sr^{2+} + 2e^- \rightleftharpoons Sr$
−2·87	$Ca^{2+} + 2e^- \rightleftharpoons Ca$
−2·714	$Na^+ + e^- \rightleftharpoons Na$
−2·37	$Mg^{2+} + 2e^- \rightleftharpoons Mg$
−2·25	$\frac{1}{2} H_2 + e^- \rightleftharpoons H^-$
−2·08	$Sc^{3+} + 3e^- \rightleftharpoons Sc$
−1·85	$Be^{2+} + 2e^- \rightleftharpoons Be$
−1·66	$Al^{3+} + 3e^- \rightleftharpoons Al$
−1·63	$Ti^{2+} + 2e^- \rightleftharpoons Ti$
−1·18	$Mn^{2+} + 2e^- \rightleftharpoons Mn$
−1·18	$V^{2+} + 2e^- \rightleftharpoons V$
−0·87	$H_3BO_3 + 3 H^+ + 3e^- \rightleftharpoons B + 3 H_2O$

* Selected from W. M. Latimer, *Oxidation potentials*, 2nd ed., Prentice Hall, Englewood Cliffs, N. J., 1952.

$E°$ (in V)	oxidized species $+ ne^- \rightleftharpoons$ reduced species
−0·86	$SiO_2 + 4 H^+ + 4e^- \rightleftharpoons Si + 2 H_2O$
−0·763	$Zn^{2+} + 2e^- \rightleftharpoons Zn$
−0·74	$Cr^{3+} + 3e^- \rightleftharpoons Cr$
−0·60	$As + 3 H^+ + 3e^- \rightleftharpoons AsH_3$
−0·53	$Ga^{3+} + 3e^- \rightleftharpoons Ga$
−0·51	$Sb + 3 H^+ + 3e^- \rightleftharpoons SbH_3$
−0·49	$2 CO_2 + 2 H^+ + 2e^- \rightleftharpoons (COOH)_2$
−0·440	$Fe^{2+} + 2e^- \rightleftharpoons Fe$
−0·41	$Cr^{3+} + e^- \rightleftharpoons Cr^{2+}$
−0·403	$Cd^{2+} + 2e^- \rightleftharpoons Cd$
−0·37	$Ti^{3+} + e^- \rightleftharpoons Ti^{2+}$
−0·342	$In^{3+} + 3e^- \rightleftharpoons In$
−0·3363	$Tl^+ + e^- \rightleftharpoons Tl$
−0·277	$Co^{2+} + 2e^- \rightleftharpoons Co$
−0·276	$H_3PO_4 + 2 H^+ + 2e^- \rightleftharpoons H_3PO_3 + H_2O$
−0·255	$V^{3+} + e^- \rightleftharpoons V^{2+}$
−0·250	$Ni^{2+} + 2e^- \rightleftharpoons Ni$
−0·136	$Sn^{2+} + 2e^- \rightleftharpoons Sn$
−0·126	$Pb^{2+} + 2e^- \rightleftharpoons Pb$
−0·15	$GeO_2 + 4 H^+ + 4e^- \rightleftharpoons Ge + 2 H_2O$
0.000	$2 H^+ + 2e^- \rightleftharpoons H_2$
+0·06	$P + 3 H^+ + 3e^- \rightleftharpoons PH_3$
+0·10	$Co(NH_3)_6^{3+} + e^- \rightleftharpoons Co(NH_3)_6^{2+}$
+0·1	$TiO^{2+} + 2 H^+ + e^- \rightleftharpoons Ti^{3+} + H_2O$
+0·141	$S + 2 H^+ + 2e^- \rightleftharpoons H_2S$
+0·15	$Sn^{4+} + 2e^- \rightleftharpoons Sn^{2+}$
+0·152	$SbO^+ + 2 H^+ + 3e^- \rightleftharpoons Sb + H_2O$
+0·153	$Cu^{2+} + e^- \rightleftharpoons Cu^+$
+0·17	$SO_4^{2-} + 4 H^+ + 2e^- \rightleftharpoons H_2SO_3 + H_2O$
+0·247	$HAsO_2(aq) + 3 H^+ + 3e^- \rightleftharpoons As + 2 H_2O$
+0·32	$BiO^+ + 2 H^+ + 3e^- \rightleftharpoons Bi + H_2O$
+0·337	$Cu^{2+} + 2e^- \rightleftharpoons Cu$
+0·36	$Fe(CN)_6^{3-} + e^- \rightleftharpoons Fe(CN)_6^{4-}$
+0·521	$Cu^+ + e^- \rightleftharpoons Cu$
+0·5355	$I_2 + 2e^- \rightleftharpoons 2I^-$
+0·559	$H_3AsO_4 + 2 H^+ + 2e^- \rightleftharpoons HAsO_2 + 2 H_2O$
+0·564	$MnO_4^- + e^- \rightleftharpoons MnO_4^{2-}$
+0·771	$Fe^{3+} + e^- \rightleftharpoons Fe^{2+}$
+0·789	$Hg_2^{2+} + 2e^- \rightleftharpoons 2 Hg$
+0·7991	$Ag^+ + e^- \rightleftharpoons Ag$
+0·80	$2 NO_3^- + 4 H^+ + 4e^- \rightleftharpoons N_2O_4 + 2 H_2O$

$E°$ (in V) oxidized species $+ne^- \rightleftharpoons$ reduced species		$E°$ (in V) oxidized species $+ne^- \rightleftharpoons$ reduced species	
+0·86	$Cu^{2+}+I^-+e^- \rightleftharpoons CuI$	+1·3595	$Cl_2+2e^- \rightleftharpoons 2\,Cl^-$
+0·920	$2\,Hg^{2+}+2e^- \rightleftharpoons Hg_2^{2+}$	+1·455	$PbO_2+4\,H^++2e^- \rightleftharpoons Pb^{2+}+2\,H_2O$
+0·94	$NO_3^-+3\,H^++2e^- \rightleftharpoons HNO_2+H_2O$	+1·50	$Au^{3+}+3e^- \rightleftharpoons Au$
+0·96	$NO_3^-+4\,H^++3e^- \rightleftharpoons NO+2\,H_2O$	+1·51	$Mn^{3+}+e^- \rightleftharpoons Mn^{2+}$
+1·0652	$Br_2+2e^- \rightleftharpoons 2\,Br^-$	+1·51	$MnO_4^-+8\,H^++5e^- \rightleftharpoons Mn^{2+}+4\,H_2O$
+1·03	$N_2O_4+4\,H^++4e^- \rightleftharpoons NO+2\,H_2O$	+1·61	$Ce^{4+}+e^- \rightleftharpoons Ce^{3+}$
+1·07	$N_2O_4+2\,H^++2e^- \rightleftharpoons 2\,HNO_2$	+1·77	$H_2O_2+2\,H^++2e^- \rightleftharpoons 2\,H_2O$
+1·19	$ClO_4^-+2\,H^++2e^- \rightleftharpoons ClO_3^-+H_2O$	+1·82	$Co^{3+}+e^- \rightleftharpoons Co^{2+}$
+1·195	$IO_3^-+6\,H^++5e^- \rightleftharpoons \frac{1}{2}I_2+3\,H_2O$	+1·98	$Ag^{2+}+e^- \rightleftharpoons Ag^+$
+1·229	$O_2+4\,H^++4e^- \rightleftharpoons 2\,H_2O$	+2·01	$S_2O_8^{2-}+2e^- \rightleftharpoons 2\,SO_4^{2-}$
+1·25	$Tl^{3+}+2e^- \rightleftharpoons Tl^+$	+2·07	$O_3+2\,H^++2e^- \rightleftharpoons O_2+H_2O$
+1·33	$Cr_2O_7^{2-}+14\,H^++6e^- \rightleftharpoons 2\,Cr^{3+}+7\,H_2O$	+2·87	$F_2+2e^- \rightleftharpoons 2F^-$

Appendix II *International atomic weights on the basis of* $^{12}C = 12 \cdot 000$ *(1969)*

The following values apply to elements as they exist in materials of terrestrial origin and to certain artificial elements[t]. When used with due regard to the footnotes, they are considered reliable to $\pm$ 1 in the last digit, or $\pm$ 3 if that digit is in small type.

Element	Atomic weight	Element	Atomic weight	Element	Atomic weight
Actinium	—	Hafnium	178·49	Praseodymium	140·9077[a]
Aluminium	26·9815[a]	Helium	4·00260[b,c]	Promethium	—
Americium	—	Holmium	164·9303[a]	Protactinium	231·0359[a,f]
Antimony	121·75	Hydrogen	1·0080[b,d]	Radium	226·0254[a,f,g]
Argon	39·948[b,c,d,g]	Indium	114·82	Radon	—
Arsenic	74·9216[a]	Iodine	126·9045[a]	Rhenium	186·2
Astatine	—	Iridium	192·22	Rhodium	102·9055[a]
Barium	137·34	Iron	55·847	Rubidium	85·4678[c]
Berkelium	—	Krypton	83·80	Ruthenium	101·07
Beryllium	9·01218[a]	Lanthanum	138·9055[b]	Samarium	150·4
Bismuth	208·9806[a]	Lawrencium	—	Scandium	44·9559[a]
Boron	10·81[c,d,e]	Lead	207·2[d,g]	Selenium	78·96
Bromine	79·904[c]	Lithium	6·941[c,d,e]	Silicon	28·086[d]
Cadmium	112·40	Lutetium	174·97	Silver	107·868[c]
Caesium	132·9055[a]	Magnesium	24·305[c]	Sodium	22·9898[a]
Calcium	40·08	Manganese	54·9380[a]	Strontium	87·62[g]
Californium	—	Mendelevium	—	Sulphur	32·06[d]
Carbon	12·011[b,d]	Mercury	200·59	Tantalum	180·9479[b]
Cerium	140·12	Molybdenum	95·94	Technetium	98·9062[f]
Chlorine	35·453[c]	Neodymium	144·24	Tellurium	127·60
Chromium	51·996[c]	Neon	20·179[c]	Terbium	158·9254[a]
Cobalt	58·9332[a]	Neptunium	237·0482[b,f]	Thallium	204·37
Copper	63·546[c,d]	Nickel	58·71	Thorium	232·0381[a,f]
Curium	—	Niobium	92·9064[a]	Thulium	168·9342[a]
Dysprosium	162·50	Nitrogen	14·0067[b,c]	Tin	118·69
Einsteinium	—	Nobelium	—	Titanium	47·90
Erbium	167·26	Osmium	190·2	Tungsten	183·85
Europium	151·96	Oxygen	15·9994[b,c,d]	Uranium	238·029[b,c,e]
Fermium	—	Palladium	106·4	Vanadium	50·9414[b,c]
Fluorine	18·9984[a]	Whosphorus	30·9738[a]	Xenon	131·30
Francium	—	Platinum	195·09	Ytterbium	173·04
Gadolinium	157·25	Plutonium	—	Yttrium	88·9059[a]
Gallium	69·72	Polonium	—	Zinc	65·37
Germanium	72·59	Potassium	39·102	Zirconium	91·22
Gold	196·9665[a]				

Footnotes

[a] Mononuclidic element.

[b] Element with one predominant isotope (about 99–100 per cent abundance).

[c] Element for which the atomic weight is based on calibrated measurements.

[d] Element for which variation in isotopic abundance in terrestrial samples limits the precision of the atomic weight given.

[e] Element for which users are cautioned against the possibility of large variations in atomic weight due to inadvertent or undisclosed artificial isotopic separation in commercially available materials.

[f] Most commonly available long-lived isotope; see *Table of selected radioactive isotopes.*[1]

[g] In some geological specimens this element has a highly anomalous isotopic composition corresponding to an atomic weight significantly different from that given.

Appendix III *Successive Ionization energies, in kJ, of the elements from hydrogen to sodium*

Element	1	2	3	4	5	6	7	8	9	10	11
H	1310										
He	2370	5250									
Li	520	7300	11 800								
Be	900	1760	14 900	21 100							
B	800	2430	3680	25 000	32 800						
C	1090	2350	4650	6180	37 800	47 600					
N	1400	2850	4520	7450	9450	53 000	64 200				
O	1310	3380	5310	7450	11 000	13 300	71 500	84 000			
F	1680	3370	6080	8400	11 000	15 200	17 800	92 000	106 000		
Ne	2080	3960	6200	9400	12 200	13 300	20 000	23 000	114 000	131 000	
Na	495	4560	6950	9550	13 400	16 600	20 100	25 500	28 900	141 000	158 000

Appendix IV   *Physical constants, conversion factors and units*

There are seven basic units in the SI system (Système Internationale d'Unités):

Quantity	Name of unit	Symbol
length	metre	m
mass	kilogramme	kg
time	second	s
electric current	ampère*	A
thermodynamic temperature	kelvin*	K
luminous intensity	candela	cd
mole	mole	mol

In addition there are a number of derived units, including the following:

Quantity	Name of unit	Symbol	Dimensions
force	newton*	N	kg m s^{-2}
energy	joule*	J	kg m^2 s^{-2}
power	watt*	W	kg m^2 s^{-3} = J s^{-1}
electrical charge	coulomb*	C	A s
potential difference	volt*	V	J A^{-1} s^{-1}
electrical resistance	ohm*	Ω	V A^{-1}
frequency	hertz*	Hz	s^{-1}
customary temperature	degree Celsius*	°C	K $-$ 273·15

*Note that units named after a person are abbreviated with a capital letter, though the unit itself must be written with a small letter if it is written out in full.

Special prefixes and symbols are used to indicate multiples and sub-multiples of the basic units, in powers of ten. The following are encountered in this book:

Multiple	Prefix	Symbol
10^6	mega	M
10^3	kilo	k
10^{-1}	deci	d
10^{-2}	centi	c
10^{-3}	milli	m
10^{-6}	micro	μ
10^{-9}	nano	n
10^{-12}	pico	p

Two common units which are frequently met in scientific literature, though they are avoided in this book, are the calorie (cal) which equals 4·184 J, and the angstrom unit (Å) which equals 10^{-10} m = 10^{-1} nm.

Some useful physical constants, expressed in SI units, are:

Quantity	Value
velocity of light, c	$2·998 \times 10^8$ m s^{-1}
Boltzmann constant, k	$1·380 \times 10^{-23}$ J K^{-1}
charge of electron, e	$1·602 \times 10^{-19}$ C
Planck's constant, h	$6·626 \times 10^{-34}$ J s
Avogadro constant, N_A	$6·022$ mol^{-1}
molar gas volume at s.t.p.	$22·4$ dm^3 mol^{-1}
or	$(2·241 \times 10^{-2}$ m^3 mol$^{-1})$
gas constant, R	$8·314$ J K^{-1} mol^{-1}
Faraday constant, F	$9·649 \times 10^4$ C mol^{-1}

Index

Page references in **bold figures** indicate items of some importance

	IA	IIA				IIIB	IVB	VB	VIB	VIIB	
1s	1 H										2 He
2s	3 Li	4 Be			2p	5 B	6 C	7 N	8 O	9 F	10 Ne
3s	11 Na	12 Mg			3p	13 Al	14 Si	15 P	16 S	17 Cl	18 Ar
4s	19 K	20 Ca			4p	31 Ga	32 Ge	33 As	34 Se	35 Br	36 Kr
5s	37 Rb	38 Sr			5p	49 In	50 Sn	51 Sb	52 Te	53 I	54 Xe
6s	55 Cs	56 Ba			6p	81 Tl	82 Pb	83 Bi	84 Po	85 At	86 Rn
7s	87 Fr	88 Ra									

Transition elements

	IIIA	IVA	VA	VIA	VIIA	VIII			IB	IIB
3d	21 Sc	22 Ti	23 V	24 Cr	25 Mn	26 Fe	27 Co	28 Ni	29 Cu	30 Zn
4d	39 Y	40 Zr	41 Nb	42 Mo	43 Tc	44 Ru	45 Rh	46 Pd	47 Ag	48 Cd
5d	57 La	72 Hf	73 Ta	74 W	75 Re	76 Os	77 Ir	78 Pt	79 Au	80 Hg
6d	89 Ac									

4f (Lanthanides): 58 Ce, 59 Pr, 60 Nd, 61 Pm, 62 Sm, 63 Eu, 64 Gd, 65 Tb, 66 Dy, 67 Ho, 68 Er, 69 Tm, 70 Yb, 71 Lu
Outer sub-shells as for Ce

5f (Actinides): 90 Th, 91 Pa, 92 U, 93 Np, 94 Pu, 95 Am, 96 Cm, 97 Bk, 98 Cf, 99 Es, 100 Fm, 101 Md, 102 No, 103 Lw
Outer sub-shells as for Pa
Outer sub-shells as for Ce & Tb

IIIA

Legend:
- Principal Shell
- Sub-shell
- Total Number of electrons required to fill each sub-shell

	1	2	3	4	5	6	7
	s	s p	s p d	s p d f	s p d f	s p d	s
	2	2 6	2 6 10	2 6 10 14	2 6 10 14	2 6 10	2